钱江源国家公园人员解说手册

Interpretation Manual for Qianjiangyuan National Park

钱江源国家公园管理局 ◆ 编

中国林业出版社
CFPH China Forestry Publishing House

图书在版编目（CIP）数据

钱江源国家公园人员解说手册 ： 汉、英 / 钱江源国家公园管理局编. -- 北京 ： 中国林业出版社， 2021.12
ISBN 978-7-5219-1544-0

Ⅰ. ①钱… Ⅱ. ①钱… Ⅲ. ①国家公园－讲解工作－开化县－手册－汉、英 Ⅳ. ①S759.992.554

中国版本图书馆CIP数据核字(2022)第001906号

中国林业出版社·自然保护分社（国家公园分社）
策划、责任编辑：肖 静

出版发行：中国林业出版社
（100009 北京西城区刘海胡同7号）
网　　址：www.forestry.gov.cn/lycb.html
印　　刷：北京雅昌艺术印刷有限公司
版　　次：2022年1月第1版
印　　次：2022年1月第1次
开　　本：787mm×1092mm 1/16
印　　张：37
字　　数：570千字
定　　价：480.00元

编辑委员会

顾　　问　汪长林　卢伦燕
主　　编　雍　怡
编　　委　方　明　钱海源　杨丹丹　刘　懿
凌芸喆　王　原　赵云鹏　郭陶然
谢惊华　胡检丽　朱建平　余建平
余顺海　汪家军　陈小南　姜伟东
图文设计　许天宇　何楚欣
翻　　译　张硕朋

前　言
PREFACE

解说国家公园，让我们从这里开始

亲爱的读者，你们好！欢迎加入钱江源国家公园环境解说团队的大家庭，和我们共同努力，向来到公园的访客传递国家公园所拥有的独特自然资源和文化遗产，及其保护工作的意义和价值。

在成为一名解说员之前，你是否和大多数人一样，对于“解说”这份工作有着各式各样的好奇或困惑呢？比如，为什么国家公园需要这个团队？他们对国家公园发挥什么特殊作用？解说员和导游有没有区别？……当然，每个人基于自己的知识储备、经验、阅历等，或多或少对“解说”会有自己的认识，但当你成为一名正式的国家公园解说员的时候，也意味着，解说成为你日常工作中需要职业化认知和掌握的一份技能。所以，让我们来一起了解，何为解说，如何才能成为一名合格的解说员。

首先，让我们先来了解什么是解说。《解说我们的遗产》一书中将解说定义为：“以信息为基础、以语言或者其他工具为载体、以启迪受众心智为目标的一种教育活动。”显然，解说不是一种简单的信息传递，它和提供信息在本质上是两回事。信息只是解说的素材，解说要做的是在访客和公园之间搭建一座沟通交流的桥梁，把公园所具备的那些抽象的、学术的、复杂的、对普通人来说难懂的内涵，用更通俗的、简洁的、有趣的、具象的（但依然是科学而准确的）但不局限于语言的丰富方式表达出来，去引发受众情感上的共鸣。因此，解说也常被认为是一门用各种方式唤起受众共鸣的艺术。而具有不同知识背景、性格特点以及个人经历的解说员，面对不同年龄、喜好以及参访预期的受众时，其解说绝没有固定的模式，而应具有丰富多元的形式。因而，每一个解说员都是一个独立的艺术家，他们赋予解说素材以生命，再用讲故事的方式将它们向受众呈现出来。

深入阅读本书之前，我们有必要对三个基本要素加以认识：解说员、访客（解说受众）和解说方案（素材）。

【解说员】千万不要小看自己，对于每个国家公园，解说员们都承担着一个无比重要和光荣的职责：面向所有访客的国家公园代言人，肩负着介绍、展示、呈现国家公园的独特价值和重要意义的职责。为此，作为解说员的你首先要比大多数人更了解你的国家公园：它成立的目标是什么？位于哪里？拥有什么资源？这些资源有什么特别之处或重要的保护价值？它的过去如何？如何得到保护？……你不仅仅是带领他们去享受国家公园的美景，而且你还将用解说把国家公园的前

世今生、资源特色和保护使命完整地呈现给公众，并激发他们的认同感和保护意愿。你在国家公园工作的经历和感悟，以及你的独特态度和理解，会让解说更具感染力。因为只有了解，才会欣赏；只有欣赏，才会关注；只有关注，才会热爱；只有热爱，才会保护。

【访客】为访客解说的前提，首先是要了解访客。解说的形式和内容如果不能与访客的参访期待、沟通习惯和认知能力相联结，那将很难有所成效。比如，如果对低龄的孩子讲述物种的科学知识，一定需要你有特别生动、直观的表达方式，往往还要辅助一些生动的说明方法，如打比方，但如果我们对大学生还在讲各种拟人化的故事，可能会遭到他们不耐烦的回应。所以，对访客了解越多，你才越能有的放矢，解说才会越具有吸引力。通常的经验是，当解说在某种程度上将展示或描述的内容与访客的个性或者经历联系起来时，访客会更愿意听，也更能感同身受，最终才会有可能达到激发听众共鸣和教育的目的。

了解受众是一项重要而必须坚持练习的工作，因为访客年龄、受教育程度、特殊的经历、此行的期待等或许可以提前收集（但并不一定每一个服务团队都能做到），但更有效的方式是在解说开始前和过程中通过观察、经验，以及有针对性的互动沟通来加以判断和甄别。很多资深的优秀解说员会提早到达集合点，与访客们进行沟通，也会在解说过程中的交流和互动中不断通过观察调整自己的解说方案。

但是也请注意，访客是通过自己的眼睛看，自己的头脑思考，而不是简单地附和和认同你。成功的解说员不是侃侃而谈自己的观点，而是为访客不断去揭开眼前所见事物和现象背后的故事和意义，并用他们自己的感受和认知去消化这一切。这一过程需要时间，也需要契机。再优秀的解说员，也不能要求你的受众记住你介绍的每个字、每句话。但是，如果有那么几个瞬间你感受到他们的情感被激发了，与你产生了共鸣，无论是惊奇还是赞赏、关切等，你的解说就向成功又迈进了一步。

在解说过程中，解说员的角色定位非常重要，这种基于知识和个人经验，针对环境、访客、时间、地点等要素而量身定制的面对面的解说服务，是任何设施解说都很难代替的。因此，作为解说活动的组织者和实施者，你需要从普通服务人员的心态转换到管理者、引领者的视角，去思考自己的角色定位：我对访客的了解如何？我希望向他们讲述什么，运用什么特殊的方式？如何评估解说的效果？从行前准备，到游线规划，再到整个活动的组织，以及最后的总结与复盘，解说员需要进行详细的目标管理和流程规划，并不断积累。

【解说方案】解说方案不是国家公园所有自然和文化资源的整合，而是其中最有代表性也最具有保护和传播价值的资源的提炼。本书的编写，基于海量背景资料的梳理、大量实地调研和潜在访客的需求分析，从而提炼出最有代表性和解说价值的主题，通过国家公园主要自然体验路线的串联，形成了一整套人员解说内容。考虑到解说员在实际解说工作中对解说素材的使用习惯，我们尽可能使用了口语化的表达方式，并在具体的解说点设计了与访客的互动环节，期望能更契合解说员的现场解说实践。而除了人员解说词和互动环节建议之外，我们在部分解说点上也标注了相应的行为注意事项、辅助道具、拓展活动和拓展知识点等内容。

解说员、访客和解说方案三者之间不是独立存在的，而是通过解说活动串联在一起的。解说从来不是简单的重复性工作，不同的季节、天气和场地条件，不同的访客类型和参访目的，甚至不同的解说员组合，都会赋予解说现场不同的使命、挑战和结果。本书结合解说员实际使用场景的需求，共制定了 5 个主题自然体验路线人员解说方案和 2 个通用型人员解说方案，可供解说员在不同场景下使用，并根据需要灵活调整与组合方案内容。照片、模型、实物等辅助道具的使用或者别出心裁的活动设计，可以让解说员的解说更加生动、有趣和吸引人；拓展知识可以作为背景知识融入解说当中；当然，最重要的是解说员对解说工作的热情。对工作的热爱可以让你在这份工作中不断成长，在一次次的活动历练中逐渐形成自己的解说风格。

最后要强调的是，这是一本永远不会定稿的书籍。解说绝没有范式和套路，每一个解说员在每一次解说中都可能有新的发现、感受和积累，并用自己的阅历和智慧，让自己的那一本解说手册不断变厚、变精彩。希望这次探索不仅能服务于各位刚刚起步的解说员，更期待你们在未来的工作中帮助我们不断完善和丰富这本书，把它延伸发展为一份解说员开展在地多元化环境解说的实用工具包。相信在未来，它也能发展成为我国国家公园体制建设中具有行业引领作用和示范效果的一套人员解说专业出版物和多功能的工具书。

注：本书所有图片（除特别标注外），由钱江源国家公园管理局、新生态工作室、城市荒野工作室、浙江大学生命科学学院、世界自然基金会和浙江野鸟会等机构提供并授权本书使用。

Interpret the national park, let's start from here

Dear interpreters, welcome to the big family of the Qianjiangyuan National Park Environmental Interpretation Team, and work with us to convey the unique natural resources and cultural heritage of the National Park, as well as the meaning and value of conservation work, to visitors.

Before becoming an interpreter, have you ever been curious or confused about the job of "interpretation" like most people? For example, why do national parks need this team? What special role do they play in national parks? Is there a difference between an interpreter and a tour guide? …Of course, everyone will more or less have their own idea of "interpretation" based on their knowledge, experience, and so on; but when you become an official national park interpreter, it also means, interpretation becomes a skill that needs professional cognition and mastery in your daily work. So, let us understand what an interpreter is and how to become a qualified interpreter.

First of all, let us understand what is interpretation? The classic book *Interpreting Our Heritage* by Freeman Tilden had already defined interpretation as: "An educational activity based on information, using language or other tools as the carrier, and enlightening the audience." Obviously, interpretation is not a simple delivery of information. They are essentially two different things. Information is only the material for interpretation. What the interpretation needs to do is to build a bridge of communication between visitors and the park, and interpret the scientific terms into understandable languages for visitors (but still scientific and accurate), not only by words but also arouse emotional resonance from the audience. Therefore, interpretation is often considered as an art that resonates with the audience in various ways. As for interpreters with different knowledge backgrounds, personalities, and personal experiences, they can have various forms of interpretation facing audiences of different ages, preferences. Therefore, each interpreter is an independent artist who gives life to the interpretation, and then presents them to the audience by telling stories.

Before using this manual, it is necessary for us to understand three basic elements: interpreters, visitors (interpretation audiences) and interpretation plans (materials).

[Interpreters] Do not underestimate yourself. For each national park, the interpreters have an extremely important and glorious responsibility: they are of the park spokespersons for all visitors, and shoulders the significant responsibility of introducing, exhibiting, and presenting the unique value of the national park. For this reason, you, as an interpreter, must first understand your park better than other people: What is the goal of its establishment? Where is it located? What resources do you have? What are the special features or important protection values of these resources? How is its past? How to protect it? ···You will not only lead them to enjoy the beauty of the park, but also use the interpretation to fully present the past and present, resource characteristics and protection mission of the national park to the public, and stimulate their sense of identity and willingness to protect the national park. Your experience and perception in the national park, as well as your unique attitude and understanding, will make the interpretation more contagious. Because only by understanding can we appreciate it; only by appreciation can we pay attention to it; only by paying attention can we love it; and only by love can we protect it.

[Visitors] The prerequisite for the interpretation of visitors is to understand the visitors first. It will be really difficult if the form and content of the interpretation cannot be connected with the visitor's expectation, communication and cognitive ability. For example, if you tell young children about the scientific knowledge of species, you must have a particularly vivid and intuitive way of expression, often with some descriptions and analogies, but if we are still telling these stories to college students, they will get bored. Therefore, the more you know about the visitor, the more you can focus and the more attractive the interpretation will be. Generally, when the interpretation connects with the visitor's personality or experience, the visitor will be more willing to listen and feel more empathetic, and finally you will be able to achieve the goal of education by stimulating the audience's resonance.

Understanding the audience is an important task that must be practiced. Although you can conduct interpretation based on visitors' age, education level, special experience, expectations of this trip, *etc.*, it is more effective to observe the audience and make targeted interactive communication. Many senior and excellent interpreters will arrive at the meeting point early to communicate with visitors, and will constantly adjust their interpretation plans through observation during the communication and interaction.

But please also note that visitors see through their own eyes and think through their own minds, rather than simply following you. Successful interpreters do not talk about their own opinions, but constantly reveal the stories and meanings behind everything they see in front of their eyes, and digest them with their own feelings and knowledge. This process takes time. It also needs opportunities. No matter how good an interpreter is, you can't ask your audience to remember every word and sentence you say. But if there are a few moments when you feel that their emotions are stimulated and resonate with you, whether it is surprise, appreciation, concern, *etc.*, then your interpretation will take another step toward success.

In the interpretation process, the status of the interpreter is very important. This kind of face-to-face interpretation service tailored to the environment, visitors, time, location, *etc.*, based on knowledge and personal experience, is irreplaceable by any interpretation facility. Therefore, as the organizer and implementer of the interpretation activity, you need to reposition yourself, changing from a normal service staff to a manager, a leader: How do I know about visitors? What do I want to tell them? What special method shall I use? How to evaluate the effect of interpretation? From pre-departure preparation, to travel planning, to the organization of the entire activity, and the final summary and review, the interpreter needs to conduct detailed target management

and process planning, and continue to accumulate experience.

[Interpretation plans] The interpretation plan is not the integration of all the natural and cultural resources in the national park, but the extraction of the most representative and most valuable resources for protection and advertising. The compilation of this manual is based on the editing and gathering of massive background information, a large number of field investigations and the analysis of the needs of potential visitors, so as to extract the most representative and interpretation themes. A complete set of interpretation is formed based on a series of main natural experience routes in the national park. We have simplified expressions as much as possible, and designed interactive links with visitors at specific interpretation points, hoping to be more in line with the interpreter's on-site interpretation practice. In addition to the interpretation and interaction suggestions, we also marked the corresponding attention, toolkit, extra activities and extra information on some interpretation points.

The interpreter, the visitor and the interpretation plan do not exist independently, but are linked together through interpretation. Interpretation is never a simple and repetitive task. Different seasons, weather and site conditions, different types of visitors and visiting purposes, and even different combinations of interpreters will give the interpreter different missions, challenges, and results. In order to enhance the practicality, this manual adopts a series of sub-volumes: 5 sub-volumes of natural experience routes and 2 general interpretation manuals to provide interpreters with clear and flexible use of interpretation programs. The design of a sub volume is also convenient for the interpreter to carry and use. The use of toolkits such as photos, models, objects, or ingenious event design can make your interpretation more vivid, interesting and attractive. Extra information can be incorporated into the interpretation as background knowledge. Of course, the most important thing is your enthusiasm for the interpretation work. This is more important than everything else. The love of work allows you to continue to grow in this job, and gradually form your own interpretation style in the experience of activities.

Finally, it should be emphasized that this is a manual that will never be finalized. There are no absolute templates and routines in interpretation. Interpreters may have new discoveries, feelings and accumulations in each of their interpretation service, and use their own experience and wisdom to make this manual thicker and more informative. We hope that this exploration will not only serve you as starting interpreters, but also look forward to your input to continuously improve and enrich this manual in the coming future, and extend it into a practical toolkit for interpreters to conduct local diversified environment interpretation. We believe that in the future, it can also become a set of interpretation manuals with industry-leading and demonstration effects in the pilot work of our country's national park system.

目 录
CONTENTS

01

概　述

Overview

1.1 国家公园的人员解说

1.1.1 人员解说的意义

面向自然保护地的环境解说，解说员的角色不再是传统旅游中的导览员或讲解员，而是最有温度、能够帮助联结公众与自然、传递国家公园资源价值的一架桥梁。在保护地大量的资源中筛选出有代表性、适宜公众去接受吸收的解说点，用解说员自身富有情感与感染力的语言面对面地去讲述这些自然故事，这是任何设施解说都无法代替的。

1.1.2 人员解说的不同形式

人员解说是指通过直接面对面向访客介绍、讲解、展示被解说对象的一种环境解说的特定形式。人员解说不仅局限于定点定路线讲解，还包括非定点解说、演讲、分享和表演，甚至可以延伸至各种由人引导开展的环境教育活动。除了人员解说，国家公园的环境解说还包括设施、媒体等其他形式。

（1）资讯服务

资讯服务是解说的一种基础形式。一般是指在入口、游客中心、景点附近的服务场所等为游客提供各类相关资讯，并为游客答疑解惑的一种解说方式。

（2）活动引导解说

指定路线的定点解说，是最传统、也最被广为熟知的一种形式。在此过程中，解说人员会带领访客有序地参访经过设计安排的指定路线上的地点、事物和现象，结合具体的场景，为访客进行解说，让访客获得实际的知识和体验。

（3）非定点解说

在具有代表性或季节性特色资源的所在地点，以特色资源解说为主要目的一种非定时、非定点、非指定访客人群的解说形式。

（4）演讲、分享和表演

由解说人员或专业人士，针对某主题进行演讲、故事分享或情景剧表演，是人员解说的进阶形式。这些解说的形式更生动、有趣，可以创造更多和受众的互动、交流，能提供更丰富的参访体验，也更容易触动访客并激发共鸣。

（5）环境教育活动

一个成熟的国家公园的人员解说系统，除了具有上述各种类型的人员解说服务外，还应该具备系统性且能针对不同主题、不同人群、不同时间地点而进行定制化设计的环境教育活动方案。

1.1 Interpretation for national parks

1.1.1 Significance of interpretation

For the environmental interpretation of protected areas, the role of the interpreter is no longer the tour guide in traditional tourism, but the first-line bridge that can help connect the public and nature and convey the value of national park resources. Selecting representative interpretation points suitable for the public to accept from the vast resources of the protected area, and telling these natural stories face-to-face in the interpreter's own emotional and infectious language is irreplaceable by any facility interpretation.

1.1.2 Forms of interpretation

Staff interpretation refers to a specific form of environmental interpretation that introduces, interprets, and displays the interpretation object directly to visitors. Staff interpretation is not limited to fixed-route interpretation, but also includes non-fixed location interpretation, speeches, sharing and performances, and can even extend to various forms of environmental education activities led by people. In addition to staff interpretations, environmental interpretations of national parks also include facilities, media and other forms.

(1) Information service

Information service is a basic function of interpretation. Generally, it refers to a way to provide tourists with all kinds of relevant information at the entrance, visitor center, and service places near scenic spots, and answer questions for tourists.

(2) Activity guide interpretation

Leading fixed interpretation routes is the most traditional and well-known form. In this process, the interpreter will lead the visitor to visit the places, things and phenomena on routes that have been designed and arranged in an orderly manner, combined with specific scenes, to interpret to the visitor, so that the visitor can obtain actual knowledge and experience.

(3) Non-fixed location interpretation

At the representative or seasonal resource locations of the park, the main purpose is the interpretation of the characteristic resources, which is a non-timely designated visitor group.

(4) Speech, sharing and performance

A speech, story sharing or melodrama performance on a topic organized by interpreters or professionals is an advanced form of staff interpretation. In addition to the more vivid and interesting form of interpretation, more importantly, it can create more interaction and communication with the audience as well as provide a richer visiting experience, and it is easier to touch visitors and stimulate resonance.

(5) Environmental education activities

A mature interpreter should have systematic environmental education curriculum activity programs designed for different themes, different groups of people, and different times and places.

1.2 编写目标

本书依托于《钱江源国家公园环境教育主题框架系统设计》《钱江源国家公园解说资源汇编报告》等内容，结合对钱江源国家公园的第一手调研资料，提取其中有代表性、重要价值并适宜面向公众进行解说、举办解说或教育活动的资源，制定了5个主题游线解说方案和2个通用型人员解说方案，共99个解说点，涵盖公园的整体概况、山系、气候、水文、植被、物种、生境、村庄、传统文化等诸多内容（图1）。

本书提供的5个主题游线解说方案，分别对应5条相对固定的自然体验路线。路线的选择结合了公园现有的步道、古道资源，根据国家公园的重要解说主题、不同访客的体验需求、重要解说资源的布局等条件，选择重点路线进行规划布局，以苏庄镇和齐溪镇为主要布局范围。每条自然体验路线都有一个与之紧密相关的解说主题和若干个相对固定的解说点。解说点的选择会综合考虑路线的基础设施、景观特色、资源情况与解说标识标牌布点等因素，其位置相对固定，每个解说点也都会有1～3个解说要点。每个解说要点下，都配有解说词，解说词的语言比较口语化，尽可能方便解说员直接调用和参考。除了主要的解说词之外，每个解说点下还配有拓展知识点、拓展活动、注意事项和辅助工具等内容，为解说员提供更多自由发挥的空间。我们并不鼓励解说员“照搬”，或者实行“拿来主义”，而是期待解说员在实际使用的过程中，能结合解说的实际场景、访客情况和自身的解说特点等，融入自己的思考，灵活调用这部分内容。

通用型人员解说是针对非游线型的解说资源，结合钱江源国家公园的参访道路系统和访客需求，以村庄传统、重要历史遗迹、重要保护资源为解说目标来选定主题解说体验点。与以固定游线、固定点位为特色的主题游线解说不同，通用型解说可适用于多个游线。通用型人员解说分通用型村庄与传统文化人员解说和生物多样性人员解说两大类。其中，通用型生物多样性人员解说共有29个解说点，涉及植物、动物和生境等与物种有关的内容，每个解说点都可以根据需要，灵活穿插在其他任何游线中使用，但最好能结合具体的场景；通用型村庄与传统文化人员解说共有17个解说点，涵盖乡土物产、乡风民俗、乡村景观等多项内容。建议解说员以村庄为基础，挑选与村落相关的一些解说资源，组成体系进行村落散点的解说。

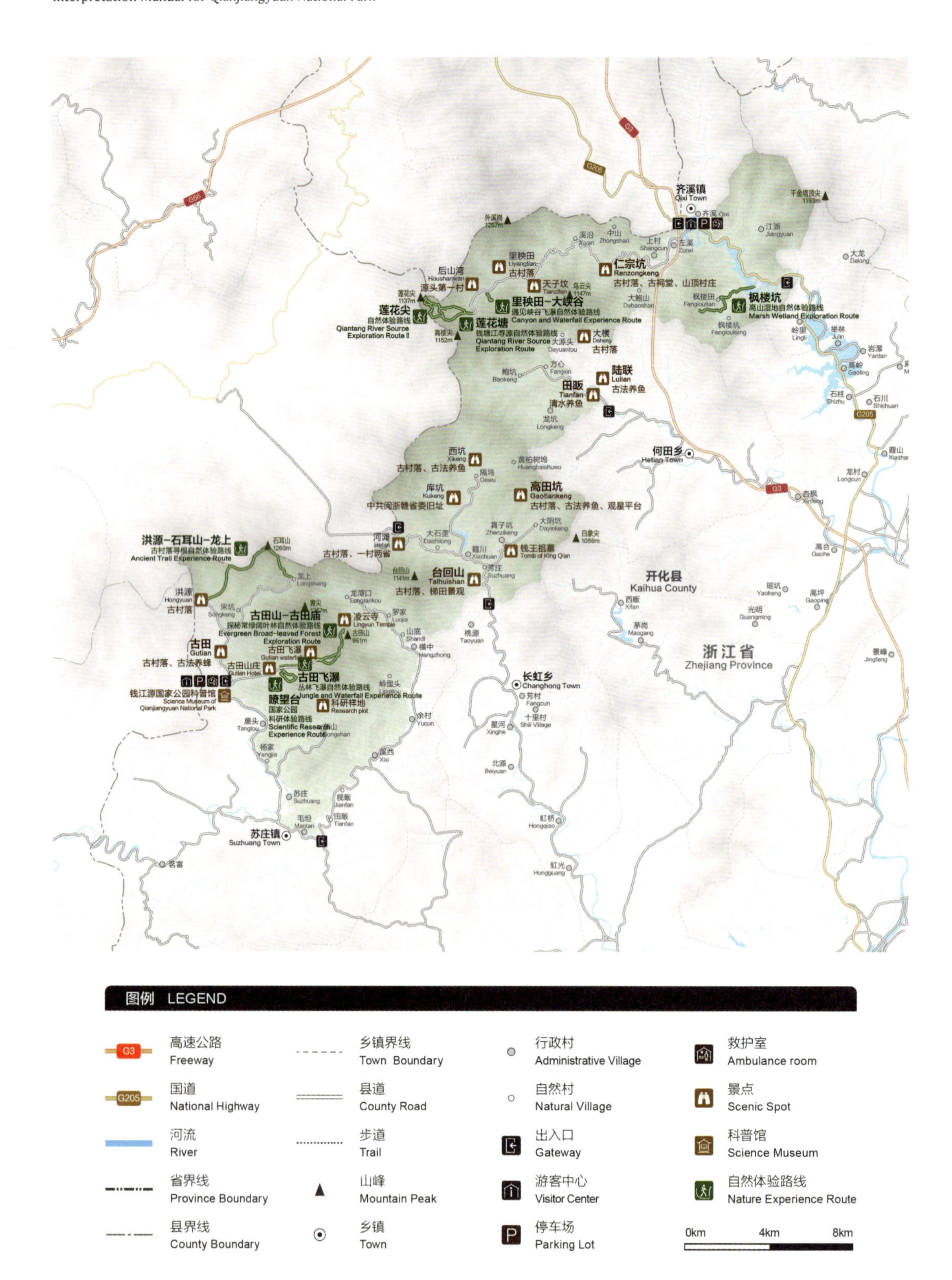

图1 钱江源国家公园人员解说总体导览图
Figure 1 Map of the nature experience routes in Qianjiangyuan National Park

1.2 Goals

This interpretation manual is based on the *Qianjiangyuan National Park Environmental Education Theme Framework System Design*, *Qianjiangyuan National Park Interpretation Resources Compilation Report* and other contents, combined with the first-hand research data of Qianjiangyuan National Park, extracting some representative, important value, and suitable for public interpretation, organize interpretation or educational activities resources from them, formulated 5 thematic tour line interpretation plans and 2 general interpretation plans, a total of 95 interpretation points, covering the entire park's overview, mountains, climate, hydrology, vegetation, species, habitats, villages, traditional culture, *etc*.

The 5 thematic tour line interpretation plans provided in this manual are based on 5 relatively fixed nature experience routes. The selection of these routes is based on the existing trails and ancient trail resources in the park. The key routes are selected for planning and layout based on the important interpretive themes of the national park, the experience needs of different visitors, and the layout of important interpretive resources. Qixi Town is the main planning area. Each nature experience route has a closely related interpretation theme and several relatively fixed interpretation points. The selection of interpretation points will comprehensively consider the route's infrastructure, landscape features, resource conditions and the placement of interpretation signs. The location is relatively fixed, and each interpretation resource will have 1-3 interpretation points. Each interpretation point is equipped with an interpretation. The language of the interpretation is more casual which makes the interpretation easier for interpreters. In addition to the main interpretation, each interpretation point is also equipped with extra information, extra activities, toolkit and attention, *etc*., to provide more space for the interpreter. We do not encourage interpreters to "copy", but expect the interpreter to integrate the actual scene into the interpretation, the situation of the visitor and the characteristics of the interpretation, and integrate your own thinking.

The general interpretation is based on non-route-oriented interpretation resources, combined with the visiting road system of Qianjiangyuan National Park and the needs of visitors, with village traditions, important historical relics, and important conservation resources as the experience goals, and the selected theme experience points. Unlike the nature experience route featuring a fixed route and a fixed location, the generale interpretation can be applied to multiple routes. General interpretation is divided into the general interpretation of biodiversity and the general interprertation of village and traditional culture. Among them, there are a total of 27 interpretation points for biodiversity, involving species-related content such as plants, animals and habitats. Each interpretation point can be flexibly integrated in any other tour lines as needed, but it is best to combine specific species. There are 15 interpretation points for village and traditional culture, covering local products, folk customs, and rural landscapes. It is recommended that the interpreters use the village as the basis to select some interpretation resources related to the village and form a system to interpret.

1.3 自然体验路线人员解说

1.3.1 自然体验路线

本书中挑选的钱江源国家公园5条主题自然体验路线分别位于公园西南部的苏庄镇古田山附近（包括S01、S02、S03共三条主题自然体验路线）和公园东北部的齐溪镇钱塘江源头地区（包括Q01和Q02两条主题自然体验路线）。每条自然体验路线根据其所在的位置、资源特色等的差异，分别对应着一个独特的主题。除此之外，每条自然体验路线的基本信息还包括长度、海拔高度、难易程度以及体验时长等（表1）。

表1　5条自然体验路线的基本信息

编号	路线名称	路线起讫点	长度（km）	最低海拔（m）	最高海拔（m）	体验时长（h）	路线难度
S01	古田飞瀑：丛林飞瀑自然体验路线	游线入口—古田飞瀑	1.6（单程）	310	466	1～1.5	大众型
S02	古田山—古田庙：探秘常绿阔叶林自然体验路线	古田山入口—古田庙—古田山出口	5.2	480	914	3～3.5	徒步型
S03	瞭望台：国家公园科研体验路线	游线入口—瞭望台—游线出口	2.1	344	610	1.5～2	专业型
Q01	莲花塘：钱塘江寻源自然体验路线	游线入口—莲花塘—游线出口	4.5	548	930	3～3.5	大众型
Q02	里秧田—大峡谷：遇见峡谷飞瀑自然体验路线	游线入口—瀑布	0.9（单程）	505	646	1～1.5	大众型

注：瞭望台步道主要供科研监测工作使用，普通游客参观需要提前预约获得许可并在解说员的带领下前往。

（1）S01 古田飞瀑：丛林飞瀑自然体验路线

这是很多访客走进钱江源国家公园的第一站。游线位于古田山入口区域。相对于其他游线，这条游线的难度较低，全程皆为宽阔的柏油马路、路程不长、坡度较缓，可以接待包括残疾人、老年人、儿童等大部分访客，也比较适宜开展一些中小规模的自然教育活动，是可接待访客量最多的一条自然体验路线，并且由于以住宿地（古田山庄）为起点，也便于安排清晨或傍晚的一些特色主题活动。在整条自然体验路线上，一共为解说员和访客安排了9个推荐解说点，为来到钱江源国家公园的访客打开认识公园的一扇窗口。

（2）S02 古田山—古田庙：探秘常绿阔叶林自然体验路线

这是认识钱江源国家公园自然生态系统原真性、完整性最重要的一条自然体验路线。古田山位于钱江源国家公园的西南侧，最高点海拔 961m，由山脚的登山口沿着古田山步道向上，可以感受这里典型的常绿阔叶林的面貌。在整条体验路线上，一共为解说员和访客安排了 12 个推荐解说点，分别从森林植被、野生动植物资源和地形、气候等角度解说古田山。

（3）S03 瞭望台：国家公园科研体验路线

这是一条以展示钱江源国家公园科研工作为主的自然体验路线。在整条自然体验路线上，一共为解说员和访客设计了 7 个推荐解说点，旨在引导访客从科学研究的视角，围绕国家公园在森林冠层、24 公顷大样地、全域覆盖的网格化监测、森林碳汇与气候变化等领域开展的研究进行系统的现场讲解，充分认识钱江源国家公园所开展的科研工作的全国性领先地位和创新实践意义。

（4）Q01 莲花塘：钱塘江寻源自然体验路线

这是以探访钱塘江源头为特色的一条自然体验路线。全程约 4.5km，以石头铺就的青石板路为主，多阶梯和小桥，路线沿着莲花溪溯源而上，从起点到终点（莲花塘），海拔高差约 300m。沿途布设有 16 个推荐解说点，主要围绕钱塘江源头的寻找和判定，钱塘江水源地与地形、森林等的关系，公园的野生动植物资源特色等角度展开解说。

（5）Q02 里秧田—大峡谷：遇见峡谷飞瀑自然体验路线

大峡谷位于齐溪镇后山湾村，这条游线路程不长，却是见证钱江源国家公园地质运动的最佳佐证。从大峡谷入口顺着溪流往上，终点为一落差超过 120m 的大瀑布。游线大约有 80% 的路程为青石板铺就的石梯，近大瀑布的一段路线为人工架设的铁梯，坡度较陡，宽仅容纳一人，行走起来略为困难。本条路线需原路返回，沿途一共为访客安排了 9 个推荐解说点，以峡谷飞瀑见证的地质运动为主题，同时还对沿途溪流中生活的鱼类、鸟类等生物以及滴水崖壁这样的典型生境进行了解说。

1.3.2 自然体验路线的分级说明

根据钱江源国家公园的解说资源特点和现状体验路线情况，选择了国家公园 8 条自然体验路线中的 5 条代表性路线作为本次人员解说的对应使用场地。这 5 条线路上的解说点基本涵盖国家公园较为重要和具有代表性的解说资源，解说员可以参照本轮自然体验路线解说点的素材内容，重新调整并组合后应用到其他路线的人员解说活动中。按道路条件情况（坡度、铺面、设施等）、步行体验、攀登难度、适宜人群等因素，钱江源国家公园的自然体验路线被划分为大众型、徒步型和专业型三个级别。本书中涉及的 5 条自然体验路线的难度等级划分及说明如表 2 和图 2 所示。

表2 钱江源国家公园自然体验路线难度分级及与本书的对应情况

	大众型	徒步型	专业型
分级说明	步道铺面平整、坡度较低或总体路线较短，指示完善清晰，沿途休息服务设施完备，行走体验轻松舒适，适合绝大多数参访者体验。	步道铺面较平整、部分路段可能为自然路面，有部分路段坡度较陡但总体路程没有明显障碍，沿途指示较为完善，有必要的休息、服务设施，路程较长，适合有一定的体力和野外徒步经验的访客。	大部分步道保持近自然或原始状态，雨天可能泥泞湿滑，部分或很多路段坡度较陡，沿途较少标识标牌指示和休息、服务设施，路线一般较长，完成有一定的难度和挑战性，适合有较丰富野外徒步登山活动经验的访客。
体验路线	S01 古田飞瀑：丛林飞瀑自然体验路线 Q01 莲花塘：钱塘江寻源自然体验路线 Q02 里秧田—大峡谷：遇见峡谷飞瀑自然体验路线	S02 古田山—古田庙：探秘常绿阔叶林自然体验路线	S03 瞭望台：国家公园科研体验路线

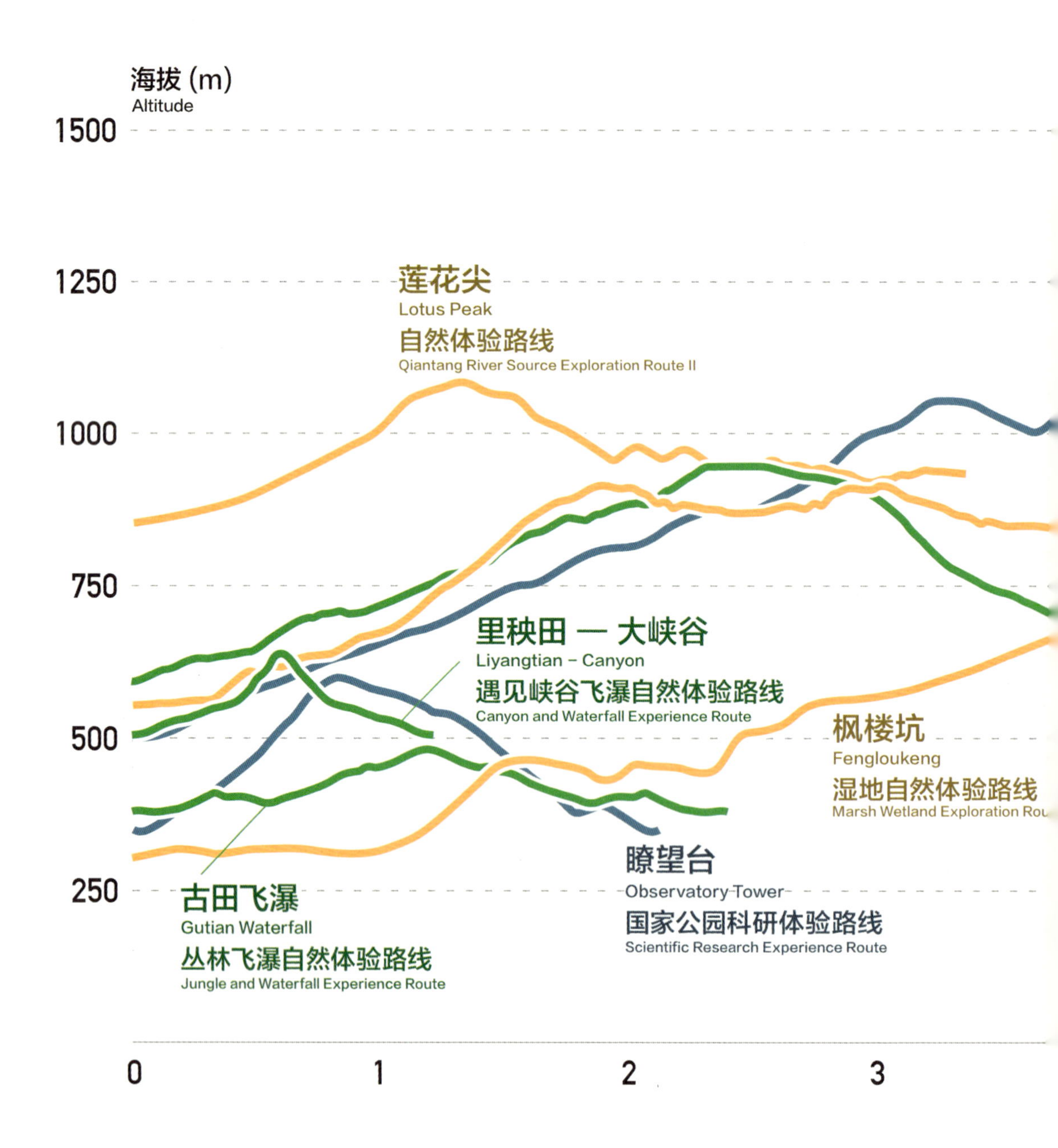

图2　各自然体验路线海拔与里程一览
Figure 2　Elevation and mileage of each nature experience route

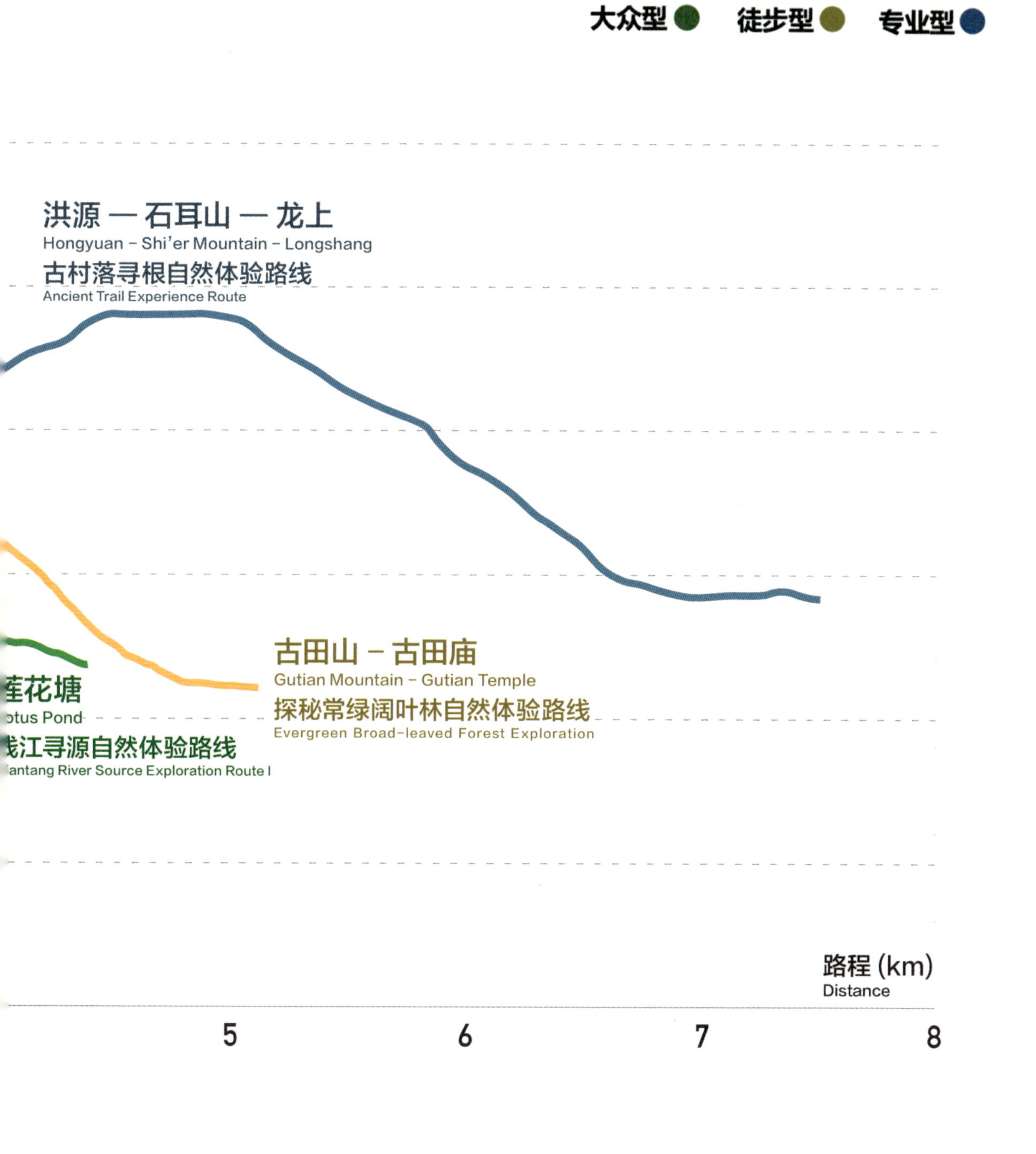
大众型
徒步型
专业型
洪源 — 石耳山 — 龙上
Hongyuan - Shi'er Mountain - Longshang
古村落寻根自然体验路线
Ancient Trail Experience Route
古田山 - 古田庙
Gutian Mountain - Gutian Temple
探秘常绿阔叶林自然体验路线
Evergreen Broad-leaved Forest Exploration
莲花塘
otus Pond
钱江寻源自然体验路线
antang River Source Exploration Route I
路程 (km)
Distance
5
6
7
8

1.3 Nature experience routes interpretation

1.3.1 Nature experience routes

The five nature experience routes of Qianjiangyuan National Park selected in this interpretation manual are located near Gutian Mountain in Suzhuang Town in the southwest of the park, including S01, S02, S03. The source area of the Qiantang River in Qixi Town in the northeast includes two nature experience routes, Q01 and Q02. Each nature experience route corresponds to a unique theme according to its location, resource characteristics, *etc*.. In addition, each nature experience route also includes basic information such as length, altitude, difficulty, and experience duration (Table 1). The respective introductions are as follows.

(1) S01 Jungle and waterfall experience route

This is the first stop for many visitors to enter the Qianjiangyuan National Park. The tour line is located in Gutian Mountain, but compared with S02, this tour line is less difficult. The whole journey is wide open road, the distance is not long, the slope is low, and it is also handicapped, elderly, children friendly, *etc*.. It is also more suitable to carry out some small and medium-scale nature education activities, that may receive the largest number of visitors. On the entire nature experience route, we arranged a total of 9 recommended interpretation points for interpreters and visitors, mainly to open a window for visitors to understand the park.

(2) S02 Evergreen broad-leaved forest exploration route

This is the most important natural exploration route for understanding the authenticity and integrity of the natural ecosystem of Qianjiangyuan National Park. Gutian Mountain is located on the southwest side of Qianjiangyuan National Park, with an altitude of 961 meters. Along the Gutian Mountain trail from the foot of the mountain, you can experience the typical evergreen broad-leaved forest here. In the entire exploration route, we have arranged a total of 12 recommended interpretation points for interpreters and visitors, interpreting Gutian Mountain from the perspectives of forest vegetation, wildlife resources, terrain, and climate.

(3) S03 Scientific research experience route

This is a nature experience route focusing on the scientific research work of Qianjiangyuan National Park. In the entire nature experience route, we have designed a total of 7 recommended interpretation points for interpreters and visitors, aiming to guide visitors from the perspective of scientific research, centering on the national park in the forest canopy, 24 hectares of large sample plots, the research carried out in the fields of grid monitoring with full coverage, forest carbon sinks and climate change, to interpret systematically on site in order to let visitors fully understand the national leading position, innovation and practical significance of the scientific research carried out by Qianjiangyuan National Park.

(4) Q01 Qiantang River source exploration route

This is a natural exploration route featuring visiting the source of the Qiantang River. The total distance is 4.5 kilometers, dominated by stone-paved roads, with many steps and small bridges, and the route traces back to the source along the Lotus Creek.

Table 1 Basic information of the five nature experience routes

Numbering	Route name	Starting and ending points of the route	length (km)	Lowest Elevation (m)	Highest Elevation (m)	Time (h)	Route difficulty
S01	Jungle and waterfall experience route	Trail Entrance – Gutian Waterfall	1.6 (one-way)	310	466	1-1.5	Public
S02	Evergreen broad-leaved forest exploration route	Trail Entrance – Gutian Temple – Trail Exit	5.2	480	914	3-3.5	Hiking
S03	Scientific research experience route	Trail Entrance – Observatory Tower – Trail Exit	2.1	344	610	1.5-2	Professional
Q01	Qiantang River source exploration route	Trail Entrance – Lotus Pond – Trail Exit	4.5	548	930	3-3.5	Public
Q02	Canyon and waterfall experience route	Trail Entrance – Waterfall	0.9 (one-way)	505	646	1-1.5	Public

Notes: The observatory tower trail is mainly used for scientific research monitoring. Ordinary visitors need to make an appointment in advance to get permission and go under the guidance of the interpreter.

From the starting point to the end point (Lotus Pond), the elevation difference is about 300 meters. There are 15 recommended interpretation points along the route, mainly focusing on the search and determination of the source of the Qiantang River, the relationship between the source of the Qiantang River and topography, forests, and the characteristics of the park's wildlife resources.

(5) Q02 Canyon and waterfall experience route

The Great Canyon is located in Houshanwan Village, Qixi Town. It is not a long route, but it is the best evidence of geological movement in Qianjiangyuan National Park. From the entrance of the Grand Canyon, follow the stream up to the end with a large waterfall with a drop of more than 120 meters. About 80% of the tour route is a stone ladder paved with bluestone slabs, and the part near the big waterfall is an artificial iron ladder. The slope is steep and only one-person wide, making it a little difficult to walk. This route is a one-way route. A total of 8 recommended interpretation points are arranged for visitors along the way, with the theme of geological movement witnessed by the waterfalls, and the fish, birds and other creatures living in the streams along the way, as well as the dripping cliffs.

1.3.2 Graded description of nature experience routes

According to the characteristics of the interpretation resources of Qianjiangyuan National Park and the current experience routes, we have selected 5 representative routes out of the 8 nature experience routes in the national park as the corresponding sites for this staff interpretation manual. The interpretation points on these 5 routes basically cover the more important and representative interpretation resources of the national park. The interpreters can refer to the material content of the interpretation points of the nature experience route in this round, re-adjust and combine them to apply to the staff interpretation activities of other routes. According to road conditions (gradient, pavement, facilities, *etc.*), walking experience, difficulty of climbing, applicable people and other factors, the natural experience route of Qianjiangyuan National Park is divided into three levels: public, hiking and professional. The difficulty levels of the 5 natural experience routes in this round of work are as shown in Table 2 and Figure 2.

Table 2 Difficulty classification of the nature experience routes of Qianjiangyuan National Park and the corresponding situation of the staff interpretation manual

	Public	Hiking	Professional
Classification descriptions	The trail is flat, with a low slope or a relatively short route, complete and clear instructions, complete rest service facilities along the way, and a relaxing and comfortable walking experience, which is suitable for most visitors	The pavement of the trail is relatively flat, some sections may be natural roads, some sections have steep slopes but there is no obvious obstacle to the overall journey, the instructions are relatively complete, there are necessary resting and service facilities along the way,and the journey time is longer, suitable for people with certain physical strength and outdoors hiking experience	Most trails remain natural or pristine. Rainy days may be muddy and slippery. Some or many sections of the road have steep slopes. There are fewer signs, resting and service facilities along the way. The routes are generally long, and there are certain difficulties and challenges in walking, suitable for visitors who have more experience in hiking and climbing in the wild
Nature experience routes	S01 Jungle and waterfall experience Route Q01 Qiantang River source exploration route Q02 Canyon and waterfall experience route	S02 Evergreen broad-leaved forest exploration route	S03 Scientific research experience route

1.4 通用型人员解说

通用型人员解说是针对非游线型的解说资源，结合钱江源国家公园的参访道路系统和访客需求，以村庄传统、重要历史遗迹、重要保护资源为体验目标，选定主题体验点。与以固定游线、固定点位为特色的自然体验路线不同，通用型人员解说可适用于多个游线。通用型人员解说分通用型生物多样性人员解说和通用型村庄与传统文化人员解说两大类。其中，通用型生物多样性人员解说共有 29 个解说点，涉及植物、动物和生境等与生物多样性有关的内容，每个解说点都可以根据需要，灵活穿插在其他任何游线中使用，但最好能结合具体的场景；通用型村庄与传统文化人员解说共有 17 个解说点，涵盖乡土物产、乡风民俗、乡村景观等多项内容。建议解说员以村庄为基础，挑选与村落相关的一些解说资源，组成体系进行村落散点的解说。

1.4.1 通用型生物多样性人员解说

通用型生物多样性人员解说是在钱江源国家公园已有的丰富的野生动植物资源的基础之上，选取了其中比较有代表性的一些动植物资源以及它们所栖息的生境进行的解说。生物多样性人员解说一共有 29 个解说点，大致可以分为植物、动物和生境解说三大类。

钱江源国家公园具有丰富的生物多样性，其中，仅高等植物就有 2000 多种。在这里，我们选取了其中比较有代表性的一些植物进行解说，既包括常绿阔叶林的典型树种木荷、甜槠等阔叶树，也包括生活在高海拔的黄山松、马尾松等针叶树；除此之外，钱江源国家公园特殊的地理环境还为许多孑遗植物或者稀有植物提供了庇护所，这里我们选取了长柄双花木、青钱柳、杜仲、南方红豆杉等具有突出个性的、辨识度高且比较有趣的物种进行解说，此外，由于钱江源国家公园气候湿润，苔藓、地衣随处可见，但对于访客来说，它们可以说是既熟悉又陌生的物种，希望通过解说能让访客重新发现和认识身边的微小之物蕴藏的大能量。

得益于复杂的地形、多样的水系、温和的气候和大面积保存完好的原真森林，钱江源国家公园为多种多样的野生动物生存提供了栖息的环境。我们分别在哺乳类、鸟类、两栖类、爬行类、鱼类和昆虫等不同的分类下，选取了一些有代表性的动物进行解说，以期待帮助访客从多个角度认识钱江源国家公园的动物。在动物的选择上会综合考虑其珍稀度、可观察性、独特性以及是否能体现公园的特色或者价值等原则。

自然界中的每一个物种都不是单独存在的，它们与环境、与其他物种之间存在着多种多样的关系。生物多样性人员解说的最后一部分，选择公园比较常见的滴水崖壁、林下阴生世界、溪流和枯枝落叶下的隐秘生境等典型的生境类型进行解说，同时也对竞争和拟态等物种间的关系进行了介绍。

小沼兰

南方红豆杉（国家一级重点保护野生植物）

蚁墙蜂（2014年在古田山发现的新物种）

浙江红花油茶（模式物种）

光唇鱼

开化青蛳

白冠燕尾

白鹇（国家二级重点保护野生动物）

黑麂（国家一级重点保护野生动物）

1.4.2 通用型村庄与传统文化人员解说

钱江源国家公园不仅拥有旖旎的自然风光，同时也保存有许多延续千百年的传统村落。自古以来，山民们靠山吃山、靠水吃水，与这片土地早已形成一种默契。他们耕种着祖辈留下来的土地，在风景和季节变换中，过着自给自足的生活。由于自然地理的天然阻隔，这些村落受到现代文明的冲击和同化较小，一些古老的生产生活方式得以存留下来，体现着当地独特的生活与朴素的价值观。这部分即从乡土美食、民间技艺、传统习俗等“事”和“物”的角度，将这些村落的独特风貌呈现出来。每个村落散点的推荐解说资源如表 3 所示。

清水养鱼　古法养蜂　九山半水半分田
香火草龙　保苗节
高田坑古村落　风干晾晒

表3 国家公园重点参访村落的人文解说要点

乡（镇）	村落名称	人文解说要点（一级）	人文解说要点（二级）
苏庄镇	古田村	01 人口汇聚之地，多元文化之乡 03 好山好水出好茶 11 依山傍水的传统村落 12 村子里最古老的居民 15 保苗节：田间的尊重与爱护	02 九山半水半分田的农业坚守 04 传统手榨茶籽油； 06 古法养蜂 09 乡愁记忆之开化气糕 14 香火草龙：颂赞自然的盛会
	唐头村	01 人口汇聚之地，多元文化之乡 12 村子里最古老的居民	02 九山半水半分田的农业坚守 13 封山育林的传统习俗 14 香火草龙：颂赞自然的盛会
	横中村	02 九山半水半分田的农业坚守 03 好山好水出好茶 12 村子里最古老的居民	08 舌尖上的开化青蛳 14 香火草龙：颂赞自然的盛会 17 在地居民的社区参与与角色转型
长虹乡	高田坑自然村（真子坑村）	07 古法养鱼 11 依山傍水的传统村落 16 守护暗夜星空	02 九山半水半分田的农业坚守 03 好山好水出好茶 05 无辣不欢与风干腌制的智慧 12 村子里最古老的居民
	台回山自然村（桃源村）	02 九山半水半分田的农业坚守 10 竹编技艺的传承 11 依山傍水的传统村落	07 古法养鱼 16 守护暗夜星空 17 在地居民的社区参与与角色转型
	库坑村	01 人口汇聚之地，多元文化之乡 17 在地居民的社区参与与角色转型	11 依山傍水的传统村落 13 封山育林的传统习俗
	河滩自然村（霞川村）	01 人口汇聚之地，多元文化之乡 17 在地居民的社区参与与角色转型	11 依山傍水的传统村落 12 村子里最古老的居民
何田乡	田畈村	02 九山半水半分田的农业坚守 07 古法养鱼 12 村子里最古老的居民	13 封山育林的传统习俗 17 在地居民的社区参与与角色转型
	陆联村	07 古法养鱼 17 在地居民的社区参与与角色转型	12 村子里最古老的居民 13 封山育林的传统习俗
齐溪镇	里秧田村	17 在地居民的社区参与与角色转型	05 无辣不欢与风干腌制的智慧 06 古法养蜂 08 舌尖上的开化青蛳
	仁宗坑村	07 古法养鱼 12 村子里最古老的居民 17 在地居民的社区参与与角色转型	02 九山半水半分田的农业坚守 03 好山好水出好茶 05 无辣不欢与风干腌制的智慧 11 依山傍水的传统村落

1.4 General interpretation

The general interpretation is aimed at a non-route-oriented interpretation resources, combined with the visiting road system of Qianjiangyuan National Park and the needs of visitors, with village traditions, important historical relics, and important conservation resources as the experience goals, and the selected theme experience points. Unlike the natural experience route featuring a fixed line and a fixed point, the general interpretation can be applied to multiple lines. General interpretation is divided into general biodiversity and general villiage and traditional culture interpretation. Among them, there are a total of 29 interpretation points for general biodiversity species interpretation, involving biodiversity-related content such as plants, animals and habitats. Each interpretation point can be flexibly interspersed in any other tour lines as needed, but it is best to combine specific ones. There are 7 interpretation points for general village and traditional culture interpretation, covering local products, folk customs, and rural landscapes. It is recommended that the interpreters use the village as the basis to select some narrative resources related to the village and form a system to interpret the scattered points of the village.

Elliot's pheasant

1.4.1 General interpretation of biodiversity

The general biodiversity interpretation is based on the abundant wild animal and plant resources in Qianjiangyuan National Park, selecting some representative animal and plant resources and the habitats in which they live. There are a total of 29 interpretation points for species interpretations, which can be roughly divided into three categories: plant, animal and habitat interpretations.

Chinese muntjacs

Qianjiangyuan National Park is rich in biodiversity. Among them, there are more than 2,000 higher plants alone. Here we have selected some representative plants for understanding. It includes broad-leaved trees such as gugertree, and sweet oachestnut, which are typical evergreen broad-leaved forests, as well as coniferous trees such as Huangshan pine and masson pine that live at high altitudes. In addition, the special geographical environments at Qianjiangyuan National Park provide the shelter for many relic plants or rare plants. Here we have selected *Disanthus*, Wheel wingnut, *Eucommia*, Chinese yew and other species with outstanding individuality, highly recognizable and more interesting species for interpretation. In addition, because of the Qianjiangyuan National Park has a humid climate, mosses and lichens can be seen everywhere, but for visitors, they can be considered as familiar and unfamiliar species. It is hoped that through interpretations, visitors can rediscover and understand the power of the tiny things around them.

Benefiting from complex terrain, diverse water systems, mild climate and large areas of well-preserved authentic forests, Qianjiangyuan National Park provides a habitat for a variety of wild animals. We have selected some representative animals to interpret from different categories such as mammals, birds, amphibians, reptiles, fish and insects, in order to help visitors understand the animals in Qianjiangyuan from multiple angles. The selection of animals will take into account their rarity, observability, uniqueness and whether it can reflect the characteristics or value of the park.

Every species in nature does not exist alone. What kind of relationship do they have with the environment and with other species? The last part of the biodiversity interpretation selects the typical dripping cliffs, the world under the trees, and streams in the park. It interprets the hidden habitats under the litter, and also briefly introduces the competitive relationship and mimicry between species.

1.4.2 General interpretation of village and traditional culture

Qianjiangyuan National Park not only has beautiful natural scenery, but also preserves many traditional villages that have lasted for thousands of years. Since ancient times, mountain people have relied on mountains and water. They have long formed a tacit understanding with this land. They cultivated the land left by their ancestors and lived a self-sufficient life in the changing landscape and seasons. Due to the natural barriers of natural geography, these villages have been less impacted by modern civilization, and some ancient production and lifestyles have survived, reflecting the uniqueness of local life and simple values. This part presents the unique features of these villages from the perspective of "things" such as local cuisine, folk skills, and traditional customs. The recommended interpretation resources for each village are as shown in Table 3.

Table 3 Basic interpretation contents of key villages visited in Qianjiangyuan National Park

Township	Village name	Recommended interpretation point (Level 1)	Recommended interpretation point (Level 2)
Suzhuang Town	Gutian Village	01 A land of immigrants and diverse cultures 03 Good environment makes perfect tea 11 Traditional villages in the mountains 12 Village sheltered by ancient trees 15 Seedling Conservation Festival	02 Agricultural development in mountain area 04 Traditional handy-made camellia oil 06 Traditional Bee-keeping 09 Taste of steamed rice cake 14 Grass Dragon Dance: A celebration of nature
	Tangtou Village	01 A land of immigrants and diverse cultures 12 Village sheltered by ancient trees	02 Agricultural development in mountain area 13 The tradition of forest conservation 14 Grass Dragon Dance: A celebration of nature
	Hengzhong Village	02 Agricultural development in mountain area 03 Good environment makes perfect tea 12 Village sheltered by ancient trees	08 Taste of Kaihua Snail 14 Grass Dragon Dance: A celebration of nature 17 Local residents engagement and role transformation
Changhong Town	Gaotiankeng Natural Village (Zhenzikeng Village)	07 Ancient fish farming 11 Traditional villages in the mountains 16 Guarding the stars and dark sky	02 Agricultural development in mountain area 03 Good environment makes perfect tea 05 Spicy, dried and preserved food 12 Village sheltered by ancient trees
	Taihuishan Natural Village (Taoyuan Village)	02 Agricultural development in mountain area 10 Inheritance of bamboo weaving skills 11 Traditional villages in the mountains	07 Ancient fish farming 16 Guarding the stars and dark sky 17 Local residents engagement and role transformation
	Kukeng Village	01 A land of immigrants and diverse cultures 17 Local residents engagement and role transformation	11 Traditional villages in the mountains 13 The tradition of forest conservation
	Hetan Natural Village	01 A land of immigrants and diverse cultures 17 Local residents engagement and role transformation	11 Traditional villages in the mountains 12 Village sheltered by ancient trees
Hetian Town	Tianfan Village	02 Agricultural development in mountain area 07 Ancient fish farming 12 Village sheltered by ancient trees	13 The tradition of forest conservation 17 Local residents engagement and role transformation
	Lulian Village	07 Ancient fish farming 17 Local residents engagement and role transformation	12 Village sheltered by ancient trees 13 The tradition of forest conservation
Qixi Town	Liyangtian Village	17 Local residents engagement and role transformation	05 Spicy, dried and preserved food 06 Traditional Bee-keeping 08 Taste of Kaihua Snail
	Renzongkeng Village	07 Ancient fish farming 12 Village sheltered by ancient trees 17 Local residents engagement and role transformation	02 Agricultural development in mountain area 03 Good environment makes perfect tea 05 Spicy, dried and preserved food 11 Traditional villages in the mountains

1.5 使用说明

不同的人员解说手册分类适用于不同的解说场景。

(1)自然体验路线解说的适用场景

适用于固定游线及固定解说点上的人员解说。主要用于对解说步道上主要设施和重要景观的解说。要求解说步道具有较好的解说设施硬件条件和解说资源。

(2)通用型生物多样性人员解说的适用场景

适用于不固定游线及资源匹配解说点上的人员解说。可适用于多个游线，配合具体的场景选择性使用，也可穿插在游线主题解说中间。

(3)通用型村庄与传统文化人员解说的适用场景

适用于不固定游线及资源匹配解说点上的人员解说。可作为游线解说的一部分，也可以单独成为一个体系，用于介绍村落散点。不固定路线，但是需对应村庄或者具体的场景。

本书内容版面的说明如图 3 所示。

01 体验路线地图

提供与该条自然体验路线有关的行进路线、解说牌、步道基础设施配备情况（如交通、休憩亭、卫生间、观景台等），以及人员解说的点位信息，并在地图上显示解说点位在路线上的相对位置。

02 路线高程示意图

结合海拔高度与步行距离，说明自然体验路线长度、坡度和路程、海拔变化等信息。

07 路线名称与所在乡镇

路线名称及所属区域。命名格式为：乡镇名 / 解说点位置 / 自然体验路线

08 解说点与位置

本条路线的解说点序号及其所处的位置。

09 解说主题、要点和场地概述

本解说点的解说目标主题、解说知识要点（通常 2~3 个，与解说词对应）及场地情况的介绍，以及本解说点在整条线路上的相对位置。

10 解说编号

本条路线的解说点编号。命名格式为：乡镇名首字母缩写 - 解说点与位置游线首字母缩写 - 解说点序号

11 地理位置

本地解说点的经纬度与海拔高度信息。
注：此处的海拔高度与实际略有出入，仅供参考。

12 图片与图解

与本解说点相关的图片与图解。

图3 《钱江源国家公园人员解说》内容版面说明

S02 古田山—古田庙：探秘常绿阔叶林自然体验路线 03

解说主题
- 主题1：点亮北纬三十度的生机绿洲
- 主题3：丛林佑护的珍贵原真
- 主题5：体制试点的创新探索

04

步道里程：往返环线约5.2km
体验时长：往返约3~3.5h
最低海拔：480m
最高海拔：914m
步道等级：徒步型
基础设施：休憩亭 / 解说牌 / 活动场地 / 卫生间 / 无餐饮

05

这是认识钱江源国家公园自然生态系统原真性、完整性最重要的一条自然体验路线。古田山位于钱江源国家公园的西南侧，最高点海拔961m，因古籍记载其山畔有田，山中深处有古森林，林中有古田庙，故名古田山。古田山步道由山脚的登山口沿着步道向上，可以感受到亚欧大陆东部典型的亚热带常绿阔叶林的风貌。

古田山的常绿阔叶林是典型的甜槠—木荷林，群落优势种突出，有明显的垂直分层。随着海拔的不断升高，访客可以逐渐感受到常绿阔叶林－常绿落叶阔叶林－针阔叶混交林的植被类型变化。到达古田庙后会发现，那里仍保留着一片形成久远的沼泽湿地。所谓古田，可能就是得名于此。

06

这条体验路线里程较长，上山途中我们多身处其间，可以重点关注沿途的植物群落和物种，而下山途中，则有多个平台可以遥望24hm² 大样地微微起伏的林冠以及远处绵延的群山，让我们从不同的角度去了解这片原始、珍贵的亚热带常绿阔叶林。

在整条体验路线上，我们一共为解说员和访客安排了12个推荐解说点，同时，根据本书提供的通用型解说方案（非定点解说方案）资料，解说员也可以在有解说价值的植物资源等停留点进行相应的引导性观察和解说。

97

03 路线编号与名称

标注路线标志性地点和解说主题，其中S为苏庄镇的首拼音字母，Q为齐溪镇的首拼音字母。

04 相关的解说主题

与钱江源国家公园解说系统规划中的五大主题相匹配，并按与游线契合的重要程度排序。

05 路线基本信息

包括里程、游览时间、海拔高差、游览难度等级和基础设施等。

06 路线整体特色介绍

包括游线的特色梳理、主要资源点介绍，以及整体解说布点情况等信息。

解说编号：SZ-GT-02 10

29°14′53.76″N，118°7′57.45″E
海拔：535m 11

12

解说季节 Interpretation Season
□ 春季 Spring □ 夏季 Summer
□ 秋季 Autumn □ 冬季 Winter
13

解说词 14

如果你的手机自带指南针功能，请大家看一下我们现在所处的位置（海拔535m，北纬29°，东经118°）。设想一下，这个点在地球上处在什么位置？当我们俯瞰整个地球时，绕着北纬30°的环形区域，又是什么样子呢？

「钱江源国家公园之于全球的意义」

其实答案就在我们常用的手机地图中。大家可以尝试打开手机自带的导航地图，然后把图层切换到卫星地图模式（或者阅读随身携带的全球北纬30°的植被分布图），找到我们现在所处的位置，然后沿着北纬30°的方向水平延伸，看一看这条线穿过哪些区域，这些区域的景观又有什么不同。

地球北纬30°的区域基本被黄色的荒漠所覆盖，唯有中国长江中下游地区的这片浓绿呈现截然不同的风貌，而钱江源国家公园正处在这片绿洲之内。

「青藏高原对东亚季风的影响」

那么这片生机绿洲是如何形成的呢？这要从4000万年前青藏高原的隆起开始讲起。如今作为“世界屋脊”的青藏高原在亿万年前曾是一片汪洋大海，因为亚欧板块和印度洋板块的挤压使得青藏高原不断抬升直至今天的高度，改变了全球的地形地貌和气候格局，对中国东部地区的影响尤为巨大。

青藏高原冬季可以阻挡来自北方的寒冷气流，夏季则截留了季风从海洋带来的暖湿气流，为中国东部地区带来了丰沛的降水，再加上这里地处亚热带，热量充足，为常绿阔叶林的生长提供了绝佳的条件，这里也因此成为地球北纬30°黄色荒漠带上独特的生机世界。

15

注意事项

第一个解说点的好坏直接影响后续访客的兴趣，所以必须让他们感受到有趣、特别，你可以：
- 借助地图或者手机APP，让访客更加直观地感受到身处何地，但前提是你对这两者都非常熟悉；
- 切换视角，从全球的北纬30°到中国的北纬30°，重要的是体现出钱江源的独特和珍贵；
- 中国北纬30°的独特环境与青藏高原的隆起带来的典型的季风气候等密不可分。

辅助道具推荐
- 标识牌：北纬三十度的生机绿洲
- 中国地形图

拓展活动推荐
- 钱江源国家公园打卡地图（P562：活动卡02）

拓展知识点 16

「世界的北纬30°」
- 在人类探究自然、认识世界的过程中，北纬30°一直是个充满了神秘和未知的地带。这里是地球至高点珠穆朗玛峰的所在地，尼罗河、幼发拉底河、长江、密西西比河等全球最重要河流的入海口，华夏文明、古印度文明、古巴比伦文明、古埃及文明均分布于此，也是百慕大三角、死海、撒哈拉沙漠等人类认知尚无法解释的神秘区域所在地。

105

13 解说方式与季节

与解说点相关的解说方式配置情况和解说季节建议。

14 主要解说词

与解说点基本对应，分为若干模块。每个模块下有一个相对完整的解说内容。解说员可以根据解说的场地、季节、访客情况等灵活使用这部分内容。

15 注意事项 / 辅助道具推荐 / 拓展活动推荐

针对解说过程中重要环节的解说技巧、道具及活动卡提示，解说员可以根据实际需要进行增减，注意需提前准备。

注：人员解说活动卡具体参考附录。

16 拓展知识点

根据本解说主题提供的拓展知识点，如专有名词解释等。

1.5 Instructions

Different staff interpretation manuals are suitable for different interpretation scenarios.

(1) Applicable scenarios for the nature experience interpretation

It is suitable for staff interpretation on fixed routes and fixed interpretation points. It is mainly used to interpret the main facilities and important landscapes along the trail. The trail should be equiped with good hardware conditions and distribution of interpretation resources.

(2) Applicable scenarios for general interpretation of biodiversity

It is suitable for staff interpretation on non-fixed tour lines and resource-matching interpretation points. It can be applied to multiple tour lines, and can be selectively used in conjunction with specific scenes. It can also be interspersed with tour line theme interpretation.

(3) Applicable scenarios for general interpretation of village and traditional culture

It is suitable for staff interpretation on non-fixed tour lines and resource-matching interpretation points. It can be used as part of the tour guide or can be a separate system to introduce the scattered spots of the village. The route is not fixed, but it needs to correspond to the village or specific scene.

The interpretation of the content layout of this manual is as shown in Figure 3.

01 Route number and name

Mark the landmark locations of the route and the theme of the interpretation, where S is the initial pinyin of Suzhuang Town and Q is the initial pinyin of Qixi Town.

02 Related interpretation themes

Match with interpretation board.

03 Basic route information

Including mileage, tour time, altitude difference, tour difficulty level and infrastructure.

06 Route name and township

Mark the route name and township.

07 Point number and location

Interpretation point number and location of this route.

08 Interpretation number

The interpretation point number of this route, and the naming format is: initical pinyin letters of township name-initial letters of tour line name-interpretation point serial number.

09 Key points and site overview

This interpretation point's interpretation target theme, extra information (usually 2－3, corresponding to the interpretation), introduction of the site situation, and the relative position of this interpretation point on the entire route.

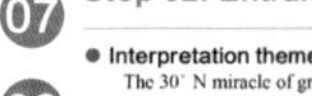

d-leaved
on route

n oasis
forest
ion and exploration

-way and 5.2 kilometers round-trip from
o the destination of Gutian Temple

rea)

/ event site / restroom / no catering

evergreen broad-leaved forest in Gutian untain is dominated by sweet oachestnut – gertree. You can see numerous sweet hestnut, gugertree, Chinese sassafras, jiang camellia, and various ferns along the . As the elevation continues to rise, you clearly feel the alteration among evergreen ad-leaved forest - evergreen-deciduous ad-leaved forest - coniferous and broad-ed mixed forest. When you reach the ian Temple, you will find that there still ains a long-standing marsh wetland. The alled Gutian (ancient field) may be named r it.

This is a relatively long experience route. On our way up the mountain, we can pay attention to the vegetation and species along the way. On our way down, there are multiple platforms that we can overlook the slightly undulating canopy of the entire 24 hectares that enables us to understand this unusual subtropical evergreen broad-leaved forest from different perspectives.

We have arranged a total of 12 recommended interpretation stops for interpreters and visitors during the entire experience route. At the same time, according to the information provided in general interpretation (non-route-oriented interpretation) of this interpretation manual, the interpreter can also find suitable plant resources to conduct corresponding guided observation and interpretation.

05

99

自然体验路线人员解说方案 Interpretation of Nature Experience Routes 02

reen broad – leaved forest exploration route

Interpretation No. SZ-GT-02

08

by the mountain gate

spective
e East Asian monsoon

oard of the trails is right in front of it

lower Yangtze River in China shows a pletely different feature, and Qianjiangyuan nal Park is right within this oasis.

act of the Qinghai–Tibet Plateau on East Asian Monsoon"

ow did this vibrant oasis form? This should back to the rise of the Qinghai – Tibet au 40 million years ago. The Qinghai – Plateau, which is now the "roof of the d", was a vast ocean hundreds of millions ars ago. The Tibetan Plateau kept rising o the squeezing of the plate, which has ged the global landform and climate n, especially on eastern China.

Qinghai – Tibet Plateau can block the cold w from the north in winter, and intercepts warm and humid airflow brought by the soon from the ocean during summer, g the southeast China in a relatively warm humid state, and create a rare continuous distribution of a large-scale subtropical evergreen broad-leaved forest in this low-elevation area, becoming a miracle of green oasis on the 30° north latitude yellow desert belt.

11

12

Attentions

The quality of the first interpretation directly affects visitors' interest for later, so you must keep them interested and special. You can:

- By using maps or mobile apps, visitors can feel where they are more intuitively, but you have to be familiar with both;
- By switching angles, from 30° north latitude around the world to 30° north latitude in China, it is important to reflect the uniqueness and preciousness of Qian Jiangyuan;
- The unique environment of 30° north latitude in China is inseparable from the rise of the Tibetan Plateau.

Recommended auxiliary facilities and toolkits

- Global map of 30° north latitude vegetation distribution
- The mobile phone built-in compass (check how to operate mainstream mobile phone brands in advance)
- Mobile phones built-in maps (Baidu, Gaode, *etc.*)
- A Chinese geomorphic model made of paper clay (highlighting the Tibetan Plateau and 30 degrees north latitude) or find a slope on site to demonstrate the blockage of the hot and cold air flow by high mountains

Recommended activities

- Task list map of Qianjiangyuan National Park (P562 ; No. 02)

Extra information

"30° north latitude of the world"

- It starts from the Sahara Desert in Africa, to the Arab region in Asia, then to the Tibetan Plateau in our country, the middle and lower Yangtze River Plain, and to the south of North America. Throughout the exploration of mankind, the 30° North latitude has always been a zone full of mystery and unknown. This is where Mt Everest stands, the worldwide essential rivers flow into the sea, as well as all the ancient civilizations and mysterious landscapes beyond our knowledge originate.

13

107

04 Introduction to the overall characteristics of the route

Including the characteristics of the tour line, the introduction of main resource points, and the overall interpretation of the distribution of points and other information.

05 Pictures and illustrations

Pictures and illustrations related to this interpretation.

10 Main interpretation

It basically corresponds to the interpretation point and is divided into several modules. There is a relatively complete interpretation under each module. The interpreter can use this part of the content flexibly according to the site, season, and visitor situation.

11 Pictures and illustrations

Pictures and illustrations related to this interpretation.

12 Attentions/Recommended auxiliary facilities and toolkits/ Recommended activities

Interpretation skills, toolkits and activity card tips during the interpretation process. Can be increased or decreased according to actual needs, and be prepared in advance.

* Please refer to the appendix for the staff activity card.

13 Extra information

Extra information related to this interpretation, such as explanation of a proper noun.

02

自然体验路线人员解说方案

Interpretation of Nature Experience Routes

2.1 S01

古田飞瀑：
丛林飞瀑自然体验路线
Jungle and waterfall experience route

古田飞瀑
Gutian Waterfall
路边平台二
Platform Number Two
种子雨收集器
Seed Rain Collector
路边平台一
Platform Number One
入口
Entrance
图例 LEGEND
钱江源国家公园
Qianjiangyuan National Park
水域
Water
等高线
Contours
步行道
Trail
车行道
Road
服务中心
Visitor Center
卫生间
Toilet
观景处
Recreation Spot
瀑布
Waterfall
设解说牌的点位
Point with Signage
设解说牌、人员解说的点位
Point with Signage and Interpretation
设解说牌、人员解说、课程的点位
Point with Signage, Interpretation and Course
0 100m 200m 300m
N
600 海拔(m)
500
400
300
200
100
路边平台一
路边平台二
古田飞瀑
(返程)
路程(km)
0 1 2 3

S01 古田飞瀑：丛林飞瀑自然体验路线

解说主题	主题 1：点亮北纬三十度的生机绿洲 主题 3：丛林佑护的珍贵原真 主题 5：体制试点的创新探索
步道里程	单程约 1.6km
体验时长	1～1.5 小时
最低海拔	310m（游线入口）
最高海拔	466m(古田飞瀑)
步道等级	大众型
基础设施	解说牌 / 活动场地 / 卫生间 / 无餐饮

作为中国首批国家公园体制试点单位，也是长江三角洲地区首个国家公园，钱江源国家公园对我国国家公园体制建设具有重要的探索和示范价值。国家公园和国家级自然保护区、森林公园、风景名胜区等不同，它是由国家批准设立，以保护具有国家代表性的大面积自然生态系统为主要目的，在我国全新设立的三级自然保护地体系中居于最高等级。这是每一个来到钱江源的访客首先需要了解的。

本条自然体验路线位于古田山脚下，是很多访客走进钱江源国家公园的第一站。本线单程约1.6km，全程均为柏油马路，路面开阔，坡度较缓，适合接待残疾人、老人、儿童等大众访客，也适宜开展一些中小规模的环境教育活动，具有较高的访客接待容量，可以作为大多数访客认识钱江源国家公园自然生态系统原真性、完整性的一个窗口。沿着步道往上，访客可以初步感受钱江源国家公园典型的常绿阔叶林生境。古田山的常绿阔叶林以甜槠、木荷等常绿阔叶树为优势种，其中混生有马尾松等针叶树。幸运的话，访客还有可能会遇到白鹇等当地的“明星”物种。

在整条自然体验路线上，一共为解说员和访客安排了 9 个推荐解说点，大部分都匹配有相应的解说牌，既可以为解说提供辅助信息，也可以帮助定位解说点。同时，根据本书提供的通用型解说方案（非定点解说方案）资料，解说员也可以在合适的植物资源等停留点进行相应的引导性观察和解说。

S01 Jungle and waterfall experience route

Themes
- 01 The 30°N miracle of green oasis
- 03 Mysterious wildlife in the forest
- 05 Scientific research innovation and exploration

Trail Length It is about 1.6 kilometers from Gutian Hotel to Gutian Waterfall
Time 1 – 1.5 hours
Lowest elevation 310 meters (entrance of the trail)
Highest elevation 466 meters (Gutian Waterfall)
Level of trail Public
Infrastructure Interpretations board / Event sites / Restrooms / No catering

As one of the first national park systems in China and the first national park in the Yangtze Delta, Qianjiangyuan National Park has important exploration and demonstration value for national park system construction. National parks are different from national-level nature reserves, forest parks, natural scenery, *etc*.. It is established by the approval of the country, with the main purpose of protecting large-scale natural ecosystems that are representative in China. And it is the highest level of China's nature reserve system. This is what every visitor who comes to the national park needs to know in the first place.

This nature experience route is located at the foot of Gutian Mountain. Starting from Gutian Hotel, it is the first stop for many visitors to enter Qianjiangyuan National Park. It is about 1.6 kilometers one way. There are wide pavements along the way with relatively smooth slopes. It can accommodate most visitors including the disabled, the elderly, children, *etc*.. It is also suitable for some small and medium-sized environmental education activities. The reception capacity is also a window for most visitors to understand the authenticity and integrity of the natural ecosystem of Qianjiangyuan National Park.

Visitors can have a taste of the typical evergreen broad-leaved forest habitat of Qianjiangyuan National Park along the trail. The evergreen broad-leaved forest of Gutian Mountain is dominated by evergreen broad-leaved trees such as sweet oachestnut-gugertree and other coniferous trees such as masson pines. If you are lucky, you may encounter local "star" species such as silver pheasants.

We have arranged a total of 9 recommended interpretation stops for interpreters and visitors during the entire experience route, most of which have matching interpretation signs, which can provide extra information for the interpretation and help locate the interpretation point. At the same time, according to the information provided in general interpretation (non-route-oriented interpretation) of this interpretation manual, interpreters can also find suitable resources to conduct corresponding guided observations and interpretations.

苏庄镇 | 古田飞瀑 | 丛林飞瀑自然体验路线　　解说编号：SZ-GTFP-01

解说点 01 集合点

29° 13′ 34.87″ N, 118° 6′ 37.61″ E
海拔：310m

● 解说主题

本条自然体验路线的总体介绍

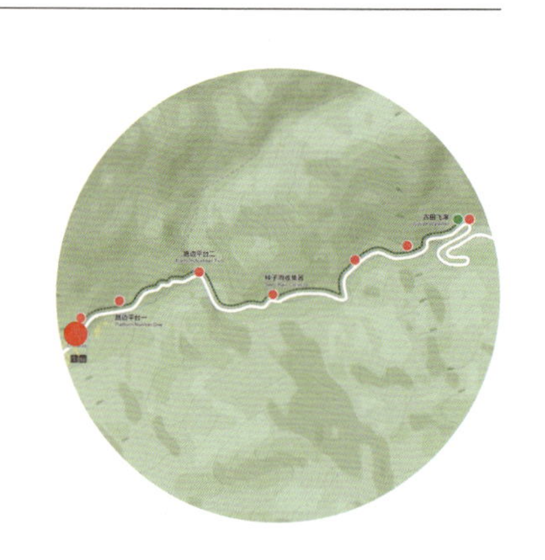

● 解说要点

1. 古田山游线的整体介绍
2. 确认听众的状态、树立自身解说员的身份

● 场地概述

入口安装有步道总体导览牌的地方

解说方式 Interpretation Style

- ☑ 人员解说 Staff Interpretation
- ☐ 教育活动 Educational Activity
- ☑ 解说牌 Interpretation Signage
- ☐ 场馆解说 Hall Interpretation

解说季节 Interpretation Season

- ☑ 春季 Spring
- ☑ 夏季 Summer
- ☑ 秋季 Autumn
- ☑ 冬季 Winter

解说词

各位朋友好！欢迎大家来到钱江源国家公园，我是公园的解说员 XXX，接下来将由我来陪伴大家一起探索这段自然体验的旅程。

我们现在所在的位置是钱江源国家公园古田山片区的主入口。在钱江源国家公园建立之前，古田山就已经是国家级自然保护区了。那么，有没有人有疑问，既然已经是国家级自然保护区，为什么还要建国家公园？国家公园和自然保护区、森林公园、风景名胜区等概念又有什么不同呢？大家可以思考一下，待会儿我们来解答这个问题。我先给大家介绍下整条自然体验路线。

今天的这条路线为往返单线，从这里到终点古田飞瀑仅1.6km 左右。该线全程就像大家现在脚下的路一样，均为宽阔的柏油马路，坡度较缓，一路我们还会走走停停，所以几乎不觉得费力。在钱江源国家公园各有特色的自然体验路线中，这是交通最为便利的一条，也是我们认识国家公园的一个重要窗口。沿途我会和大家聊一聊为什么要在这里建设国家公园，钱江源有哪些自然生态或者物种需要守护，对科学家来说公园又有哪些鲜为人知的独特价值……最后，希望大家此行结束前，能认识几种公园有代表性的动物和植物，并初步掌握识别它们的一些简单方法，如果下次有机会再见面，它们就会成为我们亲切的老朋友了！

好，闲话不多说，请大家热热身，我们准备上山喽！要提醒大家的是，虽然这段路比较宽阔平坦，但偶尔仍然会有车辆经过，大家一定要注意安全，靠路边行走。另外，由于这里终年空气湿润，路面会有些湿滑，大家要小心不要摔倒。最后，请大家千万别忘记我们是在野外！因为有很多危险是我们提前察觉不到的，所以请大家一定紧跟着我，有序地排队前进，不要独自行动，更不要脱离团队独自行动！

我们出发吧！

注：解说集合点通常设在步道起点有明显标志的集合场所，此处建议设置在本步道的总体导览牌设立处。

注意事项

请注意集合点的开场。当你出现在听众们的面前，解说员的工作已经开始了。如何吸引听众的注意力，树立自身的解说员身份，一个简短、内容明确、带有互动的开场白对解说员来说非常重要。以下几点提示可以帮你达到这一目标。

- 选择一个合适的、可以帮助大家集合并开始这次自然体验的集合地点，比如，步道起点总体导览牌边上的一块空地。
- 给自己起一个容易记忆的自然名和简短有趣的自我介绍，确保大家一开始就记住你。
- 体现你身份的制服和服饰，解说员的标志和随身的解说装备包。
- 你的介绍中一定有独特的、标识标牌上未曾涉及的内容。
- 为即将开始的自然体验之旅略作“剧透”，包括主题、时间、特别值得期待的解说点等，但不要和盘托出，而是引发大家的好奇心。
- 根据团队成员的构成，针对性地提出注意事项和安全要求。
- 记得还要提醒大家注意不采摘和带走任何公园自然物，不随便丢弃垃圾等行为准则哦。

拓展活动推荐

- 钱江源国家公园打卡地图 (P562：活动卡 02)

Suzhuang Town | Gutian Waterfall | Jungle and Waterfall Experience Route

Interpretation No. SZ-GTFP-01

Stop 01: Assembly point

Interpretation theme

General introduction to this nature experience route

Key points

1. The overall introduction of the Gutian Mountain Tour
2. Confirm the status of the audience and establish the identity as an interpreter

Site overview

Parking lot, where a general guide board for trails is installed

Hello friends! Welcome everyone to Qianjiangyuan National Park, I am the interpreter of the park, XXX, and I will be with you to explore this nature experience route.

Our current location is the main entrance to the Gutian Mountain area of the Qianjiangyuan National Park. Before the establishment of Qianjiangyuan National Park, Gutian Mountain was already a national nature reserve. So, does anyone have a question, since it is already a national-level nature reserve, why is a national park built? What are the differences between national parks and nature reserves, forest parks, and scenic spots? You can think about it, and I will answer this question later. Now, let me introduce you the whole nature experience route first.

Today's route is a round trip, about 1.6 kilometers long, and the destination is Gutian Waterfall. We have a wide asphalt road with a gentle slope during the whole trip. There are multiple stops along the way, so we won't feel tired at all. Among the unique nature experience routes of Qianjiangyuan National Park, this is the most convenient one for transportation, and it is also an important window for us to know the national park. Along the way, I will talk to everyone about why we build a national park here, what natural habitat or species Qian Jiangyuan need to protect, what are the little-known unique values of the park for scientists… Finally, I hope that before the end of this trip, you can recognize some representative animals and plants of the park, and master some simple methods to identify them. If you have the opportunity to meet again next time, they will become kind old friends !

Well, let's warm-up, we are going to hike! I need to remind you that although this part of the road is relatively wide and flat, vehicles

will still pass by occasionally. You must pay attention to your safety and walk along the roadside. Besides, due to the humid air throughout the year, the road surface is slippery, so everyone should be careful not to fall. Finally, please don't forget that we are in the wild, there are many dangers that we couldn't be aware of in advance, so please follow me, do not act alone, and do not leave the team!

Let's go!

Suggestions: The interpretation assembly point is usually set at the beginning of the trail with obvious signs. It is recommended to set it at the general guide board of this trail.

Attentions

Remember, pay attention to your debut at the assembly point. When you show up to the audience, your job as an interpreter has already begun. As an interpreter, it is very important to attract the attention of the audience and establish your own personality, such as using a short, clear, and interactive opening statement. The following tips can help you achieve this goal.

- A proper assembly place that can help gathering the crowd and start this nature experience, such as the open space next to the general guide sign at the start of the trail.
- Give yourself an name that's easy to remember as well as a short but interesting self-introduction to make sure the tourists remember you from the beginning.
- Uniforms and costumes that reflect your identity: the logo of the interpreter and supporting materials tool kit.
- There must be something unique in your introduction that isn't mentioned on the interpretation board.
- Spoil some of the upcoming natural experience trip, including the theme, time, and particularly worthy interpretation, *etc*.. in order to arouse everyone's curiosity.
- Based on specific team members, pose targeted safety requirements and Attentions.
- Don't forget to remind everyone not to pick and take away any natural objects in the park, and don't throw garbage on the ground, *etc*.

Recommended activities

- Task list map of Qianjiangyuan National Park (P562: No. 02)

苏庄镇 | 古田飞瀑 | 丛林飞瀑自然体验路线　　解说编号：SZ-GTFP-02

解说点 02 入口

29° 13′ 36.57″ N，118° 6′ 41.92″ E
海拔 314m

● 解说主题

认识钱江源国公园及其保护价值

● 解说要点

1. 认识钱江源国家公园
2. 常绿阔叶林与北纬 30° 和长江三角洲的关系

● 场地概述

古田山入口处（近古田山庄）

解说方式 Interpretation Style

☑ 人员解说 Staff Interpretation　☐ 教育活动 Educational Activity
☑ 解说牌 Interpretation Signage　☐ 场馆解说 Hall Interpretation

解说季节 Interpretation Season

☑ 春季 Spring　☑ 夏季 Summer
☑ 秋季 Autumn　☑ 冬季 Winter

解说词

各位朋友，还没有进山，大家是不是已经感受到一股原始丛林的气息迎面扑来？不知不觉间，我们就已融入山林，成了钱江源国家公园的一部分，一时间甚至忘记公园正处于繁华的长江三角洲范围内。不过解说牌上的这张地图，或许可以帮我们对钱江源国家公园的地理位置有个大体的认识。

「认识钱江源国家公园」

钱塘江自西向东在杭州湾流入东海，源头位于浙江省西部的开化县，与安徽省的休宁县、江西省的婺源县等相邻，既是三省七县、多文化交融的过渡地带，也是浙江和长江三角洲地区重要的生态屏障与水源涵养区。

钱江源国家公园总面积达 252km²，包括开化县的苏庄、长虹、何田和齐溪等 4 个乡镇，人口近万人。辖区范围整合了多个不同类型和等级的自然保护地，我们现在所在的古田山前身就是一个国家级自然保护区。

「北纬三十度的生机绿洲」

作为钱江源国家公园最具代表性的景观，这里分布着全球最为典型的亚热带低海拔地区的常绿阔叶林。它们像一块巨大的绒毯一样，覆盖着公园的大部分土地，并随着山势连绵起伏，蔚为壮观。作为访客，我们既可以身处其中，感受原始森林的静谧与幽深，也可以选择一处视野开阔的地方，从高处俯瞰这一片生机绿洲。

此外，由于现代科技的进步，我们甚至可以借助地球卫星，从地球之外，更加全方位、多角度地感受地球的整体面貌。大家的手机如果有信号，可以打开自己常用的地图软件，并切换到卫星地图模式，找到我们现在所处的位置，然后缩小比例，在一个全球视角下沿着同一纬度（水平）线分别向东和向西，看一看这条线会穿过哪些区域，这些区域又呈现怎样的自然景观。

有没有发现：这条环带区域基本被橙色和黄色的荒漠所覆盖，唯有中国长江中下游地区呈现截然不同的风貌？钱江源国家公园正是这条地球的北纬 30°“黄腰带”上的醒目的绿色“窗口”。

注意事项

第一个解说点的好坏直接影响后续访客的兴趣，所以必须让他们感受到有趣、特别。你可以：

- 借助地图或者手机 APP，让访客更加直观地感受到身处何地，但前提是你对这两者都非常熟悉；
- 切换视角，从全球的北纬 30°到中国的北纬 30°，再到长江三角洲，重要的是体现出钱江源的独特和珍贵。

辅助道具推荐

- 标识牌：长江三角洲的首个国家公园
- 标识牌：北纬三十度的生机绿洲

那么这片生机绿洲是如何形成的呢？这要从 4000 万年前青藏高原的隆起开始讲起。

如今被誉为“世界屋脊”的青藏高原在亿万年前曾是一片汪洋大海，因为地理板块的相互挤压不断抬升，形成青藏高原。它改变了全球的地形地貌和气候格局，对中国东南部地区的影响尤为巨大。

青藏高原冬季可以阻挡来自北方的寒冷气流，夏季则截留了季风从海洋带来的暖湿气流，使得中国东南部地区处于相对温暖湿润的状态，也造就了这里连续分布的大面积亚热带常绿阔叶林，因此成为北纬 30° 黄色荒漠带上独特的生机世界。

「长江三角洲最后的原真森林」

长江三角洲地区地势平坦、气候温暖湿润、河湖密布，自古以来就是我国重要的鱼米之乡和对外交往的门户。这样的地形和气候条件，也为常绿阔叶林的生长创造了极佳的条件。我国亚热带常绿阔叶林的分布区域达到 250 万 km^2，而且大都在中高海拔山地，但由于长期的人类活动影响，人口和城市迅速扩张，逐渐蚕食着原始的自然景观，仍然保持原生状态的森林几乎消失殆尽。像钱江源国家公园这样位于低海拔地区、大面积连续分布的、尚未被人类活动明显干扰的天然常绿阔叶林，在中国尤其是长江中下游地区十分罕见，因而也倍显珍贵。

拓展知识点

「国家公园的概念」

- 根据《建立国家公园体制总体方案》，国家公园是指由国家批准设立并主导管理，边界清晰，以保护具有国家代表性的大面积自然生态系统为主要目的，实现自然资源科学保护和合理利用的特定陆地或海洋区域。
- 国家公园和自然保护地、自然公园分列为我国自然保护地体系的三级，其中，国家公园是等级最高，也是最重要的自然保护地类型，属于全国主体功能区规划中的禁止开发区域，纳入全国生态保护红线区域管控范围，实行最严格的保护。而自然公园涵盖了原来的森林公园、湿地公园、风景名胜区等各种类型的保护地。

「世界的北纬 30°」

- 北纬 30° 一直是个充满了神秘和未知的地带。这里是地球至高点珠穆朗玛峰的所在地，尼罗河、幼发拉底河、长江、密西西比河等全球最重要河流的入海口，以及华夏文明、古印度文明、古巴比伦文明、古埃及文明均分布于此，百慕大三角、死海、撒哈拉沙漠等人类认知尚无法解释的神秘区域也在此。

Suzhuang Town | Gutian Waterfall | Jungle and Waterfall Experience Route

Interpretation No. SZ-GTFP-02

Stop 02: Entrance

Interpretation theme

Understand Qianjiangyuan National Park and its protection value

Key points

1. Meet Qianjiangyuan National Park

2. The relationship between the evergreen broad-leaved forest of Qianjiangyuan National Park with 30 degrees north latitude and the Yangtze River Delta.

Site overview

The entrance of Gutian Mountain (near Gutian Hotel)

Dear friends, before you enter the mountain, you have already felt the breath of a primitive jungle? Gradually, we have integrated into the mountain forest and become part of the national park, and forget that we are at the prosperous Yangtze River Delta. However, this map may help us have a general understanding of the geographical location of Qianjiangyuan National Park.

"To understand Qianjiangyuan National Park"

The Qiantang River flows from the west to the east into the East China Sea in Hangzhou Bay. The source is located in Kaihua County in the west of Zhejiang Province. It is adjacent to Xiuning County in Anhui Province and Wuyuan County in Jiangxi Province as well as an important ecological barrier and water conservation area in the Yangtze River Delta region.

The total area of the park is 252 square kilometers, involving Suzhuang, Changhong, Hetian, and Qixi in Kaihua County. Several different types and levels of nature reserves are integrated into the jurisdiction, including the Gutian Mountain National Nature Reserve, Qianjiangyuan National Forest Park, which is famous for its scientific research achievements, and the Qianjiangyuan Provincial Scenic Area, Chinese Giant Salamande Provincial Aquatic Germplasm Resource Protection Area, *etc*..

"The miracle of green oasis at 30 degrees north latitude"

As the most representative landscape of Qianjiangyuan National Park, the evergreen broad-leaved forest of the park is rolling up and down with the mountains, and it looks very spectacular from afar. The "gardener" behind the dense jungle is far at the Qinghai-Tibet Plateau. Do you know why?

You can take a look at the map of vegetation distribution at 30 degrees north latitude, or open the built-in map to satellite pattern on your phone, find where we are now, and then draw a straight line along the 30 degrees north latitude to see what areas it crosses. How is the landscape different in these areas?

We will find that this ring zone is basically covered by orange and yellow deserts, but this thick green in the middle and lower Yangtze River in China shows a completely different feature, and Qianjiangyuan National Park is the oasis on that yellow belt.

So how did this vibrant oasis form? This should trace back to the rise of the Qinghai-Tibet Plateau 40 million years ago. The Qinghai-Tibet Plateau, which is now the "roof of the world", was a vast ocean hundred of millions of years ago. The Tibetan Plateau kept rising due to the squeezing of the plate, which has changed the global landform and climate pattern, especially in eastern China. The Qinghai-Tibet Plateau can block the cold airflow from the north in winter and intercepts the warm and humid airflow brought by the monsoon from the ocean during summer, leaving the southeast China in a relatively warm and humid state, and create a rare continuous distribution of a large-scale subtropical evergreen broad-leaved forest in this low-elevation area, becoming a miracle of a green oasis on the 30 degrees north latitude yellow desert belt.

Attentions

The quality of the first interpretation directly affects visitors' interest for later, so you must keep them interested and special. You can:

- By using maps or mobile apps, visitors can feel where they are more intuitively , but you have to be familiar with both;
- Switching the angle of view, from 30° north latitude in the world to 30° north latitude in China to the Yangtze River Delta. It is important to reflect the uniqueness and preciousness of Qian Jiangyuan.

Recommended auxiliary facilities and toolkits

- Interpretation signage: The first national park in the Yangtze delta
- Interpretation signage: The 30° N miracle of green oasis

"The last primitive and indigenous forest in the Yangtze River Delta"

The Yangtze River Delta has flat terrain, warm and humid climate, and dense rivers and lakes. Since ancient times, it has been an important hometown of fish and rice production and a window to foreign exchanges. Such terrain and climatic conditions have also created excellent conditions for the growth of the evergreen broad-leaved forest. The distribution area of subtropical evergreen broad-leaved forests in China has reached 2.5 million square kilometers. However, due to the long-term human activities, the rapid expansion of population and cities has gradually eroded the original natural landscape. The forests that have remained in their original state have almost disappeared mostly in the middle and high altitude mountains. Forest like the Qianjiangyuan National Park, a natural evergreen broad-leaved forest located in a low-elevation area and continuously distributed over a large area that has not been significantly disturbed by human activities is very rare in China, especially in the middle and lower reaches of the Yangtze River.

Extra information

"Concept of National Park"

- According to the *Overall Plan for the Establishment of National Park System*, a national park refers to a specific land or ocean area that is approved for establishment and managed by the country, with clear boundaries and the main purpose of protecting a large area of the country's representative natural ecosystem, and realizing the scientific protection and rational use of natural resources.
- National parks, nature reserves, and natural parks belong to three levels of China's nature reserve system. Among them, the national park is the highest and most important type of nature protected areas. It belongs to the prohibited areas for development in the planning of the main functional areas of the country. and is under strictest protection implemented in the national ecological protection red line area. Natural parks cover various types of protected areas such as original forest parks, wetland parks, and scenic areas.

"30 degrees north latitude of the world"

- Throughout the exploration of mankind, the 30 degrees north latitude has always been a zone full of mystery and unknown. This is where Mount Everest stands, the worldwide essential rivers flow into the sea, as well as all the ancient civilizations and mysterious landscapes beyond our knowledge originate.

苏庄镇 | 古田飞瀑 | 丛林飞瀑自然体验路线　　解说编号：SZ-GTFP-03

解说点 03 路边平台一

29° 14′ 42.82″ N，118° 6′ 51.37″ E
海拔 325m

● **解说主题**

钱江源国家公园的自然环境对生物多样性的贡献

● **解说要点**

1. 地形与气候对生物多样性的影响
2. 丰富的植物区系与地理环境的关系

● **场地概述**

近瞭望台分叉口的小平台

解说方式 Interpretation Style

☑ 人员解说 Staff Interpretation　☐ 教育活动 Educational Activity
☑ 解说牌 Interpretation Signage　☐ 场馆解说 Hall Interpretation

解说季节 Interpretation Season

☑ 春季 Spring　☑ 夏季 Summer
☑ 秋季 Autumn　☑ 冬季 Winter

各位朋友，大家都是从哪里来的？来到钱江源有什么感受呢？是什么造就了钱江源国家公园这样与众不同的环境呢？我们一起通过这张钱江源国家公园的地形图来了解一下。

「古田山的地形对生物多样性的影响」

钱江源国家公园位于浙江、安徽、江西三省交界处，是典型的中山、低山丘陵区。这里山系绵延，河湖众多，仅开化县境内就有 3 条主要的山脉，分别为白际山、怀玉山和千里岗。山区不同海拔、坡度和坡向等地形地貌的组合，带来的是温度、光照、水分、养分等环境条件的差异，为多种生物的生存提供了更丰富的选择。

「古田山的气候对生物多样性的影响」

此外，气候环境也是影响物种分布和丰富程度的重要因素。从全球尺度来看，青藏高原的隆起形成了亚洲东部这一大片典型的季风气候区。每年夏季，从太平洋吹来的湿暖气流经过公园时，会被高大连绵的群山阻挡，在迎风坡产生降水；冬季，来自亚欧大陆内部的西伯利亚寒流南下，同样会受到公园高大山系的阻挡，使得这里的冬季气温明显高于长江中下游平原其他地区。

「公园丰富的植物区系与地理环境的关系」

从地理位置上来看，钱江源国家公园刚好位于我国华北和华南、华东与华中地区交界的“十字路口”，为各种植物的汇聚创造了可能；另外，钱江源国家公园复杂的山势格局塑造了多种多样的局地小气候，这里不仅生活有偏爱温凉环境的桑树、构树等来自华北植物区系的落叶阔叶树种，也有喜好湿热环境的华南植物区系植物木莲、野含笑和猴欢喜等，同时西部地区还通过一些东西走向的山岭、水系与西南、华中植物区系相连，是青钱柳、八角莲等孑遗植物传播的重要途径。

「复杂环境孕育的天然物种基因库」

钱江源国家公园 98% 以上的土地都被森林覆盖，分属钱塘江和长江两大水系，公园的森林、河谷和湿地为两千多种野生动植物提供了栖居的家园，是中国特有的世界珍稀濒危物种、国家一级重点保护野生动物白颈长尾雉、黑麂的主要栖息地，也是南方红豆杉、长柄双花木、青钱柳、杜仲等古老的孑遗植物世代生息繁衍的场所。

「国家公园科研体验路线推荐」

（在瞭望台分叉口附近空旷地停留）顺着这条路线往上，大家可以通过沿途的一些科学研究设施，如塔吊、种子雨收集器、瞭望台等，真切地感受国家公园在科研创新领域所做的一些实践和探索。感兴趣的朋友稍后一定不要错过这条路线。

Suzhuang Town | Gutian Waterfall | Jungle and Waterfall Experience Route

Interpretation No. SZ-GTFP-03

Stop 03: Roadside platform#1

● Interpretation theme

The contribution of Qianjiangyuan National Park's natual environment to its biodiversity conservation

● Key points

1. Effects of topography and climate on biodiversity in Qianjiangyuan National Park

2. The relationship between the rich flora and the geographical environment in Qianjiangyuan National Park

● Site overview

A small platform near the fork to the observatory tower

Dear friends, where do you all come from? I heard many friends from Zhejiang, Shanghai, Anhui and other places. How do you feel about coming to Qianjiangyuan? Mountainous, trees, humid, undulating terrain... so what makes the Qianjiangyuan National Park so unique? Let's take a look at this map together.

"The impact of the topography of Gutian Mountain on biodiversity"

Qianjiangyuan National Park is located at the junction of the three provinces of Zhejiang, Anhui and Jiangxi, and is a typical hilly area with moderate and low mountains. This area has a long mountain range and many rivers and lakes. There are three main mountain ranges in Kaihua County, namely Baiji Mountain, Huaiyu Mountain and Qianligang. The combination of different topography and landforms in mountainous areas, such as temperature, light, moisture, nutrients, and other environmental conditions, provides a variety of choices for the survival of various organisms.

"The impact of the climate of Gutian Mountain on biodiversity"

In addition, the climate environment is also an important factor that affects the distribution and richness of species. On a global scale, due to the uplift of the Qinghai-Tibet Plateau, a large monsoon climate region in eastern Asia is formed. In summer, when the humid heating current from the Pacific Ocean passes through the park, it will be blocked by the high mountains and generate precipitation on the windward slope. The blockade makes the winter temperature here significantly higher than other areas in the middle and lower reaches of the Yangtze River Plain. The climate is hot and humid in summer, warm in winter, distinct seasons throughout the year, and abundant rainfall provides golden climatic conditions for the growth of the subtropical evergreen broad-leaved forest.

"Relationship between park's rich flora and geographical environment"

Although Qianjiangyuan National Park is located in East China, it is a collection of typical vegetation from East, North, Central, and South China, and can be called the "Natural Museum of Plant Diversity in China". Qianjiangyuan National Park is just located at the heart of China, which has created the possibility of the gathering of various plants. In addition, the complex mountainous landscape of Qianjiangyuan National Park has shaped a variety of local microclimates. Here not only live deciduous broad-leaved tree species of the North China flora such as mulberry and fraternal trees that prefer the cool environment, but also plants of the South China flora such as ford woodlotus, wild michelia and China monkeyjoy, *etc.*. At the same time, the western region is also connected to the flora of Southwest and Central China through some east-west mountains, which provides the possibility of spreading relic plants such as wheel wingnut and dysosma.

"Gene pool bred in a complex environment"

More than 98% of the land in Qianjiangyuan National Park is covered by forests, which belongs to the two major water systems of the Qiantang River and the Yangtze River. The forests, river valleys, and wetlands of the park provide homes for more than 2,000 species of wild animals and plants, which is unique to China. The park is the main habitat of Elliot's pheasants and black muntjacs that are the world's rare and endangered species, the national first-level key protected wildlife. It is also a place where ancient relic plants such as maire yew, longstipe disanthus, wheel wingnut, and Du zhong live and reproduce for generations. Together, they have created the typical, diverse, ancient, rare and complete biological community of Gutian Mountain, which can be called "natural species gene pool".

"Recommendation of the scientific research experience route"

(Stop at the fork of the observatory tower) The typical forest vegetation and rich wild animal and plant resources of Gutian Mountain provide an ideal place for scientists to study forest ecosystems and biodiversity. Today, scientists have made rich achievements on scientific research in Gutian Mountain. The results are well-known in China and abroad. Following this route, everyone can feel some of the practice and exploration of the national park in the field of scientific research and innovation. You can look at the tower crane and feel how it could study the forest canopy 360 degrees without dead end; you can also walk into the dense forest and see how scientists record the growth process of each tree; you can also go to the observation deck at a high level, appreciate the skyline without barriers.

苏庄镇 | 古田飞瀑 | 丛林飞瀑自然体验路线　　　　解说编号：SZ-GTFP-04

解说点 04 路边平台二

29° 14′ 45.53″ N，118° 7′ 3.47″ E
海拔：345m

解说主题

从植物的花、叶、果和种子等角度认识钱江源国家公园的植物多样性

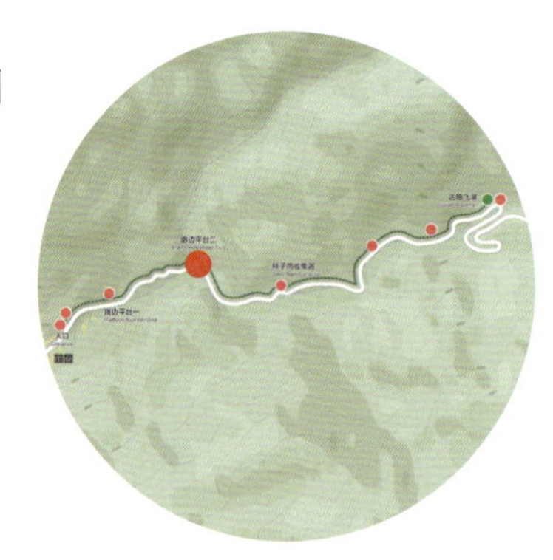

解说要点

1. 认识各种各样不同类型的花序、叶片和果实
2. 了解植物器官演化与环境的适应关系

场地概述

安装有花、叶、果相关的解说牌

解说方式 Interpretation Style

☑ 人员解说 Staff Interpretation　☑ 教育活动 Educational Activity
☑ 解说牌 Interpretation Signage　☐ 场馆解说 Hall Interpretation

解说季节 Interpretation Season

☑ 春季 Spring　☑ 夏季 Summer
☑ 秋季 Autumn　☑ 冬季 Winter

钱江源国家公园的森林里生活有数十亿棵乔木，更别说灌木和草本植物了。在我们认识它们之前，它们就是一棵树，和自然中的其他树没什么区别，但是当我们决定花点时间和精力去认识它们、和它们建立联系时，它们对我们的意义就不一样了。那么我们怎么去认识一棵树呢?

俗话说：世界上没有两片相同的叶子。植物的花、叶和果实（种子）的形状、大小、质地、排列和颜色就像植物的身份证一样，可以作为我们识别植物的依据。对于和古田山的植物打了多年交道的科研工作者来说，随地捡拾的一片干枯的叶子、一朵花或者一颗种子，他都能告诉你这属于什么植物。这样的经验并不需要什么特别的天分，却离不开长时间细心和耐心的观察与经验的积累。我们不妨从身边的这几棵树开始我们的观察实践。

木荷拥有椭圆形的大叶片，叶子表面摸起来似皮革般光滑，有一层薄薄的蜡质，可以有效地锁住水分。木荷是森林天然的“消防员”，叶片中几乎有一半都是水。不过大家不要轻易掰开木荷的叶子去尝试哦，因为木荷的汁液含有微毒素，如果不慎碰到，可能会引起皮肤的不适。此外，作为山茶家族的一员，木荷的花像山茶花一样，拥有厚实的质感和醒目的黄色花蕊。不过，大家仔细观察下木荷的花瓣，发现有什么特别的吗？木荷的 5 枚花瓣中有一枚和其他不太一样，像风衣后面的帽子一样，被称为“风帽花”。木荷的果实为咖啡色的蒴果，顶端未开裂前，中间的缝隙就像一颗五角星一样。

钱江源国家公园的森林里，经常可见散落在地上的坚果，就像动画片中经常出现的松鼠喜欢囤积的那一种，它们有可能就是壳斗科植物的果实。壳斗科是一个大家族，全世界大概有 900 多种。生活在钱江源国家公园的壳斗科植物有甜槠、苦槠、青冈等，我们怎么才能将它们区分开呢？最简单的办法就找主要特征差异。比如，甜槠和苦槠，仅一字之别，外形也极其相似，但是甜槠的壳斗和板栗相似，像个刺球，有些扎手；苦槠的壳斗是没有刺的，呈鳞片状，摸起来肉肉的。此外，我们还可以通过观察壳斗包围种子的比例来分辨不同的壳斗科植物。比如，青冈的壳斗大概只包围种子的 1/3 到 1/2，而甜槠和苦槠的壳斗却几乎将整个种子包围住了。

注意事项

- 根据现场情况，选取一些有代表性的物种进行详细的解说。
- 提前捡拾各种叶片、果实（种子），了解它们分别属于哪一类植物，背后有什么主要特点。在解说过程中才可以灵活应对、信手拈来。

辅助道具推荐

- 标识牌：一叶一世界
- 标识牌：一花一世界
- 标识牌：一果一世界
- 提前捡拾的各种叶片、果实和种子凋落物 / 标本

拓展活动推荐

- 古田山常见植物打卡 (P563：活动卡 03)
- 叶子的形态收集 (P564：活动卡 04)
- 花朵的小小收藏家 (P565：活动卡 05)
- 千奇百怪的果实 (P566：活动卡 06)

Suzhuang Town | Gutian Waterfall | Jungle and Waterfall Experience Route

Interpretation No. SZ-GTFP-04

Stop 04: Roadside platform#2

● Interpretation theme

Understanding the plant diversity of Qianjiangyuan National Park from the perspectives of flowers, leaves, fruits and seeds

● Key points

1. Meet various inflorescences, leaves and fruits
2. Understand the relationship between the evolution of plant organs and the environment

● Site overview

Equipped with interpretation board: flowers, leaves and fruits

We have just roughly estimated that there are billions of trees living in the forests of the Qianjiangyuan National Park, not to mention shrubs and herbs. Before we know them, they are just trees, not different from other trees in nature, but when we decide to spend some time and energy to recognize them and establish contact with them, they have different meanings for us. So how do we know a tree?

As the saying goes: there are no two identical leaves in the world. The shape, size, texture, arrangement, and color of the flowers, leaves, and fruits (seeds) of plants are just like the ID cards of plants and can be used as the basis for our identification of plants. For a local scientific researcher who has been dealing with plants in Gutian Mountain for many years, give him a dry leaf, a flower or a seed, and he is able to tell you what plant it is. This experience does not require any special talent, but it cannot be separated from long-term careful and patient observation as well as accumulation of experience. We may start our observation practice from the trees around us.

Gugertree has large oval-shaped leaves, and the surface of the leaves feels smooth like leather, with a thin layer of wax, which can effectively lock in moisture. gugertree is a natural "firefighter" in the forest, containing almost half of the water in its leaves. However, don't try to break the leaves, because gugertree contains toxins, which may cause skin discomfort if you accidentally touch it. In addition, as a member of the *Camellia* family, the flowers of gugertree, like tea tree flowers, have a thick pedal and striking yellow stamens, but if you carefully observe the petals of gugertree, do you find anything special? One of the five petals of gugertree is not the same as the others, like the

hat behind the coat. It is called the "hooded flower". The fruit of gugertree is a brown capsule, and before the top is cracked, the gap in the middle is like a pentagram.

In the forest of Qianjiangyuan National Park, you can often see nuts scattered on the ground, the one that squirrels love in cartoons. They may belong to the Fagaceae family. The Fagaceae family is pretty big. There are more than 900 species in the world. The plants of the Fagaceae family living in Qianjiangyuan National Park include *Castanopsis eyrei*, *Castanopsis sclerophylla*, and ring-cupped oak. So how can we distinguish them? The easiest way is to find the main feature differences. For example, *Castanopsis eyrei* and *Castanopsis sclerophylla* are different from each other, and their appearances are very similar. However, the shell of *Castanopsis eyrei* is similar to that of chestnut, feeling pinchy. The shell of *Castanopsis sclerophylla* is scale-like, feeling solid. In addition, we can also distinguish the difference by observing the proportion of the seeds surrounded by the shell. For example, the shell of ring-cupped oak only surrounds 1/3 to 1/2 of the seed, while the shells of *Castanopsis eyrei*, *Castanopsis sclerophylla* almost surround the entire seed.

Attentions

- According to the situation, select some representative species for detailed interpretation.
- Pick up the various leaves and fruits (seeds) in advance to understand what type of plant they belong to, and what are the main features behind them so that you could respond flexibly.

Recommended auxiliary facilities and toolkits

- Interpretation signage: various leaves
- Interpretation signage: colorful flowers
- Interpretation signage: tasty fruits
- Litters or specimens of various leaves and fruits (seeds) collected in advance

Recommended activities

- Common plants in Gutian Mountain (P563 : No. 03)
- Exploring different leaves (P564 : No. 04)
- Exploring different flowers (P565 : No. 05)
- Exploring different fruits (P566 : No. 06)

苏庄镇 | 古田飞瀑 | 丛林飞瀑自然体验路线　　　　解说编号：SZ-GTFP-05

解说点 05 种子雨收集器

29° 14′ 42.57″ N，118° 7′ 8.39″ E
海拔：360m

● 解说主题

钱江源国家公园的科研工作

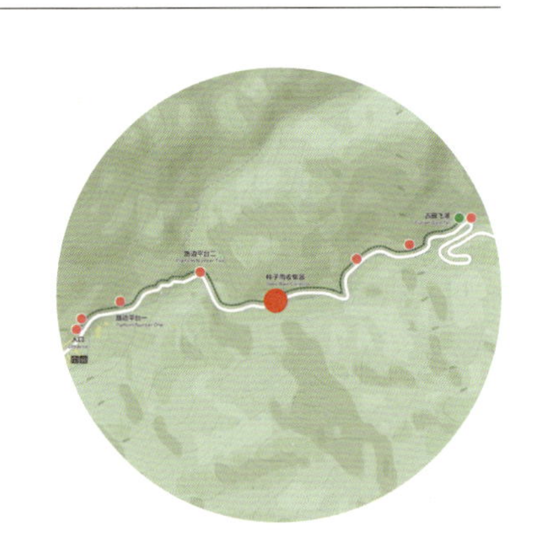

● 解说要点

1. 公园支持各种类型科研工作的开展
2. 认识种子雨收集器

● 场地概述

沿途布置有种子雨收集器的地方

种子雨收集器

解说方式 Interpretation Style

- ☑ 人员解说 Staff Interpretation
- ☑ 教育活动 Educational Activity
- ☑ 解说牌 Interpretation Signage
- ☐ 场馆解说 Hall Interpretation

解说季节 Interpretation Season

- ☑ 春季 Spring
- ☑ 夏季 Summer
- ☑ 秋季 Autumn
- ☑ 冬季 Winter

「钱江源国家公园为科学家们研究常绿阔叶林生态系统提供了场地」

各位朋友，你知道吗，钱江源国家公园平均每 2m² 的土地上，就生长有一棵树。如果按照钱江源国家公园总面积 252km² 的 90% 的森林覆盖率来算，这里至少生活有 11 亿棵树。

如果把森林比做一个大舞台的话，森林每天都在上演着死亡与新生、竞争与合作、欺骗和爱慕的故事。森林里的任何风吹草动，都有可能掀起一场“蝴蝶效应”，这就是我们常说的生态系统的“系统性”。这是经过长期演化的结果，对我们认识自然、认识森林生态系统具有极高的价值。

自二十世纪七八十年代开始，科学家们就开始关注和守护这片森林，帮助很多树木从伐木工人的刀口下逃过一劫，从而保持了最原真的模样；同时，守护森林的科学家们也得到了森林最真挚的回报。他们利用这片森林开展了多项研究工作：比如，通过对森林凋落物、种子雨等的研究，了解森林的生物多样性动态；通过对林冠层的研究，了解森林对全球气候变化的响应；以及通过红外相机，对森林里的动物活动进行实时、全方位的监测。

「认识种子雨收集器」

对于森林的研究，仅限于已有的植物是不够的，还需要了解其繁衍更替的过程，那么关于种子的探索就变得尤为重要了。不仅要了解种子的形态、结构，还要了解它们发生的时机，传播和繁衍的方式。在古田山的大样地中，布设有很多这样的方形网状设施。大家知道它是用来做什么的吗？其实这是特别用来研究种子的设施——种子雨收集器。

在一定时间和空间内，种子会大量地从母树脱落，就像下雨一样，这种现象被形象地称为“种子雨”。就好像不同的人有不同的个性一样，不同植物的种子雨发生的时间、强度及方式都有很大的差异。此外，受外部环境和气候等因素的影响，在不同地点，不同时间，甚至不同年份，种子雨的情况都有很大的变化。

因此，研究不同植物的种子雨什么时候发生、量有多大、强度如何以及种子雨是如何扩散的，对于我们认识森林生态有很重要的指导意义。

大家仔细观察一下这个“种子雨收集器”(结合现场设施，如后续移除也可准备相关的简易模型或图片)。这是一种非常简单有效的工具。它的边长和高度都可以调节，用来满足不同环境下不同研究项目的需要。古田山大样地建设以来，科研人员每周都会将每个收集器上采集的植物种子、落叶等凋落物进行回收、分类，烘干后对物种类型、生物量等指标进行记录、监测、分析和研究，探究不同树种存活和繁衍的机理，进而理解森林生态系统的存续方式等问题。

Suzhuang Town | Gutian Waterfall | Jungle and Waterfall Experience Route

Interpretation No. SZ-GTFP-05

Stop 05: Seed rain collector

Interpreration theme

Introduction of scientific research in Qianjiangyuan National Park

Key points

1. National Park's support to various scientific research work
2. Introduction of the seed rain collector

Site overview

Place with the seed rain collector

"Qianjiangyuan National Park provides a platform for scientists to study the evergreen broad-leaved forest ecosystem"

Dear friends, do you know that, in Qianjiangyuan National Park, on every 2 square meters of land, there's a tree? Let's assume, if 90% of the total area of Qianjiangyuan National Park, 252 square kilometers, is covered by trees, there are at least 1.1 billion trees living here.

If you compare the forest as a big stage, the forest is performing stories of death and rebirth, competition and cooperation, deception and admiration every day. Any wind and grass in the forest may set off a "butterfly effect", which is what we often call the "systemic" of the ecosystem. This is the result of a long-term evolution and is of great value for scientists to understand nature and forest ecosystems.

Since the 1970s and 1980s, scientists had started to research and protect this forest, not only helping many trees escape from the blade of the lumberjack, maintaining the most authentic appearance; at the same time, scientists had also received the sincere return from the forest. They have conducted a number of research work in this forest. For example, through the study of forest litter and seed rain, scientists can further understand the dynamic changes of forest biodiversity and their interrelationships; understand the response to global climate change through the study of canopy; and use a whole-national-park-wide layout infrared monitoring system to trace and record wildlife activities in the forest in real time.

"Introduction to the seed rain collector"

For forest research, it is not enough to know the existing plants, but also need to understand the process of their reproduction and replacement, so the exploration of seeds becomes particularly important.

Seed rain collector

It is necessary to understand not only the morphology and structure of seeds, but also the timing of their occurrence, the method of propagation and reproduction. In the large plot of Gutian Mountain, there are many square mesh facilities like this, do you know what it is used for? In fact, this is a special facility for seed research: seed rain collector.

In a certain time and area, a large number of seeds will fall off from the mother tree, just like rains. This phenomenon is called "seed rain". Just as different people have different personalities, the time, intensity, and method of seed rain of different plants are very different. In addition, due to factors such as the external environment and climate, the situation of seed rain varies greatly at different locations, at different times, and even in different years.

So, when does the seed rain of different plants occur? How much is it? How strong is it? And how does the seed rain spread? These questions are very important for us to understand forest ecology.

Let's take a closer look at this "seed rain collector" (It'll be the best if there is one on-site, you can also prepare related simple models or pictures). This is a very simple and effective tool. The length and height of its sides can be adjusted to meet the needs of different research projects of different species. Since the establishment of the plot, researchers have been collecting and classifying the plant seeds, fallen leaves and other litters collected on each collector every week, and record, monitor, analyze and research to explore the mechanism of survival and reproduction of different tree species, and then to understand the survival of forest ecosystems and other issues.

苏庄镇 | 古田飞瀑 | 丛林飞瀑自然体验路线　　解说编号：SZ-GTFP-06

解说点 06 孑遗植物

29° 14′ 45.77″ N，118° 7′ 21.31″ E
海拔：389m

解说主题

钱江源国家公园的孑遗植物

解说要点

1. 公园为多种孑遗植物提供了庇护所
2. 认识孑遗植物杜仲

场地概述

有杜仲生长的路边

杜仲

杜仲的“叶断丝连”现象

解说方式 Interpretation Style

☑ 人员解说 Staff Interpretation　☑ 教育活动 Educational Activity
☑ 解说牌 Interpretation Signage　☐ 场馆解说 Hall Interpretation

解说季节 Interpretation Season

☑ 春季 Spring　☑ 夏季 Summer
☑ 秋季 Autumn　☐ 冬季 Winter

「孑遗植物」

植物虽然一生中只守着一片土地，但它们却有逃离时间束缚的神奇魔法。大家知道吗？地球上最古老的树据保守估计已经 8 万岁了，它是一棵，不，是一群位于美国犹他州的杨树。大家看（展示图片），这片树林就是一棵树哦。树干看似独立，底部的根却是连在一起的，而且它已经在地球上存活了 8 万多年了，是不是非常了不起？

但是论古老，地球上还有一些植物更有发言权。它们的祖先比人类更早出现在地球上，曾经和恐龙生活在一个年代，甚至更早。一代又一代，地球经历了沧海桑田，但它们家族中古老的血液却从未改变，外表就像和化石中的祖先一个模子里刻出来的。它们就是孑遗植物。

「认识孑遗植物杜仲」

孑遗植物虽然历史久远，但是它们的样子却多姿多彩，并不呆板。请大家抬头看，我们头顶的这棵树，就是一棵孑遗植物，它的名字叫杜仲。说起杜仲，大家可能会觉得似曾相识。这是因为杜仲皮自古以来就是一味重要的中药材，同时杜仲体内含有杜仲胶，提炼萃取后是制作海底电缆的重要原料。

不过，我们怎么在野外判断一棵树是不是杜仲呢？接下来，就是见证奇迹的时刻。这是我刚从地上捡起的几片叶子，其中只有一片是杜仲叶。现在，我请几个朋友分别帮我轻轻将这些叶子扯开，看看会发生什么，其他朋友注意观察。大家发现了吗？其中有一片叶子虽被撕开，却有丝线相连，这就是杜仲叶子，而那个丝就是我们刚提到的杜仲胶。（如若不能捡到，可展示照片和录像。）

钱江源国家公园里像杜仲这样的孑遗植物一共有 21 种，比如，大家熟悉的具有扇形叶子的银杏、果实长得像一串铜钱的摇钱树——青钱柳、叶子形状似马褂的鹅掌楸以及不需要过多描述大家就非常熟悉的国家一级重点保护野生植物南方红豆杉。孑遗植物的祖先曾广泛分布在地球上很多地方，因为环境的变化，大部分都已灭绝，唯有少部分因为局地环境提供的特殊庇佑得以存活至今。生活在钱江源国家公园的这些“古老遗民”就是被这一方水土特别眷顾的幸运儿。

注意事项

- 钱江源国家公园丰富的孑遗植物得益于公园的特殊庇佑，只有保护好孑遗植物的栖息环境，它们才能更好地生息繁衍。
- 关于孑遗植物的介绍，可以参考通用型生物多样性人员解说方案中有关孑遗植物的内容。

辅助道具推荐

- 标识牌：自然界的疗愈大师
- 世界上最古老的树的照片

- 杜仲、银杏、鹅掌楸、红豆杉标本或照片

拓展活动推荐

- 孑遗植物连连看（P576：活动卡 15）

Suzhuang Town | Gutian Waterfall | Jungle and Waterfall Experience Route

Interpretation No. SZ-GTFP-06

Stop 06: Relict plants

Interpretation theme

Relict plants living in Qianjiangyuan National Park

Key points

1. Qianjiangyuan National Park provides sancutary for a variety of relict plants
2. Meet the relic plant *Eucommia ulmoides*

Site overview

The place where *Eucommia ulmoides* is growing on the roadside.

"Relict plants"

Although plants only live in a piece of land throughout their lives, they have magic that could escape from time. Do you know that the oldest tree on earth is conservatively estimated to be 80,000 years old? It is a group of poplar trees located in Utah, USA. You see (show the picture), this forest is indeed one tree. The trunks seem independent, but the roots at the bottom are connected together. It has been living on the earth for more than 80,000 years. Is it very remarkable!

But in terms of ancient origins, there are still some plants on the earth who have more say. Their ancestors appeared on this planet earlier than humans, and they have lived with dinosaurs for a generation or even earlier. From generation to generation, the earth has experienced the baptism of time, but the ancient blood in their families has never changed. Their appearance is like being carved in a mold as the ancestors in the fossils. They are relict plants.

"Du zhong (*Eucommia ulmoides*), a relic plant"

Although the relic plants have a long history, they are colorful and not dull at all. Please look up, the tree above our head is a relic plant, and its name is Du Zhong. Speaking of Du Zhong, everyone may feel familiar. This is because Du Zhong's bark has been an important traditional Chinese medicine since ancient times. At the same time, eucommia contains eucommia gum. After extraction, it is an important raw material for making submarine cables. However, how do we distinguish whether a tree is Du Zhong in the wild? Next is the moment to witness the miracle. These are the few leaves I just picked from the ground, and only one of them is Du Zhong. Now, I ask you to help me gently tear these leaves apart to see what happens? Please pay attention and do you notice? One of the leaves was torn apart, but it's silky fiber still connects. This is the Du Zhong's

leaf, and the silk is the Du Zhong's gum we just mentioned. (It is not always possible to pick up the leaves on site. It is recommended to display photos and videos.)

There are more than 30 relic plants like Du Zhong in Qianjiangyuan National Park, such as the well-known ginkgo (*Ginkgo biloba*) with fan-shaped leaves, the fruits like a string of coins wheel wingnut (*Cydocarya paliurus*), the jacket-like leaves Chinese tulip tree (*Liriodendron chinense*) and the national first-class protected plant maire yew (*Taxus wallichiana*). The ancestors of relict plants have been widely distributed in many places on earth. Most of them have been extinct due to changes in the environment. Only a few have survived because of the special protection provided by the local environment. These ancient survivors living in Qianjiangyuan National Park are the lucky ones who are especially favored by this area.

Attentions

- The rich relict plants in Qianjiangyuan National Park benefit from the special protection of the park. Only by protecting the habitat of the relict plants can they better survive and thrive.
- For the introduction of relict plants, please refer to the content about relict plants in the "General Interpretation of Biodiversity".

Recommended auxiliary facilities and toolkits

- Interpretation signage: natural remedies
- Interpretation signage: the oldest tree in the world
- Photos or specimens of some ancient relic plants

Recommended Activities

- Relic plants link game (P576 : No. 15)

苏庄镇 | 古田飞瀑 | 丛林飞瀑自然体验路线 解说编号：SZ-GTFP-07

解说点 07 明星物种

29° 14′ 46.58″ N，118° 7′ 26.2″ E
海拔：400m

解说主题

钱江源国家公园的野生动物

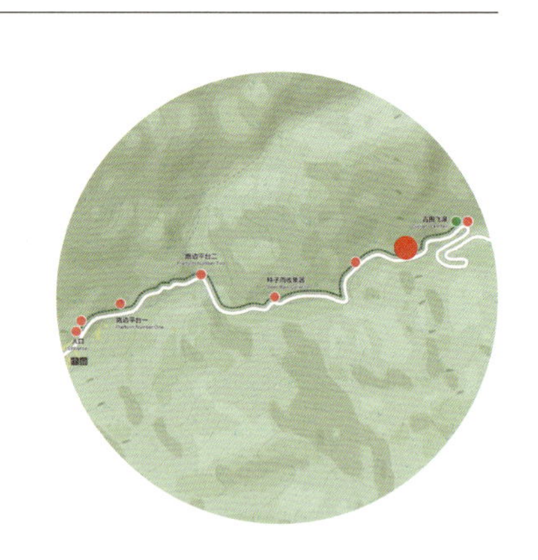

解说要点

1. 古田山为野生动物提供了栖息的家园
2. 生活在古田山的白鹇、白颈长尾雉等野生动物

场地概述

安装有解说牌：野生动物的宜居家园

白鹇
黑麂
亚洲黑熊

解说方式 Interpretation Style

☑ 人员解说 Staff Interpretation ☐ 教育活动 Educational Activity
☑ 解说牌 Interpretation Signage ☐ 场馆解说 Hall Interpretation

解说季节 Interpretation Season

☑ 春季 Spring ☑ 夏季 Summer
☑ 秋季 Autumn ☑ 冬季 Winter

「认识公园重要的野生动物」

钱江源国家公园茂密的丛林、密布的溪流孕育了丰富的植物，同时也为众多野生动物提供了理想的栖息地。我们周围的森林里就生活有白鹇、白颈长尾雉等地栖型鸟类，而在公园的更深处，还有小麂、黑麂甚至是黑熊出没。

那么，我们有多大概率可以见到它们呢？答案是非常小。这是因为这些兽类不仅数量稀少，而且生活隐蔽，警惕心强，喜欢深山老林这样人迹罕至的地方。即使是长期进行野外调研的生物学家，也只能通过声音、粪便或者脚印等间接证据窥探它们生活的蛛丝马迹。红外相机的出现，让我们得以借助现代科技对这些生活在秘境的动物有了更深一步的了解，这些动物的神秘面纱被缓缓揭开。

大家看，我手里现在就有一些红外相机拍到的照片（或视频），如黑麂、白颈长尾雉和白鹇等。我们先来认识一下它们。

黑麂又称“蓬头麂”，因头顶一丛“杀马特”红色毛发而得名。此外，黑麂的尾巴有 2/3 都是白色的，雄性还有一对外露的獠牙，这都是我们识别黑麂的重要特征。目前，全世界的黑麂加起来也不到 6000 只，其中大约有 1/10 生活在钱江源国家公园，但是由于它们生性胆小，主要生活在密林之中，而且多在晨昏活动觅食，见到它们可以说是难上加难。

相对黑麂，遇到同样生活在公园里的小麂概率会大一些。小麂全身棕褐色，体形娇小，雄性小麂长有一对角。它们和黑麂一样也是生活在深山老林之中，偶尔也会到森林边缘喝水、觅食。虽然也难得一见，但是你极有可能“听到”小麂。为什么这么说呢？这是因为小麂声音非常特别，就像狗叫一样。如果你能听到从密林中传来的这种声音，有可能是小麂发出的哦！

白鹇和白颈长尾雉是生活在林下、羽色华丽的两种地栖型鸟类，它们个头都比较大，且具有明显的外貌特征，是野外最容易识别的鸟类之一。古田山的常绿阔叶林为白颈长尾雉提供了极为适合的栖息地，是白颈长尾雉分布较为集中、数量较多的区域。但尽管如此，它们数量仍不足以达到随处可见的地步，再加上它们生性机警，且不喜鸣叫，经常隐藏在林下茂密的灌草丛中，也是极难观察到的一种鸟类。相对来说，一身白色羽毛的白鹇则时不时地出现在人们的视野之中，我们现在走的这条体验路线上就经常有朋友见到悠闲散步的白鹇夫妇。

拓展知识点

「物种共存的重要性」

- 20 世纪 90 年代，这里曾记录到我国四大猫科动物之一的云豹捕食家畜的记录，但此后再无记录。云豹是否还存在于这片山林，没有人说得清楚。但是由于人类活动的影响，作为云豹食物的小型哺乳动物的种群数量已经明显减少，同时栖息地也大面积消失，这些问题都是不争的事实。但实际上人类才是这片山林的外来者。所以大家来到这里，一定要心存敬畏之心，记住：我们是访客，它们才是这片土地的原住民。

Suzhuang Town | Gutian Waterfall | Jungle and Waterfall Experience Route

Interpretation No. SZ-GTFP-07

Stop 07: Star species

● Interpretation theme

Wildlife in Qianjiangyuan National Park

● Key points

1. The evergreen broad-leaved forest of Gutian Mountain provides a home for wildlife

2. Wild animals such as silver pheasants and Elliot's pheasants living in the evergreen broad-leaved forest of Gutian Mountain.

● Site overview

Equipped interpretation boards: a cozy home for wild animals

The dense jungles and streams of the Qianjiangyuan National Park have bred abundant plants and also provide an ideal habitat for many wild animals. There are ground dwelling birds such as silver pheasants and Elliot's pheasants in the forests around us, and in the deeper parts of the park, there are Chinese muntjacs, black muntjacs and even black bears.

How likely will you see them? The answer is very little. This is because these animals are not only rare in number but also live in deep mountains and old forests. Even biologists who have been conducting field research for a long time can only peep into the clues of their lives through indirect evidence such as sound, poop, or footprints. The development of infrared cameras has enabled us to use modern technology to have a deeper understanding of these animals living in secrets. The mystery of these animals has been slowly unveiled.

Here are some photos (or videos) taken by infrared cameras in the national park. You can see all these endangered species, such as black muntjac, Elliot's pheasant and silver pheasant. Let's get to know them first.

The black muntjac is also known as "messy muntjac". Everyone can see how the name comes from by the red hairs above their heads. In addition, two thirds of the black muntjac's tail is white, and the male has a pair of exposed fangs.

Currently, there are less than 6000 black muntjacs in the world, of which ones living in Qianjiangyuan National Park accounts for about 10%. Compared with black muntjacs, Chinese muntjacs who also live in the park are much more common. They have a tan and small

Elliot's pheasant

body. Male muntjacs have a pair of horns. If you hear a dog-like sound from a distance in the forest, it may be the Chinese muntjac!

The most well-known birds in Qianjiangyuan are the silver pheasant and the Elliot's pheasant. They all live under the forest and have gorgeous plumage, which are easier to be identified. The male bird of silver pheasant is covered with white feathers and has a long tail. It often hunts for food in the forest and has a loud voice. It is a kind of bird that can be easily observed in Gutian Mountain. You may also encounter the silver pheasants couple wandering in the forest. The Elliot's pheasant is a star species of the national park and the national first-class protected animal like the black muntjac. However, due to the scarcity, vigilance, and unwillingness to chirp, it is often hidden in the dense shrub grass under the forest, so it is not easy to be seen. The evergreen broad-leaved forest of Gutian Mountain provides a very suitable habitat for the Elliot's pheasant. It is an area where the Elliot's pheasant is concentrated with a large amount.

Extra information

"Importance of coexistence of species"

- In the 1990s, it was documented that one of the country's four major cats, snow leopards, preyed on domestic animals, but there were no records since then. No one can say for sure whether snow leopards still exist in this mountain forest. However, due to the impact of human activities, the population of small mammals that have been considered as food for snow leopards has decreased significantly along with the largely disappeared habitats. These problems are indisputable facts. But in fact, human beings are outsiders in this mountain forest. Therefore, everyone here must be in awe, remember: we are visitors, they are the native residents of this land.

苏庄镇 | 古田飞瀑 | 丛林飞瀑自然体验路线　　解说编号：SZ-GTFP-08

解说点 08 古田飞瀑

29° 14′ 51.34″ N, 118° 7′ 36.44″ E
海拔：435m

● 解说主题

古田山的水系与森林的生态功能

● 解说要点

1. 认识公园的两大水系与森林的水源涵养功能
2. 古田山作为天然的森林氧吧的评价依据

● 场地概述

古田飞瀑附近

古田飞瀑 1
喜欢瀑布溪流的小燕尾
瀑布旁的枳椇
古田飞瀑 2

解说方式 Interpretation Style

- ☑ 人员解说 Staff Interpretation
- ☑ 教育活动 Educational Activity
- ☑ 解说牌 Interpretation Signage
- ☐ 场馆解说 Hall Interpretation

解说季节 Interpretation Season

- ☑ 春季 Spring
- ☑ 夏季 Summer
- ☑ 秋季 Autumn
- ☑ 冬季 Winter

解说词

「公园的两大水系」

钱江源国家公园跨越了钱塘江和长江两大水系。钱塘江源头在公园东部的莲花塘地区，而位于公园西部的古田山的大部分水流则会顺着山势自东向西边流去，最终会汇入鄱阳湖，与长江相连。这也是我们面前这条瀑布的最终归宿。

「瀑布流水与森林的水源涵养功能」

这条瀑布的垂直落差逾 40m，常年水流不断，倾泻潭中。顺着瀑布往上走，是蜿蜒的溪流，生活经验告诉我们，森林的繁茂离不开水源的滋养，森林就像一块巨大的海绵，有着极强的吸水能力。有研究表明，单位面积森林的蓄水量比同等面积无林裸地的蓄水量多 7 倍以上。但是，森林并非只进不出，森林的水会通过蒸发、下渗、蒸腾等过程悄悄地回到环境当中。除此之外，森林还是一个巨大的“净水器”，经过森林过滤后的水源干净、污染少，甚至可以直接饮用。

「天然的森林氧吧」

古田山被称作天然的“森林氧吧”。“氧吧”这个词听起来有些养生的味道，但是却有着实实在在的评判标准。被称作“氧吧”的地方不仅空气质量要高、水质要好，还要求有较高的森林覆盖率……其中，最核心的指标是被称为“空气维生素”的负氧离子的含量。举个例子，如果你居住在城市，你们家每立方厘米空气中的负氧离子含量约为 100 个。而世界卫生组织规定，当空气中负氧离子浓度不低于 1000～1500 个 /cm^3 时，才可视为清新空气。我们这里的负氧离子含量平均值为 4500 个 /cm^3，而古田山地区监测到的最高值更达到 14.5 万个 /cm^3。大约相当于你们家的 45 倍，是世卫组织给出的清新空气标准的 3 倍还多。

「钱江源国家公园其他深度体验游线推荐」

我们今天感受到的只是公园小小的一角。接下来，无论大家想去爬山、看水、赏花、观察动物、探索原始森林的独特魅力，还是体验古村落的传统文化，钱江源国家公园都敞开怀抱欢迎大家。

注意事项

- 古田飞瀑虽是这是此行的终点站，却是访客探访公园的起点。这里既是对本游线的总结，也是抛砖引玉，为其他游线做的铺垫。

辅助道具推荐

- 标识牌：天然氧吧的最佳范本
- 标识牌：水系绵延的杰出贡献

Suzhuang Town | Gutian Waterfall | Jungle and Waterfall Experience Route

Interpretation No. SZ-GTFP-08

Stop 08: Gutian waterfall

Interpretation theme

The function of forest and water system

Key points

1. Water and forest in Gutian Mountain
2. Natural Forest Oxygen Bar

Site overview

The end of the route: Gutian Waterfall

"Two water systems in the park"

Do you know that although Qianjiangyuan National Park is located at the source of the Qiantang River, it crosses the Qiantang River and the Yangtze River? The Qiantang River is located in the east of the national park, and its source is in the lotus pond area in the east of the national park. Due to the high terrain in the middle of the park, most of the water flow in the west of the park where Gutian Mountain is located will flow from the top of the park to the west. It flows into the Poyang Lake and is connected to the Yangtze River. This is also the final destination of this waterfall in front of us.

"Water conservation function of waterfall and Forest"

The vertical drop of this waterfall is more than 40 meters, and the water flows continuously throughout the year, pouring into the pond. Going up the waterfall is a winding stream, but what about at the end of the stream? Where does the water of the stream come from? We know that forests are born of water, and the lush forests are inseparable from the nourishment of water sources, but the forests are not harvested in full. They are like a huge sponge, which has a strong water absorption capacity. Studies have shown that the water storage of a unit area of forest is more than 7 times that of the bare area of the same area without forest. In addition, the forest is a huge "water purifier", the water filtered by the forest is clean and clear, and even can be directly consumed.

"Natural forest oxygen bar"

We often say that air is colorless, tasteless, invisible and untouchable, but our senses have the intuition to judge whether the air is good or bad. There is no doubt that the air here is good, and it is called the natural "forest oxygen bar", but the "oxygen bar" can not be called at

will. It not only requires high air quality, good water quality, but also high forest coverage, but the main indicator is the content of negative oxygen ions called air vitamins. If you live in a city, your home has about 100 negative oxygen ions per cubic centimeter of air. The World Health Organization stipulates that it can be regarded as fresh air when the concentration of negative oxygen ions in the air is no less than 1000 – 1500 per cubic centimeter. What is the negative oxygen ions level where we are now? Everybody take a guess? The average value is equivalent to 45 times that of your home, more than twice the clean air standard given by WHO.

"Other in-depth experience route in Qianjiangyuan National park"

What we feel today is only a small corner of the national park, but this small corner opens a door to Qianjiangyuan National Park for us. Next, whether you want to go mountain climbing, appreciate the water, enjoy the flowers, observe the animals, explore the unique charm of the original forest, or experience the traditional culture of the ancient villages, Qianjiangyuan National Park welcomes you with open arms.

Attentions

- Although Gutian Waterfall is the end of this route, it is also the starting point of another route of an comprehensive exploration of the Gutian Mountain. This is the summary of this route, and also the initiative of the other more intensive routes.

Recommended auxiliary facilities and toolkits

- Interpretation signage: natural oxygen bar
- Interpretation signage: river source safeguards the basin

苏庄镇 | 古田飞瀑 | 丛林飞瀑自然体验路线 解说编号：SZ-GTFP-09

解说点 09 结束集散点

29° 13′ 34.87″ N，118° 6′ 37.61″ E
海拔：435m

解说主题

游线总结与分享

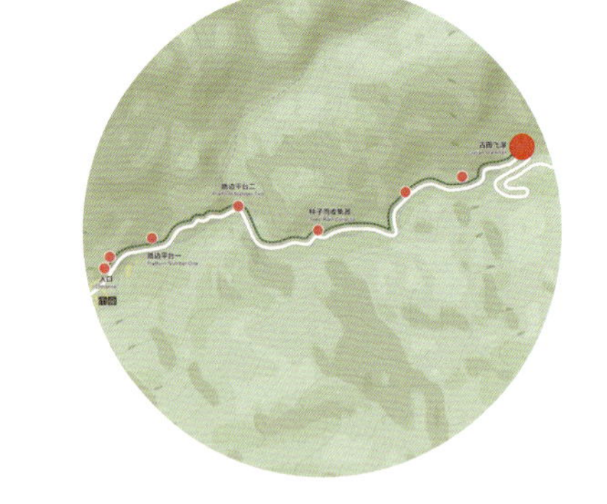

解说要点

1. 通过分享和互动的形式，带领访客进行回顾
2. 对国家公园其他参访资源的介绍

场地概述

古田飞瀑附近

解说方式 Interpretation Style

- ☑ 人员解说 Staff Interpretation
- ☐ 教育活动 Educational Activity
- ☑ 解说牌 Interpretation Signage
- ☐ 场馆解说 Hall Interpretation

解说季节 Interpretation Season

- ☑ 春季 Spring
- ☑ 夏季 Summer
- ☑ 秋季 Autumn
- ☑ 冬季 Winter

「游线总结」

各位朋友，我们此次的公园体验之路到这里就要结束了，但大家的钱江源国家公园之行才刚刚开始。在这条体验路线上，我们对钱江源作为长江三角洲首个国家公园所拥有的和守护的自然有了一个整体的印象，了解到它为多种多样的动植物提供了栖息的家园，也初步认识了一些代表性的野生动植物。同时，我们还通过种子雨收集器等了解到科学家是怎么利用这片区域开展具体的科研工作的。

「其他游线推荐」

这条路线只是抛砖引玉，接下来，钱江源国家公园还有很多美景等待着我们继续探索。如果你想深入了解古田山的森林，可以继续往上走，用脚步去丈量古田山的高度，感受更高海拔森林植被的变化；如果你想系统地了解科学家们在古田山所做的工作，可以去体验下古田山瞭望台的科研体验路线；当然，既然来到了钱江源，你一定不能错过位于莲花塘的钱塘江的寻源之旅。如果你觉得体力或者时间有限，想在最短的时间内体验最有代表性的"江源风光"，大峡谷路线会是不错的选择。如果你觉得这些路线对你来说是小菜一碟，想要挑战自己，甚至是决定可以和外界失联几小时，枫楼坑湿地自然体验路线绝对是值得一试。此外，如果你对公园的人文和历史比较感兴趣，也可以去当地的古村落走走，体验一下本地的风土民情。

「游线分享」

不知道大家一路走来，有没有在心中构建起对钱江源国家公园的初步印象呢？你心中的国家公园是什么？在结束今天古田山的自然体验行程之前，我想邀请大家一起来回顾一下沿途的所观所感。

对于整个地球而言，一花一草、一木一树都像浩瀚夜空中的星星一样，是十分渺小的，但这些小星星们汇聚一起，能发出耀眼的光芒。请大家不要忘记，作为人类，我们无时无刻不在接受这些小星星给予的光亮和温暖。

感谢大家的分享，也谢谢你们一路的支持！和你们交流也给了我很多新的启发。祝福大家未来的旅途愉快，返程平安！

注意事项

- 召集大家集合，15 分钟的小回顾，做些体力上的休整。
- 进行游线总结，并推荐其他游线。
- 通过大家的反馈，延伸启发与思考，提出回去后的行为倡议。
- 感谢大家的到访，感谢大家聆听自己的解说。

Suzhuang Town | Gutian Waterfall | Jungle and Waterfall Experience Route

Interpretation No. SZ-GTFP-09

Stop 09: Assembly ending point

● Interpretation theme

Summary and sharing

● Key points

1. Lead the visitor to review what they saw through a sharing circle
2. Introduction of other resources of the national park

● Site overview

A point near Gutian Waterfall

"Travel summary"

Dear friends, our tour today is coming to an end here, but your trip to Qianjiangyuan National Park has just begun. On this nature experience route, we have an overall impression of the nature owned and protected by Qianjiangyuan as the first national park in the Yangtze River Delta as well as a preliminary understanding of some representative wild animals and plants. At the same time, we also learned how scientists use this area to carry out specific scientific research work through the seed rain collector.

"Recommended other tours"

This route just throws a brick to attract jade. Later, there are many more beautiful scenery in Qianjiangyuan National Park waiting for us to explore. If you want to get a deeper understanding of the forests of Gutian Mountain, you can continue to go up and measure the height of Gutian Mountain with your feet and feel the changes in forest vegetation at higher elevation; if you want to systematically understand the work conducted by scientists in Gutian Mountain, you can go to experience the scientific research experience route of the Gutian Mountain Observation Platform; of course, since you have come to Qianjiangyuan, you had better not miss the source-seeking journey of Qianjiangyuan in the Lotus Pond. If you feel that your energy or time is limited, and want to experience the most stunning scenery in the shortest time, the Grand Canyon will be a good choice. If you think these routes are a piece of cake for you, you want to challenge yourself, or even decide you can lose contact for a few hours, the Fenglou Pit Wetland Nature Experience Route is definitely worth trying. In addition, if you are interested in the humanities and history of the park, you can also go to the ancient villages here to experience the local customs.

Attentions

- Gather everybody, organize a 15-minute quick review, and take a break.
- Summarize the route and recommend other routes.
- Lighten inspiration and propose environmental protection strategy.
- Thank them all for visiting and listening to your interpretation. (Suggest to promote the national park to those around them after going back.)

"Travel sharing"

I don't know if you have built up your initial impression of the Qianjiangyuan National Park? What is the national park in your heart? Before ending today's Gutian Mountain Nature Experience route, I would like to invite everyone to review your feelings along the way.

For the entire planet, flowers, grasses and trees are like tiny stars in the vast night sky, but when these little stars gather together, they can emit dazzling light. Please don't forget that as human beings, we are benefiting from the light and warm we received from these "little stars" all the time.

Thank you for sharing, and for your support along the way! Communicating with you has given me a lot of new inspiration. I wish you all a happy future journey and a safe return home!

S02 古田山—古田庙：探秘常绿阔叶林自然体验路线

Evergreen broad-leaved forest exploration route

古田庙湿地
Gutian Temple Wetland
杉木林
China Fir Forest
山脊线
Ridgeline
休憩亭四
Lounge Number Fo
休憩亭三
Lounge Number Three
休憩亭二
Lounge Number Two
溪流倒木
Fallen Trees over the Stream
休憩亭一
Lounge Number One
潜龙池
Qianlong Pond
入口
Entrance
岔路
Forked Road
观景平台二
Landscape Platform Number Two
观景平台一
Landscape Platform Number One
木荷林
Gugertree Forest
往古田山庄方向
图例 LEGEND
钱江源国家公园
Qianjiangyuan National Park
水域
Water
等高线
Contours
步行道
Trail
车行道
Road
卫生间
Toilet
观景处
Recreation Spot
休憩亭
Rest Lounge
寺庙
Temple
设解说牌的点位
Point with Signage
设解说牌、人员解说的点位
Point with Signage and Interpretation
设解说牌、人员解说、课程的点位
Point with Signage, Interpretation and Course
0 100m 200m 300m
N
海拔（m）
1000
900
800
700
600
500
休憩亭一
休憩亭二
休憩亭三
休憩亭四
古田庙湿地
休憩亭四
观景平台一
观景平台二
路程（km）
0 1 2 3 4 5 6

S02 古田山—古田庙：探秘常绿阔叶林自然体验路线

解说主题 主题 1：点亮北纬三十度的生机绿洲
主题 3：丛林佑护的珍贵原真
主题 5：体制试点的创新探索

步道里程 往返环线约 5.2km
体验时长 往返约 3～3.5h
最低海拔 480m
最高海拔 914m
步道等级 徒步型
基础设施 休憩亭 / 解说牌 / 活动场地 / 卫生间 / 无餐饮

这是认识钱江源国家公园自然生态系统原真性、完整性最重要的一条自然体验路线。古田山位于钱江源国家公园的西南侧，最高点海拔 961m，因古籍记载其山畔有田，山中深处有古森林，林中有古田庙，故名古田山。古田山步道由山脚的登山口沿着步道向上，可以感受到亚欧大陆东部典型的亚热带常绿阔叶林的风貌。

古田山的常绿阔叶林是典型的甜槠 – 木荷林，群落优势种突出，有明显的垂直分层。随着海拔的不断升高，访客可以逐渐感受到常绿阔叶林 – 常绿落叶阔叶林 – 针阔叶混交林的植被类型变化。到达古田庙后会发现，那里仍保留着一片形成久远的沼泽湿地。所谓古田，可能就是得名于此。

这条体验路线里程较长，上山途中我们多身处其间，可以重点关注沿途的植物群落和物种，而下山途中，则有多个平台可以遥望 24hm² 大样地微微起伏的林冠以及远处绵延的群山，让我们从不同的角度去了解这片原始、珍贵的亚热带常绿阔叶林。

在整条体验路线上，我们一共为解说员和访客安排了 12 个推荐解说点，同时，根据本书提供的通用型解说方案（非定点解说方案）资料，解说员也可以在有解说价值的植物资源等停留点进行相应的引导性观察和解说。

S02 Evergreen broad-leaved forest exploration route

Theme
- 01 The 30°N miracle of green oasis
- 03 Mysterious wildlife in the forest
- 05 Scientific research innovation and exploration

Trail Length It is about 2.5 kilometers one-way and 5.2 kilometers round-trip from the foot of Gutian Mountain to the destination of Gutian Temple (Lingyun Temple)

Time 3 – 3.5 hours round trip

Lowest elevation 480 meters

Highest elevation 914 meters (Gutian Temple area)

Level of trail Hiking

Infrastructure Lounge / interpretation board / event site / restroom / no catering

This is the most essential nature experience route to understand the authenticity and integrity of the natural ecosystem of Qianjiangyuan National Park. Gutian Mountain is located on the southwest side of the national park with an elevation of 961 meters. Gutian Mountain got its name from the ancient books, documenting that there are fields by its hillside, with ancient forests deep in the mountains, and the Gutian Temple is hidden in the forest. Walking along the Gutian Mountain trail from the bottom, you can feel the typical evergreen broad-leaved forest in Eastern Asia and Europe.

The evergreen broad-leaved forest in Gutian Mountain is dominated by sweet oachestnut – gugertree. You can see numerous sweet oachestnut, gugertree, Chinese sassafras, Zhejiang camellia, and various ferns along the way. As the elevation continues to rise, you can clearly feel the alteration among evergreen broad-leaved forest – evergreen-deciduous broad-leaved forest – coniferous and broad-leaved mixed forest. When you reach the Gutian Temple, you will find that there still remains a long-standing marsh wetland. The so-called Gutian (ancient field) may be named after it.

This is a relatively long experience route. On our way up the mountain, we can pay attention to the vegetation and species along the way. On our way down, there are multiple platforms that we can overlook the slightly undulating canopy of the entire 24 hectares that enables us to understand this unusual subtropical evergreen broad-leaved forest from different perspectives.

We have arranged a total of 12 recommended interpretation stops for interpreters and visitors during the entire experience route. At the same time, according to the information provided in general interpretation (non-route-oriented interpretation) of this interpretation manual, the interpreter can also find suitable plant resources to conduct corresponding guided observation and interpretation.

苏庄镇 | 古田山—古田庙 | 探秘常绿阔叶林自然体验路线　　　　解说编号：SZ-GT-01

解说点 01 集合点

29°14′53.76″ N，118°7′57.45″ E

海拔：530m

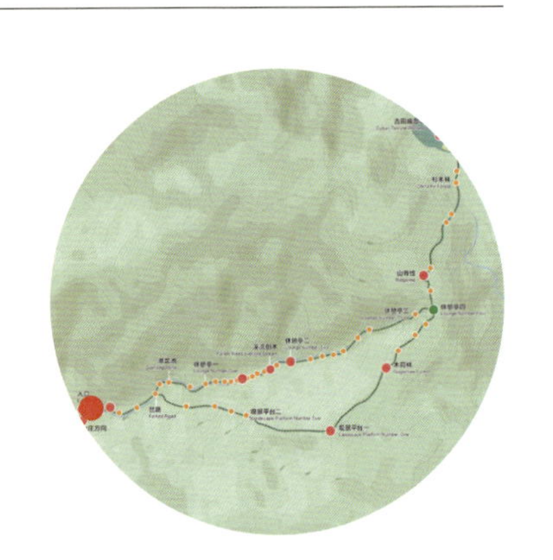

● 解说主题

本条自然体验路线的总体介绍

● 解说要点

1. 古田山游线的整体介绍
2. 确认听众的状态、确立解说员的角色定位

● 场地概述

入口山门外的集合点，门前有步道总体导览牌

解说方式 Interpretation Style

☑ 人员解说 Staff Interpretation　☐ 教育活动 Educational Activity

☑ 解说牌 Interpretation Signage　☐ 场馆解说 Hall Interpretation

解说季节 Interpretation Season

☑ 春季 Spring　☑ 夏季 Summer

☑ 秋季 Autumn　☑ 冬季 Winter

各位朋友好！欢迎大家来到钱江源国家公园，我是公园的解说员 XXX，接下来将由我陪伴大家一起探索这段自然的旅程。

现在我们脚下的土地就是古田山了，大家知道古田山在钱江源国家公园的哪个方位吗？我们不妨一起来看一看标识牌上的这张地图。古田山位于钱江源国家公园的西南侧，但是细心的朋友可能注意到了，古田山其实是一片很大的区域，一共有 4 条自然体验路线。接下来我们要走的这条，是最能帮助我们全面地认识和探索古田山常绿阔叶林的一条自然体验路线。

大家知道古田山为什么叫“古田”吗？这其实要和我们今天所要挑战的最高点有关。我们今天行程的最高处有一座凌云寺，也叫古田庙。这个“古”呢，来源于山林中郁郁葱葱的参天古木，“田”则是因为这里虽然保持自然原真，却不乏人类足迹，曾有农家的水田菜畦点缀山间。古田庙前就有一片“有故事的”水田。至于故事是什么，到了山顶，我再讲给大家听。

好，现在大家热热身，我们准备上山喽！我们从这里上山，最终还会回到这里，全程大约需要 3h，我们要到达的最高点比我们现在的位置海拔高 400m，沿途有不少比较陡的路段，步道设施也比较原生态，对大家来说需要一定的体力和耐力，会是个不小的挑战。但大家也不用担心，我会安排好休息和停顿的时间。如果你们累了，也可以随时提醒我。

随着海拔的升高，森林植被也在发生着变化：从四季常绿的常绿阔叶林到有季相变化的常绿落叶阔叶林、针阔叶混交林，最后是耐寒的针叶林。沿途，我们不仅可以观赏到青冈、木荷、甜槠、马尾松等高大的乔木，聆听小鸟与溪流的欢歌，说不定还会遇到当地的明星物种白鹇、小鹿等。

要提醒大家的是，无论上山还是下山，大家都要注意安全。请紧跟我，有序地排队前进。山里的信号可能不太好，大家千万不要独自行动，更不要奔跑追逐，一定要注意安全！接下来，让我们出发吧！

注：解说集合点通常设在步道起点有明显标志的集合场所，此处建议设置在本步道的总体导览牌设立处。

注意事项

记住，请注意集合点的开场。当你出现在听众们的面前，解说员的工作已经开始了。如何吸引听众的注意力，树立自身的解说员身份，一个简短、内容明确、带有互动的开场白对解说员来说非常重要。以下几点提示可以帮你达到这一目标。

- 选择一个合适的、可以帮助大家集合，并开始这次自然体验的集合地点，比如，步道起点总体导览牌边上的一块空地。
- 给自己起一个容易记忆的自然名和简短有趣的自我介绍，确保大家一开始就记住你。
- 体现你身份的制服和服饰，解说员的标志和随身的解说装备包。
- 你的介绍中一定有独特的、标识标牌上未曾涉及的内容。
- 为即将开始的自然体验之旅略做“剧透”，包括主题、时间、特别值得期待的解说点等，但不要和盘托出，而是要引发大家的好奇心。
- 根据团队成员的构成，针对性地提出注意事项和安全要求。
- 记得还要提醒大家注意不采摘和带走任何公园自然物，不随便丢弃垃圾等行为准则哦。

辅助道具推荐

- 游线总体导览牌

Suzhuang Town | Gutian Mountain—Gutian Temple | Evergreen broad—leaved forest exploration route

Interpretation No. SZ-GT-01

Stop 01: Assembly point

● Interpretation theme

General introduction to this nature experience route

● Key points

1. Overall introduction of Gutian Mountain Route
2. Make sure the audience is listening and establish the identity as an interpreter

● Site overview

The assembly point is outside the entrance of the mountain gate, there is a general guide board in front of the gate

Nice to meet you everybody. Welcome to the Qianjiangyuan National Park. I am the interpreter of the national park XXX, and I will explore this natural journey with you together.

Now the Gutian Mountain is right under our feet. Does anybody know which direction Gutian Mountain is located? Let's take a look at this guide board together. Gutian Mountain is located on the southwest side of Qianjiangyuan National Park, but mindful friends may notice that Gutian Mountain is a large area with a total of four routes. What we are experiencing today is the first one: the nature experience route, where the evergreen broad-leaved forest can be explored. Friends with good endurance can continue to challenge other routes later.

Do you know why Gutian Mountain is called "Gutian"? This actually has to do with the highest point we are going to challenge today. There is a Lingyun Temple at the top of our destination, also called the Gutian Temple. The word "Gu" means old, which is derived from the lush towering ancient trees in the mountain. "Tian", also known as "fields", comes from the fact that although the mountain remains natural, it still has human footprints with farmland paddy fields often bordering the mountains. In front of Gutian Temple, there is a paddy field with a story behind it. As for the story, I will tell you when we reach the top of the mountain.

Well, now warms up everyone! it's time for hiking! This route will take about 3 hours and the elevation will increase about 400 meters. There are many steep sections along the way while the trail facilities are also relatively primal. It might pose some challenge to your strength and endurance

As the elevation increases, we will see different forest landscapes: from evergreen broad-leaved forest to evergreen-deciduous broad-leaved forest as well as coniferous and broad-leaved mixed forest, and finally to coniferous forest. Along the way, we have the opportunity to see various plants such as ring-cupped oak, gugertree, sweet oachestnut, and Masson pine, and we might encounter local animals as well. Of course, there must be the chorus of birds, the singing of creeks, and everyone's laughter.

Along the way. I must remind you that everyone must pay attention to your own safety, whether it's going up or down the mountain. Please follow me closely but orderly. The signal in the mountain may not be very good. Don't act alone! Don't run! Be careful!

Let's go!

Notes: Interpretation assembly point is usually at the start of the trail with a clearly marked meeting place. It is recommended to set it next to the general guide board of this trail.

Attentions

Remember, pay attention to your debut at the assembly meeting point. When you show up to the audience, your job as an interpreter has already begun. As an interpreter, it is very important to attract the attention of the audience and establish your own personality, such as using a short, clear and interactive opening statement. The following tips can help you achieve this goal.

- A proper assembly place that can help gathering the crowd and start this natural experience, such as the open space next to the general guide sign at the start of the trail.
- Give yourself an name that's easy to remember as well as a short but interesting self-introduction to make sure the tourists remember you from the beginning.
- Uniforms and costumes that reflect your identity: the logo of the Interpreter and supporting materials tool kit.
- There must be something unique in your introduction that isn't mentioned in the ID card.
- Spoil some of the upcoming nature experience routes, including the theme, time, and particularly worthy interpretation, *etc*. in order to arouse everyone's curiosity.
- Based on specific team members, pose targeted safety requirements and precautions.
- Don't forget to remind everyone not to pick and take away any natural objects in the park, and don't throw garbage on the ground, *etc*..

Recommended auxiliary facilities and toolkits

- Overall guide board of the route

苏庄镇 | 古田山—古田庙 | 探秘常绿阔叶林自然体验路线 解说编号：SZ-GT-02

解说点 02 山门入口步道

29°14′53.76″ N，118°7′57.45″ E
海拔：535m

解说主题

北纬三十度的生机绿洲

解说要点

1. 公园的常绿阔叶林在全球的意义
2. 常绿阔叶林的形成与青藏高原和东亚季风的关系

场地概述

入口山门，门前有步道的总体导览牌

解说方式 Interpretation Style

- ☑ 人员解说 Staff Interpretation
- ☐ 教育活动 Educational Activity
- ☑ 解说牌 Interpretation Signage
- ☐ 场馆解说 Hall Interpretation

解说季节 Interpretation Season

- ☑ 春季 Spring
- ☑ 夏季 Summer
- ☑ 秋季 Autumn
- ☑ 冬季 Winter

如果你的手机自带指南针功能，请大家看一下我们现在所处的位置（海拔 535m，北纬 29°，东经 118°）。设想一下，这个点在地球上处在什么位置？当我们俯瞰整个地球时，绕着北纬 30° 的环形区域，又是什么样子呢？

「钱江源国家公园之于全球的意义」

其实答案就在我们常用的手机地图中。大家可以尝试打开手机自带的导航地图，然后把图层切换到卫星地图模式（或者阅读随身携带的全球北纬 30° 的植被分布图），找到我们现在所处的位置，然后沿着北纬 30° 的方向水平延伸，看一看这条线穿过哪些区域，这些区域的景观又有什么不同。

地球北纬 30° 的区域基本被黄色的荒漠所覆盖，唯有中国长江中下游地区的这片浓绿呈现截然不同的风貌，而钱江源国家公园正处在这片绿洲之内。

「青藏高原对东亚季风的影响」

那么这片生机绿洲是如何形成的呢？这要从 4000 万年前青藏高原的隆起开始讲起。如今作为“世界屋脊”的青藏高原在亿万年前曾是一片汪洋大海，因为亚欧板块和印度洋板块的挤压使得青藏高原不断抬升直至今天的高度，改变了全球的地形地貌和气候格局，对中国东部地区的影响尤为巨大。

青藏高原冬季可以阻挡来自北方的寒冷气流，夏季则截留了季风从海洋带来的暖湿气流，为中国东部地区带来了丰沛的降水，再加上这里地处亚热带，热量充足，为常绿阔叶林的生长提供了绝佳的条件，这里也因此成为地球北纬 30° 黄色荒漠带上独特的生机世界。

注意事项

第一个解说点的好坏直接影响后续访客的兴趣，所以必须让他们感受到有趣、特别，你可以：

- 借助地图或者手机 APP，让访客更加直观地感受到身处何地，但前提是你对这两者都非常熟悉；
- 切换视角，从全球的北纬 30° 到中国的北纬 30°，重要的是体现出钱江源的独特和珍贵；
- 中国北纬 30° 的独特环境与青藏高原的隆起带来的典型的季风气候等密不可分。

辅助道具推荐

- 标识牌：北纬三十度的生机绿洲
- 中国地形图

拓展活动推荐

- 钱江源国家公园打卡地图（P562：活动卡 02）

拓展知识点

「世界的北纬 30°」

- 在人类探究自然、认识世界的过程中，北纬 30° 一直是个充满了神秘和未知的地带。这里是地球至高点珠穆朗玛峰的所在地，尼罗河、幼发拉底河、长江、密西西比河等全球最重要河流的入海口，华夏文明、古印度文明、古巴比伦文明、古埃及文明均分布于此，也是百慕大三角、死海、撒哈拉沙漠等人类认知尚无法解释的神秘区域所在地。

Suzhuang Town | Gutian Mountain–Gutian Temple | Evergreen broad–leaved forest exploration route

Interpretation No. SZ-GT-02

Stop 02: Entrance of the trail by the mountain gate

Interpretation theme

The 30° N miracle of green oasis

Key points

1. The significance of protection from a global perspective
2. The impact of the Qinghai – Tibet Plateau and the East Asian monsoon

Site overview

Entrance of the mountain gate, the general guide board of the trails is right in front of it

If your mobile phone comes with a compass function, please take a look at our current location: (520 meters above sea level, 29° north latitude, 118° east longitude). Imagine where this point is on the earth; when we look down on the whole planet, what the area circling the 30° north latitude looks like.

"The significance of Qianjiangyuan National Park to the world"

You can take a look at this map of vegetation distribution at 30° north latitude, or open the Baidu or Gaode map that comes with your phone, and then switch the layer to satellite pattern, find where we are now, and then draw a straight line along the 30° north latitude to see what areas it crosses.

How is the landscape different in these areas? Have you noticed that the area of 30° north latitude of the earth is basically covered by yellow deserts, but this thick green in the middle and lower Yangtze River in China shows a completely different feature, and Qianjiangyuan National Park is right within this oasis.

"Impact of the Qinghai–Tibet Plateau on the East Asian Monsoon"

So how did this vibrant oasis form? This should trace back to the rise of the Qinghai – Tibet Plateau 40 million years ago. The Qinghai – Tibet Plateau, which is now the "roof of the world", was a vast ocean hundreds of millions of years ago. The Tibetan Plateau kept rising due to the squeezing of the plate, which has changed the global landform and climate pattern, especially on eastern China.

The Qinghai – Tibet Plateau can block the cold airflow from the north in winter, and intercepts the warm and humid airflow brought by the monsoon from the ocean during summer, leaving the southeast China in a relatively warm and humid state, and create a rare continuous distribution of a large-scale subtropical evergreen broad-leaved forest in this low-elevation area, becoming a miracle of green oasis on the 30° north latitude yellow desert belt.

Attentions

The quality of the first interpretation directly affects visitors' interest for later, so you must keep them interested and special. You can:

- By using maps or mobile apps, visitors can feel where they are more intuitively, but you have to be familiar with both;
- By switching angles, from 30° north latitude around the world to 30° north latitude in China, it is important to reflect the uniqueness and preciousness of Qian Jiangyuan;
- The unique environment of 30° north latitude in China is inseparable from the rise of the Tibetan Plateau.

Recommended auxiliary facilities and toolkits

- Global map of 30° north latitude vegetation distribution
- The mobile phone built-in compass (check how to operate mainstream mobile phone brands in advance)
- Mobile phones built-in maps (Baidu, Gaode, *etc.*)
- A Chinese geomorphic model made of paper clay (highlighting the Tibetan Plateau and 30 degrees north latitude) or find a slope on site to demonstrate the blockage of the hot and cold air flow by high mountains

Recommended activities

- Task list map of Qianjiangyuan National Park (P562：No. 02)

Extra information

"30° north latitude of the world"

- It starts from the Sahara Desert in Africa, to the Arab region in Asia, then to the Tibetan Plateau in our country, the middle and lower Yangtze River Plain, and to the south of North America. Throughout the exploration of mankind, the 30° North latitude has always been a zone full of mystery and unknown. This is where Mt Everest stands, the worldwide essential rivers flow into the sea, as well as all the ancient civilizations and mysterious landscapes beyond our knowledge originate.

苏庄镇 | 古田山—古田庙 | 探秘常绿阔叶林自然体验路线　　解说编号：SZ-GT-03

解说点 03 原真森林

29° 14′ 56.09″ N，118° 8′ 4.58″ E
海拔：576m

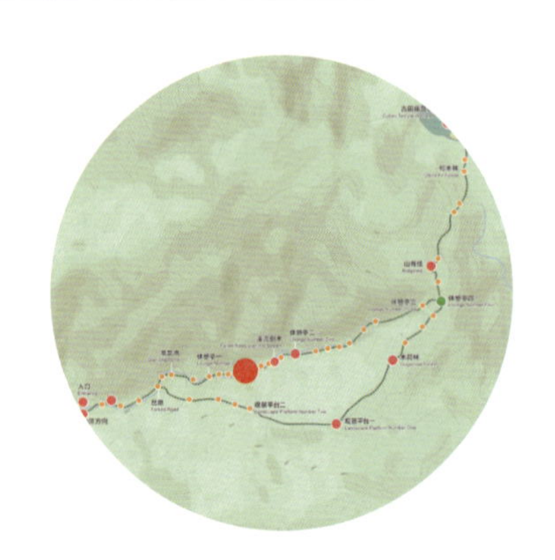

解说主题

长江三角洲最后的原真森林

解说要点

1. 长江三角洲最后的原真森林
2. 天然的物种基因库

场地概述

林荫道上，可以观察到常绿阔叶林

解说方式 Interpretation Style

☑ 人员解说 Staff Interpretation　☐ 教育活动 Educational Activity
☑ 解说牌 Interpretation Signage　☐ 场馆解说 Hall Interpretation

解说季节 Interpretation Season

☑ 春季 Spring　☑ 夏季 Summer
☑ 秋季 Autumn　☑ 冬季 Winter

解说词

各位朋友，我们现在身处的这片密林，就是古田山典型的常绿阔叶林。这片森林又被称为“长江三角洲最后的原真森林”。为什么这么说呢？

「长江三角洲最后的原真森林」

大家知道，长江三角洲从古至今都是我国经济最发达、人口最密集、现代化程度最高的地区之一。相应的，这里也成了受人类活动影响而改变最多的地区之一。钱江源恰好位于这个巨大又热闹的城市群的边缘。但是请大家再看看现在我们身边漫山遍野的浓绿森林，这难道不是大自然对钱江源特别的眷顾吗？

常绿阔叶林在全球均有分布，但由于东亚季风的影响，这里的常绿阔叶林生长尤其茂盛，生物多样性丰富，被称作“典型常绿阔叶林”。我国南方各省也均分布有常绿阔叶林，但是像古田山这样位于低海拔地区、大面积连续分布的情况，是非常少见的。

「天然的物种基因库」

古田山的森林生态系统约占公园总面积的93%，这里有高等植物2245种，其中不乏一些古老的孑遗植物，中国的特有种和稀有种。比如，秋冬开花的长柄双花木、“摇钱树”青钱柳、国家一级重点保护野生植物南方红豆杉等。它们世世代代在这里生息繁衍，共同造就了古田山典型、多样、古老、稀有和完整的生物群落，堪称“天然的物种基因库”。

注意事项

- 从长江三角洲经济发展地区切入，围绕繁华与原真的对比，说明钱江源常绿阔叶林的脆弱与珍贵；
- 最好找到一片你认识的常绿阔叶林，选择一片开阔的地方，从身边的植物开始，介绍这里的植物多样性。

辅助道具推荐

- 标识牌：长三角的首个国家公园
- 标识牌：原真森林的顶梁柱
- 古田山常绿阔叶林的航拍图

拓展活动推荐

- 古田山常见植物打卡 (P563：活动卡03)

拓展知识点

「稀有种的“庇护所”和安乐园」

- 中国科学院等机构在古田山设立24ha大样地系统研究当地生物多样性及保护价值，研究发现大量稀有种分布，如三峡槭、厚壳树、大果卫矛、四照花和红紫珠等，数量占总物种数的37.1%。这些稀有或濒危植物分布区狭窄，个体数稀少，大多数零星分布于各类植物群落中，仅有少数能作为优势种或次优势种形成特定的群落，我们将其称为“稀有植物群落”。

Suzhuang Town | Gutian Mountain－Gutian Temple | Evergreen broad－leaved forest exploration route

Interpretation No. SZ-GT-03

Stop 03: Primitive and indigenous forest

● Interpretation theme

The important value of Gutian Mountain forest

● Key points

1. The last primitive and indigenous forest in the Yangtze River Delta
2. Natural species gene pool

● Site overview

On the way where you can observe the evergreen broad-leaved forest

Dear friends, the lush forest right in front of us is a typical evergreen broad-leaved forest in the Gutian Mountain. This forest has been called the last primitwe and indigenous forest in the Yangtze River Delta. Why do you say that!

"The last primitive and indigenous forest"

As we all know, the Yangtze River Delta has been one of the most economically developed, densely populated, and modernized areas in China since ancient times. In other words, it is the most affected ecosystem by human activities as well, and Qian Jiangyuan is surrounded by this huge and lively urban agglomeration.

Now please take a second look at the lush green forests around us. Isn't this a special blessing from nature?

Evergreen broad-leaved forests are distributed all over the world. However under the influence of the East Asian monsoon, the evergreen broad-leaved forests here are lush and rich in biodiversity, and are called "typical evergreen broad-leaved forests". Although they are also distributed in the southern provinces of China, it is very rare that evergreen broad-leaved forests are widely distributed in such a low-elevation area continuously like the Gutian Mountain.

"Natural species gene pool"

The forest ecosystem of Gutian Mountain occupies about 93% of the national park and is home to more than 2,000 species. These creatures have lived and thrived here from generation to generation, and together generate the typical, diverse, ancient, rare and complete biological community of Gutian Mountain, known as "natural species gene pool".

Attentions

- Starting from the economic development of the Yangtze River Delta, focusing on the contrast between urbanization and nature, indicating how vulnerable and precious the Qianjiang Yuan evergreen broad-leaved forest is;
- It's best to find an evergreen broad-leaved forest you know, choose an open ground, and start with the plants around you to introduce their diversity.

Recommended auxiliary facilities and toolkits

- Global distribution map of forest vegetation (highlight evergreen broad-leaved forest)
- The geographical location of the Yangtze River Delta
- Aerial image of the evergreen broad-leaved forest of Gutian Mountain

Recommended Activities

- Common plants in Gutian Mountain (P563 : No. 03)

Extra information

"Shelter and paradise for rare species"

- Chinese Academy of Sciences and other institutions have set up a large plot of 24 hectares in Gutian Mountain to systematically study the local biodiversity and conservation value, and unveiled a large number of rare species distributions, such as *Acer wilsonii*, *Ehretia acuminata*, *Euonymus myrianthus*, Chinese dogwood, and *Callicarpa rubella*, which accounts for 37.1% of the total species. These rare or endangered plants have a narrow distribution area with few individuals. Most of them are scattered in various plant communities. Only a few can form specific communities as dominant or subdominant species. We call them "rare plant communities".

苏庄镇 | 古田山—古田庙 | 探秘常绿阔叶林自然体验路线　　解说编号：SZ-GT-04

解说点 04 倒木

118°8′17.92″ E，29°14′58.92″ W
海拔：588m

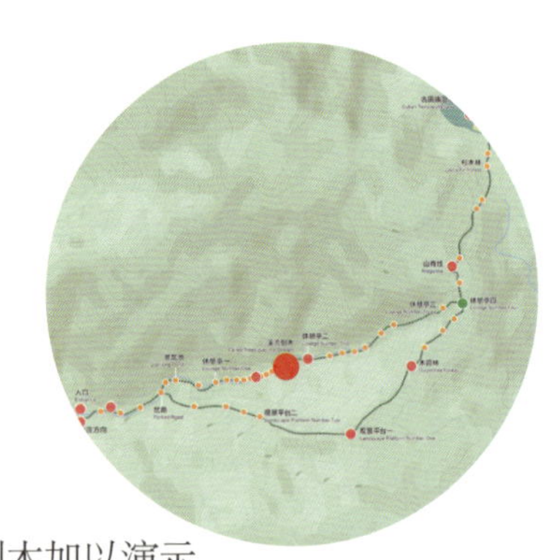

解说主题

认识倒木、林窗与森林群落更新的关系

解说要点

1. 认识倒木对于森林更新的意义
2. 认识林窗对森林生态系统更新的意义

场地概述

上山途中，可观察溪流倒木景观，但最好能找到林中的倒木加以演示

解说方式 Interpretation Style

- ☑ 人员解说 Staff Interpretation
- ☐ 教育活动 Educational Activity
- ☑ 解说牌 Interpretation Signage
- ☐ 场馆解说 Hall Interpretation

解说季节 Interpretation Season

- ☑ 春季 Spring
- ☑ 夏季 Summer
- ☑ 秋季 Autumn
- ☑ 冬季 Winter

「倒木对于森林更新的意义」

行走在森林中，除了各种高大的乔木之外，我们经常还会看到一些倒下的树木。它们有的是树木到了一定年纪后自然死亡的结果，也有可能是因为遭昆虫蛀蚀、雷击、台风、火灾、泥石流等自然灾害引起的林木死亡现象。

倒木虽死不衰。大家可以凑近观察下这个倒木，看看除了木头之外，有什么别的发现没？这里有……

任何生命都有自己的极限，树木也是。它们到了一定的年纪以后，会衰老、死亡。但倒木并非单纯的树木死亡事件，还是森林生态系统自动更新过程中非常重要的一个环节。树木的倒下，不仅可以为林下的幼苗让出更多的生存空间，同时它们本身也还在继续参与着森林生态系统的物质循环。倒下的树木是各种昆虫、真菌和微生物的温床。这批小小的居民承担着分解者的角色，可以将倒木转化成养分，进入土壤并再次参与到自然界的物质循环中去。

「认识林窗与森林生态」

倒木的另一个重要的贡献，就是形成“林窗”，这是森林实现自我更新的一种重要的生态学现象。原始森林植物生长十分茂盛，一般林下空间很难有充足的光照，但是在有些地方，森林仿佛被从天而降的巨石砸穿了一样，留下一处垂直的空隙，使得光照可以直接照射到地面，这种现象即是林窗。

林窗的形成还要从几颗种子开始。在古田山阳光雨露的滋润下，沉睡的种子挣脱大地母亲的怀抱，迅速发芽，长成小树苗、大树。它们根与根相连，枝与枝相依，不浪费每一份阳光，但这样有一个坏处就是树林下面新生的小树苗因得不到充足的光照而难成活。照这样下去可不行，生命只有不断更迭才能常生常新。

聪明的大自然懂得用简单的办法，借助自然死亡的老树来解决问题，倒伏的大树可以在森林茂密的林冠中顶上打开一扇“窗”，让阳光照射到林下。这样藏在土壤种子库中休眠的种子接收到了阳光的讯号，便开始萌发、生长，取代死去的树木，使森林得以生机永续。

注意事项

- 选取典型的倒木进行解说，注意引导访客主动观察倒木与环境的关系，而不仅仅是倒木本身。比如，大树倒下后倒木上有什么变化，环境发生了什么变化，和周围有什么区别？

辅助道具推荐

- 标识牌：朽木与新生
- 倒木上生长的真菌照片
- 倒木与昆虫照片
- 林窗照片

Suzhuang Town | Gutian Mountain–Gutian Temple | Evergreen broad–leaved forest exploration route

Interpretation No. SZ-GT-04

Stop 04: Fallen trees over the stream

Interpretation theme

Understand the relationship between fallen trees, forest gaps and communities

Key points

1. The significance of fallen tree for forest regeneration
2. Understand forest gaps and forest ecology

Site overview

On the way up the mountain, where visitors are able to observe the fallen trees over the stream

"The significance of fallen trees for forest regeneration"

While walking in the forest, in addition to all kinds of tall trees, we may also see some fallen trees right in or beside the trail. Some of them fell down after their natural death at a certain age, while others might die because of natural disasters such as lightning, typhoon, fire, debris flow and insect pests.

However, the fallen trees always pregnant another lively world. Let's look closely at this mysterious wood, what else can you find?

Every life has its limits, and so do trees. When they reach a certain age, they become old and die. However, falling down is not a simple signal of a tree's death. It's also a very important cradle of a restarted regeneration of an ecosystem. When trees fall down, they will not only create room and light for seedlings, but also continue to contribute to the material cycle of the forest ecosystem. These small residents play the role of decomposers, transforming fallen trees into nutrients in the soil, and once again participating in the cycle of nature.

"Getting to know forest gaps and forest ecology"

The falling trees has another contribution to the forest, that is the creation of "forest gap". In Chinese, we call it window of the forest. This is a very important ecological phenomenon of forest self-regeneration. Primal forest plants grow so abundantly that there is little light underneath, but in some places, the forest seems to have been hit by a big stone, leaving a vertical gap that allows light to reach the ground directly. This phenomenon is called "forest gap".

The story begins with a few seeds. Moisturized by the sunshine and dewdrops of Gutian

Mountain, the hibernating seeds broke free from mother earth, and quickly germinated and grew into small saplings then big trees, roots by roots, branches by branches. One disadvantage is that the young saplings beneath the forest can hardly survive because of insufficient sunlight. If then, the circle of life can't sustain like that.

Our smart mother nature applies an easy solution to let the naturally died trees to solve the problem. The fallen trees can open a "window" in the thick forest canopy to let the sunlight shed to the ground. In this way, the dormant seeds hidden in the soil can receive the signal from the sunlight, then start to germinate and grow, replacing the dead trees, so that the forest can live forever.

Attentions

- Find if there is a typical fallen tree to make your interpretation tangible. Remind the visitors to pay special attention to the relation between fallen tree and the environment, instead of just the tree. Such as what has happened to the tree after its falling down? What happens to the environment, and what is the difference between it and its surroundings?

Recommended auxiliary facilities and toolkits

- Interpretation signage: death and rebirth
- Photos of fallen tree with fungi or insects
- Photos of forest gap

苏庄镇 | 古田山—古田庙 | 探秘常绿阔叶林自然体验路线　　解说编号：SZ-GT-05

解说点 05 休憩亭二

29° 14′ 56.09″ N，118° 8′ 4.58″ E
上山第二个休憩亭　海拔：636m

● 解说主题

认识常绿落叶阔叶混交林（过渡带）

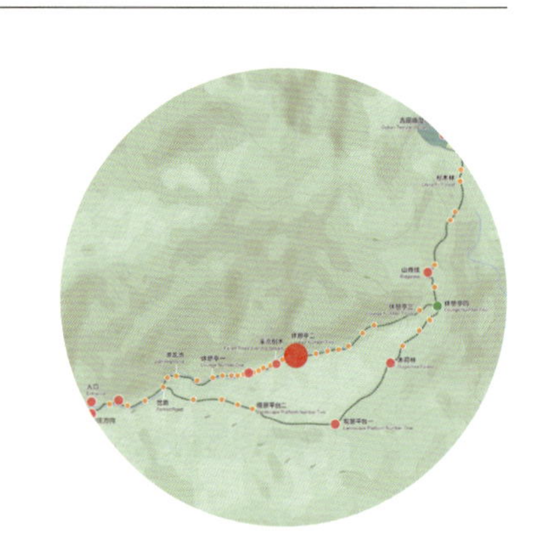

● 解说要点

1. 认识常绿落叶阔叶混交林及其代表植物群落
2. 认识公园丰富的植物区系

● 场地概述

上山途中，可观察常绿落叶阔叶混交林景观

解说方式 Interpretation Style

☑ 人员解说 Staff Interpretation　☐ 教育活动 Educational Activity
☑ 解说牌 Interpretation Signage　☐ 场馆解说 Hall Interpretation

解说季节 Interpretation Season

☑ 春季 Spring　☑ 夏季 Summer
☑ 秋季 Autumn　☑ 冬季 Winter

现在请大家仔细观察一下我们身边的植物，大家有没有什么新发现？和我们刚上山时看到的植物有什么不同吗？

「认识常绿落叶阔叶混交林与代表性稀有植物群落」

各位朋友，我们现在来到的是以常绿落叶阔叶混交林为主的区域。什么是常绿落叶阔叶混交林呢？从名字中就可看出，它是指由四季常绿和秋冬落叶的阔叶树种混合组成的森林类型。和山脚相比，这里的温度、湿度、光照等环境条件随着海拔升高发生了变化，一些耐寒的落叶树种便取代了对热量需求较高的常绿树种。

古田山的常绿落叶阔叶混交林，根据局部地形、气候等条件的差异，又会有不同的组合，比如，适合生长在沟谷的南酸枣－紫楠林和稀有植物群落披针叶茴香－紫茎林，以及生长在山坡的短柄枹－细叶青冈林。我们现在所处的就是XXX林（根据实际情况解说）。

「森林过渡带的家族成员」

请问大家都是来自哪里？你平时有注意观察过身边的植物吗？无论你从哪里来，你都有可能在这里找到熟悉的植物朋友。这是因为钱江源国家公园虽然地处华东，却汇集了来自华东、华北、华中和华南等全国各地的典型植被，堪称“中国植物多样性的天然博物馆”。

从地理位置上来看，钱江源国家公园刚好位于我国华北和华南，华东与华中地区交界的“十字路口”，为各种植物的汇聚创造了可能；另外，钱江源国家公园复杂的山地地形地貌塑造了多种多样的局地小气候，这里不仅生活有偏爱温凉环境的桑树、构树等来自华北植物区系的落叶阔叶树种，也有喜好湿热环境的华南植物区系植物木莲、含笑和猴欢喜等。同时，西部地区还通过一些东西走向的山岭、水系，与西南、华中植物区系相连，是青钱柳、八角莲等孑遗植物传播的重要途径。

注意事项

- 这里只提供了常绿落叶阔叶混交林的简单介绍，具体到物种可参看“通用型生物多样性解说方案”的内容。
- 植物区系的内容最好能够结合地图解说。

辅助道具推荐

- 标识牌：常绿落叶阔叶混交林——过渡地带的森林变化
- 标识牌：一园一世界

拓展活动推荐

- 古田山常见植物打卡（P563：活动卡 03）

Suzhuang Town | Gutian Mountain－Gutian Temple | Evergreen broad－leaved forest exploration route

Interpretation No. SZ-GT-05

Stop 05: Lounge No. 2

Interpretation theme

Understand the evergreen-deciduous broad-leaved mixed forest (Transition Zone)

Key points

1. Understand the evergreen-deciduous broad-leaved mixed forest and representative plant communities

2. Understand the family members of the forest transition zone

Site overview

On the way up the mountain, where visitors are able to observe the evergreen-deciduous broad-leaved mixed forest landscape

Now please take a closer look at the plants around us. Have you noticed anything new? Is there any difference between the plants we saw here and by the mountain foot?

"Understand evergreen-deciduous broad-leaved mixed forest and representative rare plant communities"

Dear friends, we are now in an area dominated by evergreen-deciduous broad-leaved mixed forest. What is evergreen-deciduous broad-leaved mixed forest? As you can see from the name, it refers to a forest type that is a mixture of evergreen in all seasons and deciduous broad-leaved tree species. The environmental conditions here such as temperature, humidity, and light have changed with the rise of the elevation compared with the mountain foot, and some hardy deciduous tree species have replaced evergreen trees with higher temperature requirements.

The evergreen-deciduous broad-leaved mixed forest in Gutian Mountain has different combinations based on different terrains, such as the Nepali hog plum－*Phoebe sheareri* forest and the rare plant community Chinese anise－Chinese stewartia forest suitable for growing in the ravine as well as the *Quercus glandulifera*－ring-cupped oak forest growing by the hillside. We are now in the XXX forest (interpret base on the actual condition).

"Family members of the forest transition zone"

Where does everybody come from? Have you paid attention to the plants around you? No matter where you are from, you may find familiar plant friends here. Although Gutian Mountain is located in the subtropical zone, it

also contains plants in East, South, Central, and North China.

But having said that, the most typical vegetation here is still some species of evergreen broad-leaved forest. For example, the family of Fagaceae, which produces nuts, the aromatic genus *Cinnamomum*, which can extract essential oils, and the magnificent family of Magnoliaceae.

In addition, there are the genus *Distylium* of family Hamamelidaceae; the holly genus of the holly family; the *Elaeocarpus* and *Sloanea* genera of the Elaeocarpus family; the *Camellia* genus, the *Adinandra* genus, the *Ternstroemia* genus and the *Cleyera* genus of the Theaceae family, and the *Ardisia* genus; *Colicwood* genus of the Myrsinaceae family; *Symplocos* genus of the Symplocos family and so on. All of them are common species of broad-leaved forest.

Attentions

- The introduction to the evergreen and deciduous broad-leaved mixed forest is brief. For specific species, please refer to the “General Interpretation of Biodiversity”.
- As for the content of flora, it is better to interpret with the map.

Recommended auxiliary facilities and toolkits

- Interpretation signage: Deciduous and evergreen broadleaved forest: transition zone
- Interpretation signage: a collective display of floras throughout the country

Recommended Activities

- Common plants in Gutian Mountain (P563 : No. 03)

苏庄镇 | 古田山—古田庙 | 探秘常绿阔叶林自然体验路线　　解说编号：SZ-GT-06

解说点 06 休憩亭四

29° 14′ 56.09″ N，118° 8′ 4.58″ E
上山第四个休憩亭　海拔：860m

解说主题

认识针叶阔叶混交林（攀向高处的群落变化）

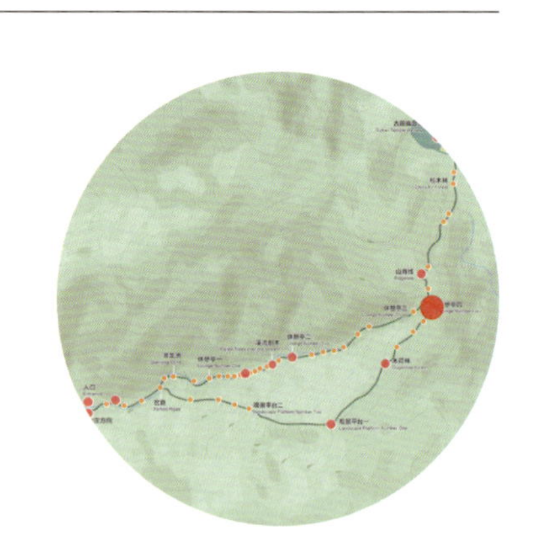

解说要点

1. 认识针叶阔叶混交林与海拔和植被演替的关系
2. 认识针叶阔叶混交林常见的树种

场地概述

上山途中，可观察针叶阔叶混交林景观

解说方式 Interpretation Style

☑ 人员解说 Staff Interpretation　☑ 教育活动 Educational Activity
☑ 解说牌 Interpretation Signage　☐ 场馆解说 Hall Interpretation

解说季节 Interpretation Season

☑ 春季 Spring　☑ 夏季 Summer
☑ 秋季 Autumn　☑ 冬季 Winter

有朋友在上山的途中已经注意到了，随着山体的爬升，这里针叶树的数量和种类越来越丰富，按照我们刚刚经过的常绿落叶阔叶混交林的命名规则，大家可以试着给这一片森林命个名。没错，是针叶阔叶混交林。

「认识针叶阔叶混交林与海拔和植被演替的关系」

在垂直方向上，针叶阔叶混交林一般位于常绿阔叶林之上海拔较高的区域，而从群落发展的角度来看，则代表了植被演替的一个阶段。大家知道什么是植被演替吗？植被演替，是指随着时间的推移，一种植物群落被另一种植物群落逐渐改变或替代的过程。

现在我们所在的针叶阔叶混交林就属于本地植被演替的中期阶段。更早的时候，这里是马尾松等针叶树种的天下，它们是早期的“拓荒者”。随着马尾松群落的不断壮大，森林越来越茂密，越来越阴郁，导致林下环境由于缺少足够的光照，不再适合马尾松幼苗的生长，而一些耐阴的阔叶树种，如木荷、甜槠等，则占据了马尾松林下的空间，因其生长速度快、叶片宽阔、树形高大，慢慢取代了后者，直至占据整片森林，最后演替成相对稳定的常绿阔叶林顶极群落。

「认识针叶阔叶混交林中常见的树种」

古田山针阔叶混交林中的植物种类丰富，组合多样。其中最常见的是以马尾松为针叶树种，加上多种阔叶树组成的植被群落，如马尾松－石栎－青冈林，马尾松－甜槠林，木荷－马尾松－甜槠林。

古田山有一处特别的针叶阔叶混交林：南方红豆杉－甜槠－木荷针叶阔叶混交林，位于我们西北方的洪源自然村附近。这片巨大的南方红豆杉林组成的群落，在古田山区域仅发现一处，属于稀有植物群落，也是当地人世代守护的结果。

注意事项

- 介绍针阔叶混交林代表的演替阶段，说明植被演替的含义。
- 列举代表树种与珍稀植物，如南方红豆杉群落。
- 涉及具体物种时，可参考“通用型人员解说方案——生物多样性”的相关内容。

辅助道具推荐

- 标识牌：针阔叶混交林——攀向高处的森林变化
- 本地植被演替卡片
- 捡拾的马尾松凋落物（松针、球果）
- 南方红豆杉物种照片

拓展活动推荐

- 古田山常见植物打卡（P563：活动卡 03）

Suzhuang Town | Gutian Mountain—Gutian Temple | Evergreen broad—leaved forest exploration route

Interpretation No. SZ-GT-06

Stop 06: Lounge No. 4

Interpretation theme

Identify mixed coniferous and broad-leaved forests (community changes as elevation increases)

Key points

1. Understand the relationship between coniferous and broad-leaved mixed forests as well as elevation and vegetation succession

2. Understand the common tree species in coniferous and broad-leaved mixed forests

Site overview

On the way up the mountain, where visitors are able to observe the coniferous and broad-leaved mixed forests landscape

Some friends may have noticed on our way up the mountain that as elevation increases, the number and types of conifers here are getting richer. Based on the naming rules of the evergreen-deciduous broad-leaved mixed forest we just passed, you can try to guess the name of this forest. Yes, it is coniferous and broad-leaved mixed forest.

"Understand the relationship between coniferous and broad-leaved mixed forests as well as elevation and vegetation succession"

In a vertical direction, coniferous and broad-leaved mixed forests are generally located above evergreen broad-leaved forests at higher elevations; and from the perspective of community development , it represents one of the stages of vegetation succession. Does anybody know what vegetation succession is? Vegetation succession refers to the process of gradually replacing one plant community with another plant community over time.

The coniferous and broad-leaved mixed forest we are looking at now belongs to the mid-term stage of local vegetation succession. In the early stage, here was the world of Masson pine and other coniferous trees. They were the early "pioneers". As the Masson pine community continued to grow, the forest became denser and more gloomy, resulting in a lack of sufficient sunlight, no longer suitable for the growth of masson pine seedlings, and some shade-tolerant broad-leaved tree species, such as gugertree, Sweet oachestnut, *etc*., take the opportunity to enter the masson pine forest because of their fast growth, wide leaves, and tall trees, slowly replacing the latter and

eventually occupying the entire forest, and finally succeeded as a relatively stable top community of evergreen broad-leaved forest.

"Understand the common tree species in coniferous and broad-leaved mixed forests"

The Gutian Mountain coniferous and broad-leaved mixed forests are rich in plant species with diverse combinations. Among them, the most common is Masson pine as a coniferous species, plus various broad-leaved tree vegetation communities, such as Masson pine – Japanese oak – ring – cupped oak forest, Masson pine – sweet oachestnut forest, gugertree – masson pine – sweet oachestnut forest.

There is a special coniferous and broad-leaved mixed forest in Gutian mountain: Chinese yew – sweet oachestnut – gugertree broad-leaved mixed forest, located near Hongyuan Natural Village to the northwest of us. This huge community of Chinese yew forest is only one found in the Gutian Mountain area. It is a rare plant community and is the result of the protection from the local people for generations.

Attentions

- Introduce the successional stages of coniferous and broad-leaved mixed forest and explain the meaning of vegetation succession.
- Examples of representative tree species and rare plants: Chinese yew.
- When introducing the specific species, you can refer to the "General Interpretation of Biodiversity".

Recommended auxiliary facilities and toolkits

- Interpretation signage: coniferous and broadleaved forest—forest succession
- Photos with different distribution of vegetation types as the elevation changes
- Forest litter of Masson Pine find on-site (pine needles, cones)
- Photos of Chinese yew species

Recommended activities

- Common plants in Gutian Mountain (P563 : No. 03)

苏庄镇 | 古田山—古田庙 | 探秘常绿阔叶林自然体验路线　　解说编号：SZ-GT-07

解说点 07 山脊线

118°8′39.05″ E，29°15′9″ W
山脊线　海拔：895m

● 解说主题

平凡汇聚的博大家庭

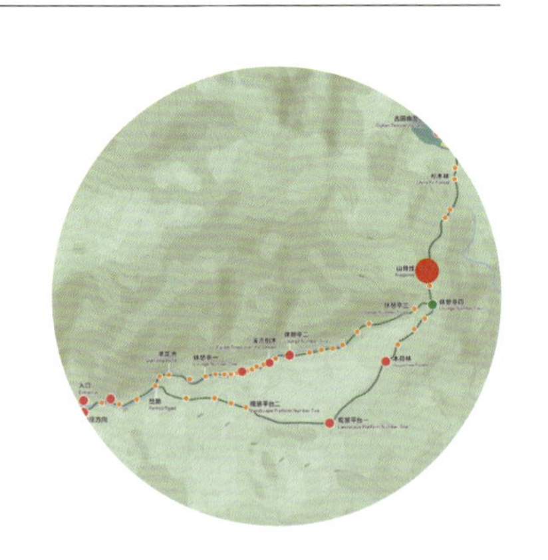

● 解说要点

1. 古田山的优势种群
2. 森林中的垂直结构

● 场地概述

古田山的山脊位置，是整条步道海拔最高的位置

解说方式 Interpretation Style

☑ 人员解说 Staff Interpretation　☐ 教育活动 Educational Activity
☑ 解说牌 Interpretation Signage　☐ 场馆解说 Hall Interpretation

解说季节 Interpretation Season

☑ 春季 Spring　☑ 夏季 Summer
☑ 秋季 Autumn　☑ 冬季 Winter

解说词

「古田山的优势种群」

如果让你选择一种最常见的植物代表古田山，大家觉得会是什么呢？如果从分布的广泛性和集中程度来看，甜槠、木荷等乔木一定是不二之选。沿途我们如果留心的话，就会发现地上掉落的甜槠子是不少浙江人童年的美味回忆。

有朋友会说，仅仅因为盛产甜槠就称为代表树种是不是有点主观？不，我可是有客观证据的。在古田山 24ha 的大样地中，生活着数以万计的植物。最擅长计算的科学家发现，甜槠不仅在数量上占据优势，胸高断面积也排在首位。换句话说，它们不仅多，而且壮。而第二壮的是木荷。这兄弟俩成了这片林子当之无愧的统治者。

「森林中的垂直结构」

植物群落会随着时间的变化而不断演替，最终达到一种相对稳定的阶段，被称为森林顶极群落，这也是各种植物对环境利用的最优化阶段，拥有典型的垂直分层结构。

作为森林的主宰，乔木理所当然地占据森林的高位，这一层被称为乔木层。乔木之下，灌木见缝插针，形成灌木层。灌木之下，是比较低矮的草本层以及最底部的由苔藓和地衣等植物组成的地被层。

除了植物外，动物也会有分层，大家可以留心观察一下，看看会有什么发现。原来，植物的分层是垂直结构的，而动物、微生物等的分层主要依赖于植物分层。环境条件越丰富，群落的层次就越多，层次结构就越复杂。环境条件差，层次就少，层次结构也就越简单。

注：古田山的山脊附近，是整条步道海拔最高的位置。附近较多木荷树，植被有一定林相的变换。

注意事项

- 解说的时候最好配有实景；
- 走到林下，结合实景，给访客展示森林的垂直分层结构。

辅助道具推荐

- 沿途捡拾的甜槠、木荷、青冈的叶片标本和果实凋落物

拓展知识点

「胸高断面积」

- 指乔木主干离地表面 1.3m 处胸高处的横截面面积。

Suzhuang Town | Gutian Mountain－Gutian Temple | Evergreen broad－leaved forest exploration route

Interpretation No. SZ-GT-07

Stop 07: Ridgeline

- **Interpretation theme**
  Gathering of an ordinary family
- **Key points**
  1. Dominant population in Gutian Mountain
  2. Vertical structure in the forest
- **Site overview**
  The ridge of Gutian Mountain which is the highest elevation of the whole trail

"Dominant species of Gutian Mountain"

If you were to choose one of the most common plants to represent Gutian Mountain, what do you think it would be? From the perspective of distribution and concentration, sweet oachestnut and gugertree must be the best choice. If we pay attention to it along the way, we will find the sweet oachestnut seeds falling down, which is a delicious memory of many Zhejiang persons about their childhood.

Some may argue that it is a bit subjective to call it a representative tree species simply because it is rich in sweet oachestnut. No, I have objective evidences. Tens of thousands of plants live in the large plot of 24 hectares in Gutian Mountain. Scientists who are most good at calculations find that not only do they have an advantage in quantity, but also in the DBH (diameter at breast height) area. In other words, they are not only numerous, but also plump. And the second most abundant is gugertree. The two brothers deservedly became the rulers of this forest.

"Vertical structure in the forest"

The plant community will succeed over time and eventually reach a relatively stable stage that is called the forest top community, It is also the most optimal stage for various plants to use the environment, with a typical vertical layered structure.

As the master of the forest, the arbor naturally occupies the high position of the forest. This layer is called the arbor layer. Under the arbor, the bushes form a shrub layer. Under the shrub, there is a relatively low herbaceous layer and the bottom layer composed of moss and lichen and other plants.

In addition to plants, animals also have layers. Take a closer observation and see what you find. It turns out that the layering of plants is a vertical structure, while the layering of animals, microorganisms, fungi, *etc*. mainly depends on the layering of plants. The richer the environmental conditions, the more layers there are in the community, and the more complex the hierarchy is. Poor environmental conditions mean fewer levels and a simpler hierarchy.

Notes: The ridge position of Gutian Mountain is the highest elevation of the whole trail. There are many gugertree nearby, and the vegetation has a certain change of forest phase.

Attentions

- It's best to have real scenes when interpreting.
- Walk down to the forest and combine the actual scene to show visitors the vertical layered structure of the forest.

Recommended auxiliary facilities and toolkits

- Pick up leaf specimens and fruit litters of sweet oachestnut, gugertree, and ring-cupped oak along the way.

Extra information

"DBH area"

- "DBH area" refers to the area of the height of the tree trunk 1.3 meters above the ground.

苏庄镇 | 古田山—古田庙 | 探秘常绿阔叶林自然体验路线　　解说编号：SZ-GT-08

解说点 08 古田庙湿地

29° 15′25.91″ N，118° 8′45.13″ E
古田庙　海拔：840m

解说主题

山间湿地的保护与恢复

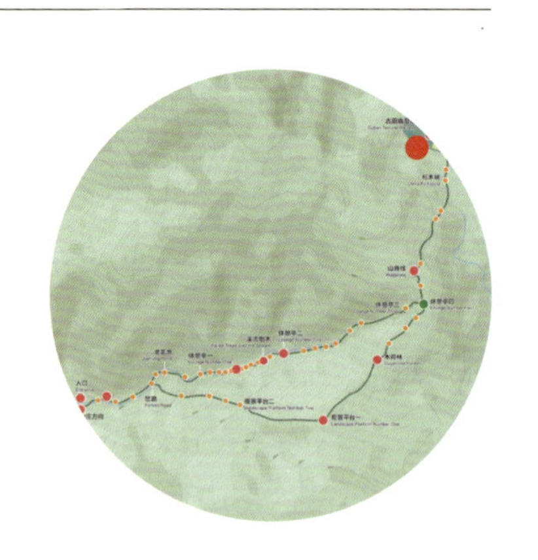

解说要点

1. 沼泽湿地的概念与形成原因
2. 沼泽湿地的意义与守护

场地概述

古田庙的前场空地水田边

解说方式 Interpretation Style

☑ 人员解说 Staff Interpretation　☑ 教育活动 Educational Activity
☑ 解说牌 Interpretation Signage　☐ 场馆解说 Hall Interpretation

解说季节 Interpretation Season

☑ 春季 Spring　☑ 夏季 Summer
☑ 秋季 Autumn　☑ 冬季 Winter

各位朋友，大家一路辛苦啦！我们现在所处的位置是此行的最高处——古田庙。大家待会儿可以在这里休息停留，各处走走，看看远处的山林，头顶的蓝天，放松下身心。但在这之前呢，我想请大家在眼前这片水田边稍作停留。为什么要在这里停住呢？有的朋友可能会说："这不就是一块普普通通的水田吗？公园里到处都是。"其实不然，这可是一片"有故事的"水田。

「沼泽湿地的概念与形成原因」

这里过去是一片由常年积水而形成的天然沼泽湿地。什么是沼泽湿地呢？和其他湿地相比，沼泽湿地普遍有明显的草根层，能像海绵一样吸收水分，对涵养水源、调节径流有非常重要的意义。沼泽湿地仅占浙江省湿地总面积的万分之二（0.02%），是浙江省非常珍贵且稀缺的湿地类型。

另外，不知道大家有没有注意到，古田山有些地名，如莲花塘、仁宗坑、枫楼坑都有一个"坑"或"塘"字，正是沼泽湿地在这里存在过的证明，而湿地形成的秘密也藏在这些字里。无论是"坑"或"塘"，都是因为地质原因形成的山间凹地，可以汇聚水源，同时这里底部平坦开阔，非常适宜造屋屯田，往往成为先民们定居安家的优先选择地点。

「沼泽湿地的意义与守护」

如今，原始自然湿地的状态几乎都已难觅踪影，很多在历史上被开发成为农田，随着人口迁出，村落缩小而逐渐被荒废。在我国的其他地方，许多自然湿地通过建立保护区或自然公园被有效地保护下来，但仍有众多未受到应有的重视。钱江源国家公园所保护的这些沼泽湿地，对于整个浙江省的湿地保护都具有一定意义。

如何借助国家公园发展的契机，让这些位于钱塘江源头，蕴含珍贵生态服务功能的湿地发挥其原有的功能，是我们今天面临的挑战，也是眼前这片"古田"留给我们的思考。

注意事项

- 到达山顶，先请访客休息，再进行解说。
- 从视线引导，观赏的角度开始，由过去到当下，解说沼泽湿地的概念。
- 由湿地的消失提升至守护的思考与倡议。

辅助道具推荐

- 标识牌：守护珍贵的沼泽湿地
- 标识牌：白际山脉的支撑与佑护

Suzhuang Town | Gutian Mountain–Gutian Temple | Evergreen broad–leaved forest exploration route

Interpretation No. SZ-GT-08

Stop 08: Gutian Temple wetland

● Interpretation theme

Protection and restoration of mountain wetlands

● Key points

1. The concept and formation of marsh wetlands
2. The significance and protection of marsh wetlands

● Site overview

The open field in front of the Gutian Temple

Good job, everybody! We finally reached the highest point of today's trip, Gutian Temple. Everyone can take a break and stay here for a while. You are welcome to walk around, take a look at the distant forest, the blue sky above, and relax. But before that, I would like to invite you to stop by the paddy field in front of you for a moment. Why is it here? Some may argue that this is just a normal paddy field and there are paddys like that all over Gutian Mountain. In fact, this is a "story" paddy field.

"The concept and formation of marsh wetlands"

Here was a natural marsh wetland formed by perennial water accumulation. What is a marsh wetland? Compared with other wetlands, marsh wetlands generally have a distinct grass root layer, which can absorb water like a sponge and is very important for conserving water sources and regulating runoff. Marsh wetlands only account for two ten-thousandths (0.02%) of the province's total wetland area, making it a very precious and scarce type of wetland in Zhejiang Province.

In addition, I don't know if you have noticed the names of some places in Gutian Mountain, for instance, Lotus Pond, Renzong Pit, Fenglou Pit. They all have the word "pit" or "pond", which proves that the marsh wetland existed here. The secret is also hidden in these words. Whether it is a "pit" or a "pond", it is a mountain concave formed by geological reasons, which can gather water sources. At the same time, the bottom here is flat and open, which is very suitable for building houses and farming at ancient times.

"The significance and protection of marsh wetlands"

Nowadays, pristine natural wetlands are almost impossible to find. Many areas that have been developed into farmland in history have gradually become deserted as

the population moves out. In other parts of our country, many natural wetlands have been effectively conserved by establishing protected areas or natural parks, but there are still many that have not received the attention they deserve. These marsh wetlands protected by Qianjiangyuan National Park have certain significance for the protection of wetlands in Zhejiang Province.

How to take advantage of the development of national parks to let these wetlands with precious ecological significance play their original functions is the challenge we are facing today, and it is also the thinking left to us by the "gutian" in front of us.

Attentions

- When reach the top of the mountain, let the visitors take a break before interpretation.
- Starting from the perspective of nature appreciation, explain the concept of marsh wetlands from the past to the present.
- Propose the sense of protection due to the disappearing of wetland.

Recommended auxiliary facilities and toolkits

- Interpretation signage: precious wetlands in the mountain
- Interpretation signage: shelter and nourishing of the Baiji Mountain

苏庄镇 | 古田山—古田庙 | 探秘常绿阔叶林自然体验路线　　解说编号：SZ-GT-09

解说点 09 古田庙

29°14′56.09″ N，118°8′4.58″ E
海拔：845m

● 解说主题

认识针叶林（适应高海拔的佼佼者）

● 解说要点

1. 认识针叶林在整个山地垂直植被分布中的位置
2. 认识针叶林代表树种：马尾松、黄山松、杉木

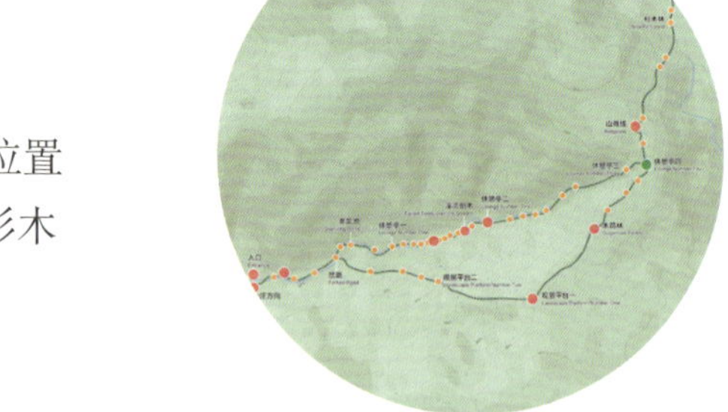

● 场地概述

古田庙，可观察典型针叶林景观

解说方式 Interpretation Style

☑ 人员解说 Staff Interpretation　☐ 教育活动 Educational Activity
☑ 解说牌 Interpretation Signage　☐ 场馆解说 Hall Interpretation

解说季节 Interpretation Season

☑ 春季 Spring　☑ 夏季 Summer
☑ 秋季 Autumn　☑ 冬季 Winter

各位朋友，欢迎来到适应高海拔的佼佼者——针叶树的世界。到此为止，我们已经见识了四种比较典型的植被群落。它们分别是什么，大家还记得吗？没错，从山脚的常绿阔叶林开始，到常绿落叶阔叶混交林，再到针叶阔叶混交林，最后是我们现在所身处的针叶林。

「认识针叶林在整个山地垂直植被分布中的位置」

在上一个点我们已经了解到马尾松等针叶树种是森林植被演替早期的“拓荒者”，这是从时间上来看的。而在垂直方向上，针叶林通常位于海拔较高的地方。

大家知道为什么吗？（估计访客会提到温度变化。）

是的，因为随着海拔的上升，温度不断降低，土壤有机质分解慢，养分相对贫乏，对温度、养分等需求较高的阔叶树种逐渐被耐寒、耐贫瘠的针叶树种所取代。马尾松的松针即是这一适应的见证。

「认识针叶林代表树种：马尾松、黄山松、杉木」

马尾松、黄山松、杉木等是钱江源国家公园针叶林中的代表树种。如果用三个短语概括的话，分别是“先头部队”马尾松，“登高能手”黄山松和“优质建材”杉木。

首先来看看马尾松。作为拓荒者，马尾松不仅具有探索家的大无畏风范，还非常大公无私。它们的存在为更多新物种和植株个体在群落中的生存提供资源和空间，并推动群落向地带性常绿阔叶林演替。

黄山松大家应该比较熟悉了，但黄山松不只是黄山才有哦。古田山就有大片的黄山松林，它们所处的海拔较马尾松更高，能适应更恶劣的生长环境，常常在石缝、峭壁中傲然独立，是当之无愧的“登高能手”。

我们要介绍的第三种针叶树“杉木”，因其枝干挺拔、生长迅速、易于种植，但同时又具有耐腐蚀、耐白蚁蛀蚀等特点，是我国秦岭以南，长江中下游地区最主要的用材树种之一，自古以来即被广泛用作各种建筑材料，尤其是建筑的精细结构部件。

注意事项

- 带领访客系统整理一下整个森林植被随着海拔的变化而变化的规律；
- 从针叶林与环境的关系出发，得出针叶林在植被演替和垂直变化角度生长环境的共性；
- 介绍三种针叶树的时候，可以参考“通用型人员解方案——生物多样性”解说资料。

辅助道具推荐

- 标识牌：针叶林——适应高海拔的佼佼者
- 古田山植被垂直分布图
- 马尾松、黄山松、杉木的照片或者实物标本

拓展活动推荐

- 古田山常见植物打卡（见P563：活动卡03）

Suzhuang Town | Gutian Mountain–Gutian Temple | Evergreen broad–leaved forest exploration route

Interpretation No. SZ-GT-09

Stop 09: Lounge Gutian Temple

Interpretation theme

Understand the coniferous forest (the leader in adapting to high elevation)

Key points

1. Understand the location of coniferous forest in the vertical vegetation distribution throughout the mountain

2. Understand the representative tree species of coniferous forest: Masson pine, Huangshan pine, Chinese fir

Site overview

Near the Gutian temple, where visitors are able to observe the typical coniferous forests landscape

Look around us, aren't we now surrounded by conifers? Welcome to the world of typical coniferous trees in Gutian Mountain. So far, we have seen four typical vegetation communities. Does anybody remember who they are? That's right, starting from the evergreen broad-leaved forest at the foot of the mountain, to the evergreen-deciduous broad-leaved mixed forest, then the coniferous and broad-leaved mixed forest, and finally to the coniferous forest here.

"Understand the position of coniferous forests in the vertical distribution of vegetation throughout the mountain"

At the previous stop, we have learned that coniferous trees such as Masson pine are forest planters who were succeeded as early "pioneers". This is from the perspective of time. In the vertical direction, coniferous forests are usually located at higher elevations.

Do you know why? (Usually visitors will mention temperature changes)

Yes, because as the elevation increases, the temperature continues to decrease, the organic matter in soil decomposes more slowly, and the nutrients are relatively inadequate. Broad-leaved tree species with high demands for temperature and nutrients are gradually replaced by cold-resistant and barren-resistant coniferous tree species. The pine needles of Masson pine are the evidence to this adaptation.

"Recognize the representative tree species of coniferous forest: Masson pine, Huangshan pine, Chinese fir"

Masson pine, Huangshan pine, Chinese fir, *etc.*

are the representative tree species in the coniferous forest of Qianjiangyuan National Park. If summed up in three phrases, they can be "pioneer" Masson pine, "climbing expert" Huangshan pine and "high-quality building material" Chinese fir.

First look at Masson pine. As a pioneer, Masson pine is not only a dauntless explorer, but also very selfless. Their existence provides resources and space for the survival of more new species and individual plants in the community, and promotes the succession of the community to zonal evergreen broad-leaved forest.

Everyone is probably familiar with Huangshan pine, but Huangshan pine is not unique to Huangshan. Gutian Mountain has a large area of Huangshan pine forest. They are located at a higher elevation than Masson pine, and the growth environment is even worse. They are often proudly standing in stone cracks and cliffs. They are well-deserved "climbing experts".

The third kind of conifer we are going to introduce is Chinese fir. Because of its upright branches, rapid growth, and easy planting, but at the same time it has the characteristics of corrosion resistance and termite resistance. So it becomes one of the main tree species in the south of the Qinling Mountains and the middle and lower Yangtze River that has been widely used in various building materials since ancient times, especially the fine structural parts of buildings.

Attentions

- Systematically sort out the pattern of the entire forest vegetation changing with elevation.
- Starting from the relationship between coniferous forests and the environment, the commonness of the growth environment of coniferous forests in the vegetation succession and the change in vertical directions.
- When introducing three types of conifers, you can refer to the "General Interpretation of Biodiversity".

Recommended auxiliary facilities and toolkits

- Interpretation signage: coniferous forest—mountaintop dominants
- Map of the vertical vegetation zonation along the altitudinal gradient
- Pictures or specimens of coniferous trees

Recommended activities

- Common plants in Gutian Mountain (P563 : No. 03)

苏庄镇 | 古田山—古田庙 | 探秘常绿阔叶林自然体验路线　　解说编号：SZ-GT-10

解说点 10 木荷林

29°15′0.14″ N，118°8′34.64″ E
海拔：850m

● 解说主题

认识常绿阔叶林

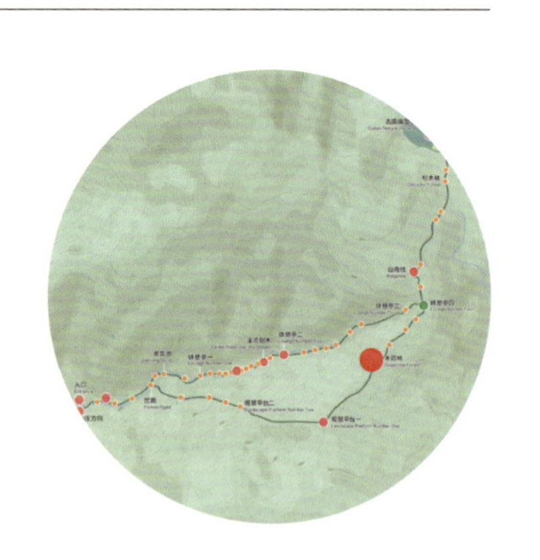

● 解说要点

1. 常绿阔叶林的概念
2. 生物学中的“密度制约”法则

● 场地概述

下山途中，可观察常绿阔叶林林冠景观以及林下生境

木荷林　木荷的花　木荷凋落的花朵与果实

解说方式 Interpretation Style

☑ 人员解说 Staff Interpretation　☐ 教育活动 Educational Activity
☑ 解说牌 Interpretation Signage　☐ 场馆解说 Hall Interpretation

解说季节 Interpretation Season

☑ 春季 Spring　☑ 夏季 Summer
☑ 秋季 Autumn　☑ 冬季 Winter

各位朋友，上山的途中我们已经见到了四种森林类型，现在就来考考大家：我们现在所在的森林应该属于哪种类型呢？非常棒！是常绿阔叶林。

「常绿阔叶林的概念」

大家是怎么判断出来的？通过植物的叶片、植物种类、海拔等？看来大家已经基本掌握森林分类的技巧了。没错，常绿阔叶林，顾名思义，就是四季常绿的森林，从严格意义上来说，它特指在温暖湿润的亚热带气候条件下，由常绿树种组成的森林植被。不过值得注意的是，四季常绿不等于不落叶哦，常绿阔叶林只不过是“冬天不会落叶的森林”。

「“密度制约”法则」

古田山自然保护区的森林中还分布着以甜槠、木荷、青冈等为代表的上百种不同的植物，它们的习性各不相同，是如何共存于同一个群落的呢？

这就不得不提起植物之间相互制衡的一套奇妙的法则，叫作“密度制约”。它指同一种植物物种当密度较大时，会主动抑制自己的种群增长，为多种植物共存提供空间。密度制约也是森林中各种物种共存和生物多样性维持的重要机制。

注意事项

- 结合一路的所见所闻，让访客能多参与到本次旅途中来；
- 以提问方式让公众思考物种之间如何共存，引入密度制约的概念。

辅助道具推荐

- 标识牌：一叶一世界
- 古田山常绿阔叶林的航拍图

拓展知识点

「密度制约效应」

- 研究发现，密度制约效应主要发生在物种生活史的早期阶段，如种子和幼苗阶段。这一时期物种更容易受专一性天敌的侵害而出现较高的死亡率。
- 密度制约发生在土壤深处植物的根茎间。植物幼苗所受到的密度制约与植物累积病原真菌和外生菌根真菌的速度紧密相关。一个树种受到密度制约影响越大，该树种就越容易累积病原真菌，幼苗死亡率也会升高；反之，受密度制约限制较小的物种，则能较快累积外生菌根真菌。这两种因素往往共同作用，决定不同物种的具体分布密度。

Suzhuang Town | Gutian Mountain－Gutian Temple | Evergreen broad－leaved forest exploration route

Interpretation No. SZ-GT-10

Stop 10: Evergreen broad-leaved forest

Interpretation theme

Understand evergreen broad-leaved forest

Key points

1. The concept of evergreen broad-leaved forest
2. The "density restriction" rule

Site overview

On the way down the mountain, where visitors are able to observe the canopy landscape of the evergreen broad-leaved forest and the area under the forest habitat

Dear friends, we have seen four types of forests on the way up the mountain. Now let's have a little quiz. What type of forest are we at? Great! It is an evergreen broad-leaved forest.

"The concept of evergreen broad-leaved forest"

How can we distinguish them? By plant leaves, plant species, elevation, *etc*.? It seems that everyone has mastered the basic skills of forest classification. Yes, evergreen broad-leaved forests, as the name implies, are evergreen forests in all seasons. Strictly speaking, they refer specifically to forest vegetation gradually formed by evergreen tree species under warm and humid subtropical climate conditions. It is worth noting that evergreens in all seasons doesn't mean their leaves never fall. Evergreen broad-leaved forests are nothing more than "forests that do not deciduous in winter".

"The law of 'density restriction'"

Hundreds of different plants represented by sweet oachestnut, gugertree, ring-cupped oak, *etc*. are distributed in the forest of Gutian Mountain Nature Reserve. How are they able to exist within the same community?

This has to mention a wonderful set of rules for balances between plants, called "density restriction", which means that when the density of the same plant species is getting too high, it will actively inhibit the growth of its own population and provide space for the coexistence of multiple plants. Density restriction is also an important mechanism for the coexistence of various species in the forest and the maintenance of biodiversity.

Attentions

- Encourage visitors to participate more in this journey combining what they see and hear along the way.
- Introduce the concept of density restrictions and raise questions to let tourists think about how species coexist together.

Recommended auxiliary facilities and toolkits

- Interpretation signage: various leaves
- Aerial view of evergreen broad-leaved forest in Gutian Mountain.

Extra information

"Typical cases of density restriction law"

- The density restriction effect mainly occurs in the early stages of the life history of the species such as the seed and seedling stage, because this period is more susceptible to mortality caused by the invasion of specific natural enemies.
- Density restriction occurs deep in the soil, between the roots and stems of plants.During tree growth, the rate at which plants accumulate pathogenic fungi and ectomycorrhizal fungi is related to the intensity of density restriction of their seedlings: the greater the density restriction, the more likely the tree species to accumulate pathogenic fungi, thus increasing the mortality of seedlings; in the contrast, species that are less restricted by density can accumulate ectomycorrhizal fungi faster. These two factors often work together to determine the specific distribution density of different species.

苏庄镇 | 古田山—古田庙 | 探秘常绿阔叶林自然体验路线　　解说编号：SZ-GT-11

解说点 11 观景平台

29° 14′ 49.96″ N，118° 8′ 22.52″ E
可以远眺山间的区域　海拔：833m

解说主题

白际山脉

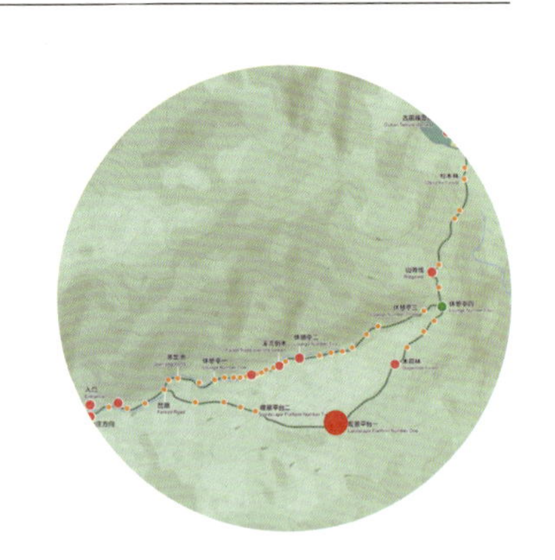

解说要点

1. 白际山脉的地理位置
2. 白际山脉对钱江源国家公园的影响

场地概述

环线下山途中的观景平台，可远眺白际山脉

解说方式 Interpretation Style

- ☑ 人员解说 Staff Interpretation
- ☐ 教育活动 Educational Activity
- ☑ 解说牌 Interpretation Signage
- ☐ 场馆解说 Hall Interpretation

解说季节 Interpretation Season

- ☑ 春季 Spring
- ☑ 夏季 Summer
- ☑ 秋季 Autumn
- ☑ 冬季 Winter

各位朋友，此刻，在我们眼前的是连绵的群山，我们往往在赞叹壮阔景象的同时却叫不出山脉的名字。

「白际山脉的地理位置」

现在，就让我们来登高望远，认识一下钱江源所处的山脉吧。我们来看一下这幅地图（拿出手中的白际山脉走势图），开化县有3条主要的山脉，分别是白际山、怀玉山和千里岗，其中，与钱江源国家公园关系最密切的是白际山脉。

白际山脉山势高峻，自东北向西南延伸，跨越串联了安徽、浙江、江西三个省份，在公园境内则依次跨过齐溪、何田、长虹、苏庄诸乡镇，连绵46km。

「白际山脉对钱江源国家公园的影响」

我们已经知道，由于青藏高原和东亚季风的影响，形成了我国东部温润多雨的气候。回到我们所在的钱江源国家公园，这里的气候其实也和山地息息相关。

每年夏季，从太平洋吹来的湿暖气流经过白际山脉时，会被高大连绵的群山阻挡，在迎风坡产生降水；冬季，来自亚欧大陆内部的西伯利亚寒流南下，同样会受到白际山脉的阻挡，使得这里的冬季气温明显高于长江中下游平原其他地区，形成了夏季炎热潮湿，冬季温暖，全年四季分明，雨量充沛的气候特点，为亚热带常绿阔叶林的生长提供了最适宜的气候条件。

注意事项

- 选择一个可以远眺连绵山峰的视角；
- 让访客建立起对钱江源国家公园山脉的整体印象；
- 让访客理解白际山脉对钱江源气候的影响。

辅助道具推荐

- 标识牌：一山一世界

Suzhuang Town | Gutian Mountain－Gutian Temple | Evergreen broad－leaved forest exploration route

Interpretation No. SZ-GT-11

Stop 11: Landscape platform No. 1

● Interpretation theme

Baiji Mountain

● Key points

1. Geographical position of Baiji Mountain
2. Impact of Baiji Mountain on Qianjiangyuan National Park

● Site overview

Landscape platform on the way down the mountain

My friends, at this moment, in front of us, there are rolling mountains. While we often admire the magnificent scenery, we often cannot tell the name of the mountains.

"Geographical location of Baiji Mountain"

Today, let's get to know the mountains of Qianjiangyuan. Let's take a look at this map (Take out the Baiji Mountain trend map). There are three main mountains in Kaihua County, namely Baiji Mountain, Huaiyu Mountain and Qianligang, of which the most closely related to Qianjiangyuan National Park is Baiji Mountain, which distributes from northeast to southwest, connecting the provinces Anhui, Zhejiang and Jiangxi for more than 46 kilometers.

"Impact of Baiji Mountain on Qianjiangyuan National Park"

We already know that the influence of the Qinghai－Tibet Plateau and the East Asian monsoon formed a warm and rainy climate in eastern China. Back to the Qianjiangyuan National Park where we are now, the climate here is actually closely related to the mountains. Every summer, when the humid heating current from the Pacific Ocean passes through the Baiji Mountain, it will be blocked by the high mountains and generate precipitation on the windward slope.

The blockade makes the winter temperature here significantly higher than other areas in the middle and lower reaches of the Yangtze River Plain. The climate is featured by hot and humid summer, warm winter, distinct seasons throughout the year, and abundant rainfall, providing perfect climatic conditions for the growth of subtropical evergreen broad-leaved forest.

Attentions

- Choose an angle from which everyone can overlook the rolling mountains.
- Let visitors build an overall impression of the mountains of Qianjiangyuan National Park.
- Let visitors understand the influence of the Baiji Mountains on the climate of Qianjiangyuan.

Recommended auxiliary facilities and toolkits

- Interpretation signage: diverse vegetation varied with elevation

苏庄镇 | 古田山—古田庙 | 探秘常绿阔叶林自然体验路线 解说编号：SZ-GT-12

解说点 12 结束集散点

29° 14′ 53.76″ N，118° 7′ 57.45″ E
海拔：525m

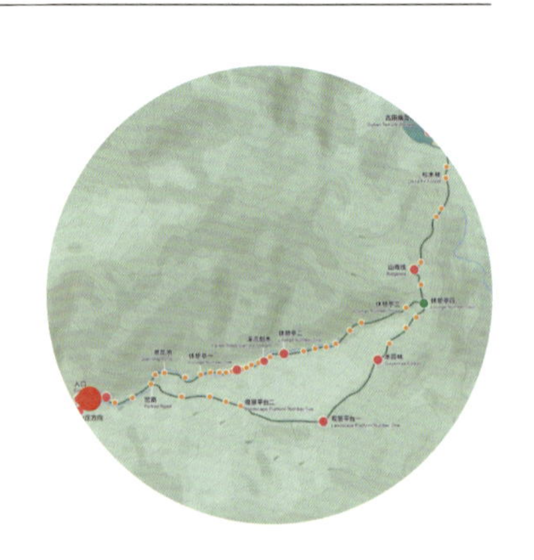

解说主题

游线总结与其他游线推荐

解说要点

1. 通过分享的形式，带领访客进行回顾
2. 对国家公园其他参访资源的介绍

场地概述

出发集合点或结束点

解说方式 Interpretation Style

- ☑ 人员解说 Staff Interpretation
- ☐ 教育活动 Educational Activity
- ☑ 解说牌 Interpretation Signage
- ☐ 场馆解说 Hall Interpretation

解说季节 Interpretation Season

- ☑ 春季 Spring
- ☑ 夏季 Summer
- ☑ 秋季 Autumn
- ☑ 冬季 Winter

「游线总结」

各位朋友，我们这条线的体验之旅到这里就接近尾声了，如果让大家用一句话、一个短语或者一个词概括本次旅程，你脑海中浮现的文字会是什么呢？

对我来说，是森林，呈现自然状态的，未被人干扰的森林。一路走来，我们认识了各种各样的植被类型，从山下到山上，从常绿阔叶林到针叶林，从木荷到马尾松，同时，我们还向森林学到了一些自然的智慧，如林窗效应、密度制约法则等。

当我们赞叹这些美景的时候，不要忘记在现代文明的冲击下，古田山这片小小的绿色方舟仍以她最原真的模样展现在世人面前，是多么不容易。

「其他游线推荐」

除了我们走的这条游线之外，钱江源还有很多地方值得大家去探索。你可以化身科学家，在古田山的瞭望台自然体验路线，进行一次科学考察之旅；也可以去古村落走走，体验一下当地的风土民情。当然，既然来到了钱江源，你一定不能错过莲花塘的钱塘江寻源之旅。如果你想在最短的时间内，体验最精致的风景，大峡谷会是不错的选择。

「游线分享」

在结束今天古田山的自然体验行程之前，我想邀请大家一起来回顾一下沿途的所观所感。我这里有 3 个小问题想问问大家。

第一个问题：随着海拔的升高，我们依次观察到了集中分布的植被类型，分别是什么？（答：常绿阔叶林、常绿落叶阔叶林、针叶阔叶混交林、常绿针叶林）

第二个问题：和钱江源国家公园联系最密切的山脉是哪座？（答：白际山脉）

最后一个问题：如果让你从此次徒步中选择一个点和朋友、亲人分享，你会选择什么呢？可以用关键词概括下吗？

感谢大家的分享，也谢谢你们一路的聆听、参与和支持！和你们交流给了我很多新的启发。希望大家未来的旅途愉快，返程平安！

注意事项

- 召集大家集合，15min 的小回顾，做些体力上的休整；
- 推荐其他游线；
- 通过大家的反馈，延伸启发与思考，提出回去后的行为倡议；
- 感谢大家的到访，感谢大家聆听自己的解说（请大家回去后向身边人宣传国家公园）。

Suzhuang Town | Gutian Mountain—Gutian Temple | Evergreen broad—leaved forest exploration route

Interpretation No. SZ-GT-12

Stop 12: Assembly ending point

- **Interpretation theme**

  Summary and sharing, and recommendation for other routes

- **Key points**

  1. Lead visitors to review what they saw through a sharing circle
  2. Introduction of other tourist resources of the national park

- **Site overview**

  Departure meeting point or ending point

"Travel summary"

Dear friends, our walk is coming to an end. What will be the word or phrase that appears in your mind to conclude today's trip?

For me, it is the forest. Along the way, we have known a variety of forest types, from the mountain to the mountain, from the evergreen broad-leaved forest to the evergreen coniferous forest, from the gugertree to the masson pine. At the same time, we also learned some ancient wisdom from the forest, such as forest gap effect, density restriction law, *etc*..

When we admire these beautiful sceneries, don't forget how difficult it is for the small green ark of Gutian Mountain to be displayed in the most authentic way under the impact of modern civilization.

"Recommended other routes"

In addition to this route, Qian Jiangyuan has many other places worth exploring. You can be like a scientist and go to the observation platform at Suzhuang Town, and enjoy a scientific investigation tour; you can also go to the ancient village to experience the local customs. Of course, since you are here, you can't miss the source-seeking journey of Qianjiangyuan in Lianhua Pond. If you want to experience the most exquisite scenery in the shortest time, the Grand Canyon is a good choice. Along the way, you can also experience the wetland natural experience route of Fenglou Pit.

"Travel sharing"

Before ending today's Gutian Mountain Nature Experience journey, I would like to invite you to review what you saw and your feelings along the way. I have three little questions here that I want to ask everyone.

The first question: as the elevation increases,

what are the concentrated forest communities we have observed? (evergreen broad-leaved forest, evergreen deciduous broad-leaved forest, coniferous and broad-leaved mixed forest, and evergreen coniferous forest)

Second question: which mountain range is most closely related to Qianjiangyuan National Park? (Baiji Mountain)

Last question: if you were to choose a stop from this walk and share it with friends and relatives, what would you choose? could you summarize it by keywords?

Thank you for sharing, and thank you for your support along the way. Communicating with you also inspired me a lot. I wish you all a pleasant journey in the future and a safe return home!

Attentions

- Gather everybody, organize a 15-minute quick review, and take a break.
- Recommend other routes.
- Lighten inspiration and propose environmental protection strategy.
- Thank them all for visiting and listening to your interpretation. (Please popularize the Qianjiangyuan National Park after return home.)

S03 瞭望台：国家公园科研体验路线

Scientific research experience route

500m
450m
400m
350m
入口
Entrance
7
1
气象站
Meteorological Station
2
古田山庄
Gutian Hotel
3
4
5
瞭望台
Observatory Tower
6
N
图例 LEGEND
钱江源国家公园
Qianjiangyuan National Park
水域
Water
等高线
Contours
步行道
Trail
车行道
Road
服务中心
Visitor Center
餐厅
Restaurant
救护室
Ambulance Room
卫生间
Toilet
观景处
Recreation Spot
设解说牌的点位
Point with Signage
设解说牌、人员解说的点位
Point with Signage and Interpretation
设解说牌、人员解说、课程的点位
Point with Signage, Interpretation and Course
0
200m

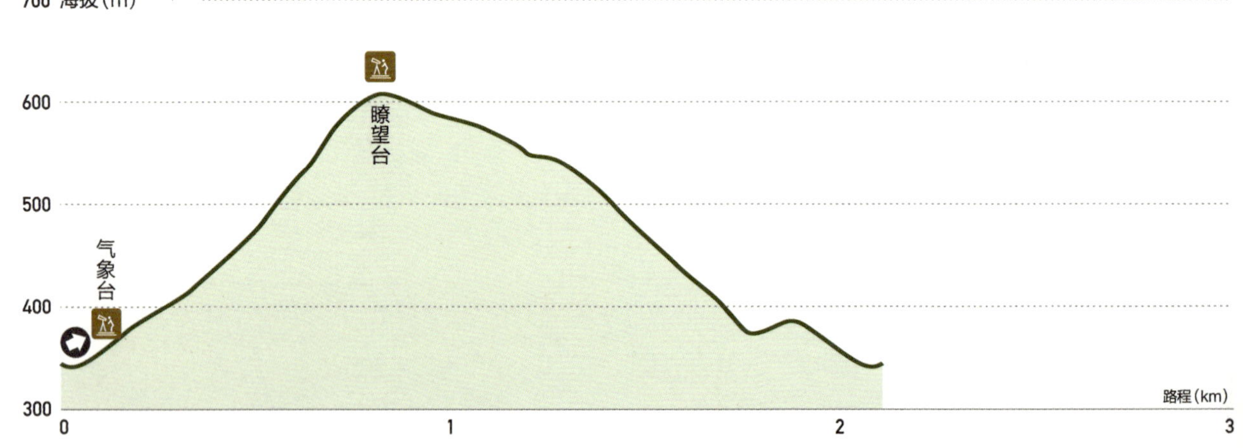

700 海拔（m）
600
500
400
300
瞭望台
气象台
路程（km）
0
1
2
3

S03 瞭望台：国家公园科研体验路线

解说主题 主题 1：点亮北纬三十度的生机绿洲
主题 3：丛林佑护的珍贵原真
主题 5：体制试点的创新探索

步道里程 往返环线约 2.1km
体验时长 往返环线 1.5～2h
最低海拔 344m
最高海拔 610m
步道等级 专业型
基础设施 解说标牌 / 活动场地 / 卫生间 / 无餐饮

这是以展示钱江源国家公园科研工作为主要目的的一条自然体验路线，围绕国家公园在森林冠层、24ha 大样地、全域覆盖的网格化监测、森林碳汇与气候变化等领域开展的研究进行的系统现场讲解，旨在引导访客从科学研究的视角，充分认识钱江源国家公园所开展的科研工作的全国性领先地位和创新及实践意义。

这是一条面向预约访客，特别是相关行业专业从业者的自然体验路线，毗邻大样地、森林塔吊等相关研究区域和设施。瞭望台则提供了登高观察和整体性认识科研工作的场所。整条步道不向普通访客自由开放，而需先通过公园预约平台的相关资质审核并完成必要预约流程，而后才能在解说人员的带领下进入。

在整条自然体验路线上，我们一共为解说员和访客设计了 7 个推荐解说点，同时，根据本书提供的通用型人员解说方案（非定点解说方案）资料，解说员也可以在合适的动植物资源停留点进行相应的引导性观察和解说。

S03 Scientific research experience route

Theme
- 01 The 30° N miracle of green oasis
- 03 Mysterious wildlife in the forest
- 05 Scientific research innovation and exploration

Trail Length about 2.1 kilometers round trip
Time 1.5 – 2 hours round trip
Lowest elevation 344 meters
Highest elevation 610 meters
Level of trail Professional
Infrastructure Interpretation board / event site / restroom / no catering

This is a nature experience route focusing on the research work of Qianjiangyuan National Park. The systematic on-site interpretations are conducted on the research of forest canopy, 24 hectares of the Gutian Plot, gridded infrared monitoring system, forest carbon sinks and climate change. It is aimed to guide visitors to understand the national leading position as well as the innovation and practical significance of the scientific research carried out by Qianjiangyuan National Park.

This is a nature experience route for reserved visitors, especially professional practitioners in related industries. It is next to the large plot, forest tower cranes and other related research areas and facilities. The observation platform provides a place for overlooking and overall understanding of scientific research work. The entire trail is not freely open to ordinary visitors, but needs to pass the relevant qualification review through the park reservation platform and complete a necessary reservation process, before entering the park guided by the interpreter.

We have arranged a total of 7 recommended interpretation stops for interpreters and visitors during the entire experience route. At the same time, according to the information of general interpretation (non-targeted interpretation plan) provided in this interpretation manual, interpreters can also find suitable resources to conduct corresponding guided observations and interpretations.

ZOOMLION

苏庄镇 | 瞭望台 | 国家公园科研体验路线　　　　解说编号：SZ-LWT-01

解说点 01 集合点

29° 13′ 34.87″ N，118° 6′ 37.61″ E
海拔：328m

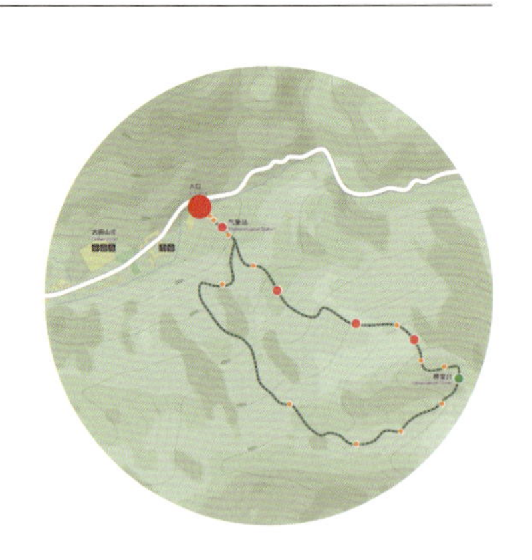

● 解说主题

本条自然体验路线的总体介绍

● 解说要点

1. 整条游线的总体介绍
2. 确认听众的状态，树立自身解说员的身份

● 场地概述

集合点可以选择在古田山庄

解说方式 Interpretation Style

- ☑ 人员解说 Staff Interpretation
- ☐ 教育活动 Educational Activity
- ☑ 解说牌 Interpretation Signage
- ☐ 场馆解说 Hall Interpretation

解说季节 Interpretation Season

- ☑ 春季 Spring
- ☑ 夏季 Summer
- ☑ 秋季 Autumn
- ☑ 冬季 Winter

解说词

各位朋友，欢迎大家来到钱江源国家公园，我是公园的解说员 XXX。

今天，我们要参访的自然体验路线是一条综合展示钱江源国家公园科研工作的引领性、创新性和实践意义的主题路线。各位通过预约来参访这条路线，大部分是对国家公园的科研有兴趣，或是来自相关领域的专业人士。我们也期待今天的行程会帮助大家更系统、生动地理解钱江源国家公园在科研领域的杰出成就。

钱江源国家公园最具有保护价值的自然资源就是我们身处的这片亚热带常绿阔叶林。也许很多人会觉得，这样的森林景观并不罕见，但在我国经济最发达的长江三角洲地区能保留这样一片大面积连续分布的低海拔的原始常绿阔叶林却非常不易。那么，这里是何以被保护下来的呢？

中华人民共和国成立初期，开化县以林业为支柱产业，在古田山建了伐木场。1973 年，当时的杭州大学（今浙江大学）生物系教授诸葛阳带队来到开化县考察，发现古田山的原始森林，并意识到其重要的保护价值，马上写信给当时的开化县革命委员会，呼吁把这片森林保护起来。他的建议得到了相关部门的重视，很快在 1975 年成立古田山省级自然保护区。随后，诸葛阳教授、郑朝宗教授等老一辈科学家带领考察和研究团队对古田山的动植物等自然资源进行了系统全面的考察，进一步确定了这里的保护价值。

2001 年，古田山国家级自然保护区成立。时至今日，来自浙江大学、中国科学院、北京大学等众多科学研究团队在古田山开展工作，古田山产出的科研成就与培养的专业人才在全国名列前茅。古田山也已成为国际知名的亚热带森林生物多样性和生态学研究的专业平台。

接下来，我们就循着科学家们的足迹，近距离了解和感受一下这里的科研工作，亲自体验科学家们是怎么进行野外调查和研究的，比如，测量一棵树的胸径，调查一块样方，布置红外相机，搭设种子雨收集器等，我们还有机会了解人类是怎么去探究高耸入云的林冠层的奥秘的。

有没有很期待呢？接下来就让我们出发吧！

注意事项

记住，请注意集合点的开场。当你出现在听众们的面前，解说员的工作已经开始了。如何吸引听众的注意力，树立自身的解说员身份，一个简短、内容明确、带有互动的开场白对解说员来说非常重要。以下几点提示可以帮你达到这一目标。

- 选择一个合适的可以帮助大家聚焦这次步道体验的集合地点，比如，步道导览牌边上。
- 体现你身份的制服、解说员的标志和便携的解说装备包。
- 你的介绍中一定有独特的、导览牌上未曾涉及的内容。
- 根据团队成员的构成，针对性地提出步道过程中的注意事项和建议。

辅助道具推荐

- 游线总体导览牌

拓展活动推荐

- 钱江源国家公园打卡地图（P562：活动卡 02）

Suzhuang Town | Observatory Tower | Scientific Research Nature Experience Route

Interpretation No. SZ-LWT-01

Stop 01: Assembly meeting point

Interpretation theme

General introduction to this nature experience route

Key points

1. Overall introduction of this route
2. Make sure the audience is listening and establish the identity as an interpreter

Site overview

Assembly meeting point can be set at the Gutian Mountain Villa

Dear friends, welcome everyone to the Qianjiangyuan National Park. I am your interpreter XXX.

The nature experience route we are going to visit today is a theme route that comprehensively demonstrates the leading, innovative and practical significance of the scientific research work of Qianjiangyuan National Park. I know everyone here booked this route by appointment, and most of you are interested in the scientific research of the national park or professionals in related fields. We really look forward to today's journey, which will help you understand the outstanding achievements of Qianjiangyuan National Park in scientific research more systematically and vividly.

The natural resource of this subtropical evergreen broad-leaved forest we are at has the most conservation value in Qianjiangyuan National Park. Many people may think that such a forest landscape is not uncommon, but it is very difficult to retain such a large area of continuous evergreen broad-leaved forest in a low-elevation area in China's most developed Yangtze River Delta region. So how is it preserved here?

In the early days of the founding of the People's Republic of China, Kaihua County built a logging field in Gutian Mountain with forestry as its pillar industry. In 1973, Zhuge Yang, a professor of the Department of Biology of Hangzhou University (now Zhejiang University), led a team to investigate in Kaihua County, discovered the original forest of Gutian Mountain, and realized its important protection value. He immediately wrote to the Kaihua County Revolutionary Committee to call for the protection of this forest. His proposal then received the attention from the relevant departments, and soon the Gutian Mountain Provincial Nature

Reserve was established in 1975. Later, professors such as Professor Zhuge Yang and Professor Zheng Chaozong led an investigation and research team to conduct a systematic and comprehensive investigation of the natural resources like animals and plants in Gutian Mountain, and further confirmed the protection value here.

The Gutian Mountain National Nature Reserve was established in 2001. Today, many scientific research teams from Zhejiang University, the Chinese Academy of Sciences, Peking University and other scientific research teams are working in Gutian Mountain. The scientific research achievements and professional talents produced in Gutian Mountain are among the best in the country. In this case, Gutian Mountain has also become an internationally renowncd profcssional platform for subtropical forest biodiversity and ecology research.

Next, we will follow the footprints of scientists and take a close look at the scientific research work here, and experience how scientists conduct field investigations and research in person, for example, measuring the diameter of breast height of a tree, investigating a sample, setting up infrared cameras and seed rain collectors. We also have the opportunity to understand how humans have explored the mystery of the forest canopy that is soaring into the clouds.

Are you looking forward to it? Let's go!

Attentions

Remember, pay attention to your debut at the assembly meeting point. When you show up to the audience, your job as an interpreter has already begun. As an interpreter, it is very important to attract the attention of the audience and establish your own personality, such as using a short, clear and interactive opening statement. The following tips can help you achieve this goal.

- A proper assembly place that can help gathering the crowd and start this nature experience, such as the open space next to the general guide sign at the start of the trail.
- Uniforms and costumes that reflect your identity: the logo of the interpreter and supporting materials tool kit.
- There must be something unique in your introduction that isn't mentioned in the guide board.
- Based on specific team members, pose targeted safety requirements and precautions.

Recommended auxiliary facilities and toolkits

- Overall guide board of the route

Recommended activities

- Task list map of Qianjiangyuan National Park (P562 : No. 02)

苏庄镇 | 瞭望台 | 国家公园科研体验路线

解说编号：SZ-LWT-02

解说点 02 塔吊

29°14′39.25″ N，118°6′53.18″ E

海拔：338m

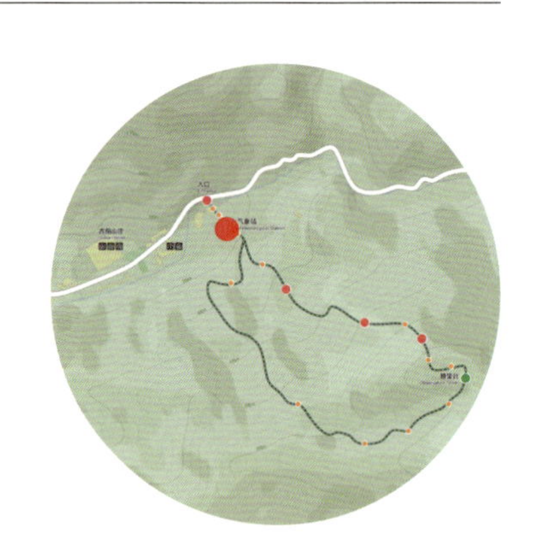

解说主题

林冠层的研究

解说要点

1. 林冠层的研究方法与塔吊研究的优势
2. 古田山的塔吊研究

场地概述

气象站附近，可观察远处塔吊的地方

解说方式 Interpretation Style

☑ 人员解说 Staff Interpretation　☐ 教育活动 Educational Activity

☑ 解说牌 Interpretation Signage　☐ 场馆解说 Hall Interpretation

解说季节 Interpretation Season

☑ 春季 Spring　☑ 夏季 Summer

☑ 秋季 Autumn　☑ 冬季 Winter

「林冠层的研究方法」

各位朋友，说到森林的生物多样性，我们往往想到的是走进森林内部进行调查研究。其实高大的树冠似乎才是森林生物量最集中，生物多样性最高的区域，但它们又往往“高高在上”，遥不可及。如果我们想研究树冠，大家觉得可以有哪些办法？（大家畅所欲言）

为了探究林冠层，历史上科学家曾采用过单绳攀爬技术，或借助空中走廊、热气球和林冠筏等各种方法。尽管这些方法为科学家研究林冠层提供了多种可能，但仍然有很大的局限性。大家可以想象一下，想要爬到几十米甚至更高的树冠层难度较高，本身是非常危险的，这种危险不仅来自脱离地面的高度，而且还有蛇、马蜂和蚂蚁等生物的威胁；此外，在攀爬的过程中，一些附生在树干上的植物、树木的枝干、栖息在树上的昆虫和鸟类等都可能会受到侵扰；最后，对于科学研究来说，这种方法效率低，对林冠层研究的深度、广度和持续度都会是很大的制约。

我们面前这个大块头——塔吊的应用，则是对上述传统方法局限性的突破。大家可能在城市的建筑工地见到过类似的高大的吊车，它可通过调节高度和臂展范围精确地投放物体或者把人送到相应的位置，就像这样（结合塔吊模型演示）。横杆环绕一圈正好组成一个立体的圆柱空间，理论上科学家可以进入这个圆柱体内的所有区域。

「古田山的塔吊研究」

古田山塔吊是中国七大林冠塔吊之一，独立高度和半径都是60m，大约相当于20层楼高。你可以想象吗？在这样一个巨大的圆柱形空间里，你可以到达你想去的任何一处地方。塔吊的优势不言而喻。

和传统的方法相比，借助塔吊的研究不仅更全面、细致，而且科研人员也可以精确地把监测和研究仪器投放到森林的大部分地方进行持续的观测，同时，借助塔吊进入林冠中的指定位置，把对森林的干扰和破坏降到最低。

古田山塔吊是组成我国生物多样性监测网络林冠生物多样性监测网络的一个重要节点，也是目前全球19个林冠层监测平台之一。它为监测高大乔木的生理生态、CO_2动态和气候变化、生物多样性等问题提供了重要的科研支撑。

注意事项

- 结合森林的垂直分层，让访客思考不同分层森林研究方法的差异。
- 讲解塔吊的工作原理以及优势。
- 适当补充林冠层的内容。

辅助道具推荐

- 标识牌：塔吊——揭秘森林冠层

拓展活动推荐

- 塔吊设计图 / 模型制作（P585：活动卡24）

拓展知识点

「林冠层的概念与研究意义」

- 什么是林冠层呢？顾名思义，就是森林的最顶层。这里是整个森林中与外界环境相互作用最直接和最活跃的界面，对气候变化和人类的干扰十分敏感，同时这里也是老鹰、寄生植物等众多生物的家园，极大地丰富了地球上的生物多样性。在全球变暖和人类活动干扰的背景下，森林生态系统一旦被破坏，首当其冲的就是林冠层。
- 林冠层虽然重要，却因为难以借助常规方法进入，一直是植物生态学研究最少的领域之一，因而又被称作“最后的生物学前沿”和“地球第八大洲”。

Suzhuang Town | Observatory Tower | Scientific Research Nature Experience Route

Interpretation No. SZ-LWT-02

Stop 02: Tower Crane

Interpretation theme

Study on the forest canopy

Key points

1. Research methods of forest canopy and advantages of tower crane research
2. Tower crane research of Gutian Mountain

Site overview

The place near the meteorological station, where visitors are able to visit the tower crane in the distance

"The research methods of the forest canopy and the advantages of tower crane research"

Dear friends, is it enough for us to understand the trees in the forest if we simply know the methods of positioning, listing, measuring diameter of breast height, *etc*.? Apparently not. Tall canopies seem to be the areas with the most concentrated biomass and the highest biodiversity, but they are often "out of reach". If we want to study the canopy, what methods do you think can be done?

Indeed, in history, scientists have used various methods such as freehand climbing, using ropes, ladders, hot air balloons or helicopters to explore the forest canopy, but these methods will have some limitations. Let's take tree climbing as an example. On one hand, it is difficult for climbers to reach the canopy of tens of meters or higher, which is quite dangerous. On the other hand, it may accidentally damage the plant's branches or insects' habitats during the climbing. In addition, for scientific research, this method is inefficient and unable to complete a comprehensive and systematic monitoring and research.

The big guy in front of us is the tower crane. Its application is a breakthrough in the limitations of the above traditional methods. You may have seen a similar tall crane at a construction site in the city. It can accurately drop objects or send people to the corresponding position by adjusting the height and arm span, just like this (show with the help of the model of tower crane). The bar wraps-around to form a three-dimensional cylindrical space. In theory, scientists can enter all areas in this cylinder.

"Tower crane research of Gutian Mountain"

Gutian Mountain Tower Crane is one of the seven largest canopy tower cranes in China. Its independent height and radius are both 60 meters, which is equivalent to about 20 floors. You can imagine: standing on the top of a 20-story building, this building is cylindrical with the same radius and height, you can enter any room of this building freely, which is the range of a forest that a tower crane can accurately study.

Compared with traditional methods, tower crane research is more comprehensive and meticulous. Researchers can accurately place monitoring and research instruments in most parts of the forest for continuous observation. At the same time, tower cranes can be used to enter designated locations in the canopy. Forest disturbance and destruction are also minimized.

The research scope of the Gutian Mountain Tower Crane covers 1.13 hectares of typical mid-subtropical low-elevation evergreen broad-leaved forest. It provides important scientific research support for monitoring the physiology of tall trees, CO_2 dynamics, climate change, biodiversity and other issues.

Attentions

- Combined with the vertical stratification of the forest, visitors are able to think about the differences between different stratified forest research methods.
- Explain the mechanism and advantages of tower crane.
- Appropriately supplement the content of the forest canopy.

Recommended auxiliary facilities and toolkits

- Interpretation signage: tower crane—revealing the canopy

Recommended activities

- Design drawing creation or model making of tower crane (P585: No. 24)

Extra information

"The concept and research significance of the forest canopy"

- What is the forest canopy? As the name suggests, it is the top of the forest. This is the most direct and active layer in the entire forest that interacts with the external environment. It is very sensitive to climate change and human interference. It is also home to many creatures such as eagles and parasitic plants, which greatly enriches the biological diversity on earth. Under the background of global warming and human interference, once the forest ecosystem is destroyed, the forest canopy bears the brunt.
- Although the forest canopy is important, it has been one of the least studied areas of plant ecology because it is difficult to access by conventional methods, so it is also called "the last biological frontier" and "the eighth continent of the earth".

苏庄镇 | 瞭望台 | 国家公园科研体验路线　　解说编号：SZ-LWT-03

解说点 03 24hm² 大样地

29° 14′ 45.56″ N, 118° 6′ 59.29″ E
海拔：400m

● 解说主题

24hm² 大样地动态监测研究

● 解说要点

1. 认识样方与 24hm² 大样地
2. 24hm² 大样地的科学研究工作

● 场地概述

刚开始进入林地的场地

解说方式 Interpretation Style

- ☑ 人员解说 Staff Interpretation
- ☐ 教育活动 Educational Activity
- ☑ 解说牌 Interpretation Signage
- ☐ 场馆解说 Hall Interpretation

解说季节 Interpretation Season

- ☑ 春季 Spring
- ☑ 夏季 Summer
- ☑ 秋季 Autumn
- ☑ 冬季 Winter

「样方是科学研究的代表性区域」

东亚地区的常绿阔叶林因为受东亚季风的影响，形成了独有的特征，被称作“典型常绿阔叶林”。在这里的研究，对于中国、东亚乃至整个世界，都具有代表性意义。

但是野外情况复杂多样，科学家要研究一片森林，不可能穷尽其中的每一棵树，每一片土地，而必须选择一些有代表性的区域作为样本，这就是样方。

什么是样方呢？样是样本，方是方形的区域。所以，样方顾名思义就是包含各种研究样本的一片方形的区域。

最早的样方面积的大小由研究对象决定。一般草本植物的样方最小，大概相当于四个人手拉手伸展开来的面积，灌木其次，乔木的较大，有 100～200m²，甚至更大，相当于一般三居室户型面积的 1～2 倍。

「认识 24hm² 大样地」

随着科学研究的不断深入，我们对于认识生态系统整体性与动态变化的需求越来越强烈，同时随着研究的深入和研究技术方法的精进，样方的精度已经不能满足需要，更大面积的样地就产生了。样地越大，我们得到的数据也会更准确，更能反映真实的情况。

对面的这片森林，就是中国科学院植物研究所联合浙江大学等单位于 2004 年建成的一块总面积 24hm² 的大型样地，相当于将近 34 个标准足球场面积（105m × 68m）。这么大规模的样地，在全球都相当罕见。这里是我国森林生物多样性监测网络最早建立的五大样地之一，也是世界热带森林研究中心监测网络的重要组成部分。

「24hm² 大样地的科学研究工作」

如果我们把这片样地比作一个村子的话，树木就是这里的常住民。你要做好科学研究，就需要认识每一个居民，并且定期对其进行调查访问。所以，身为一个科学家，需要做的工作有哪些呢？

首先，你要了解样地内每一棵树“住”在哪儿，记录下它们的定位。

注意事项

- 尽量结合实际情况，选择一个角度，让访客不仅可以从宏观的角度看到这个样地，还能近距离观察到一棵树上的标记。

辅助道具推荐

- 标识牌：24 公顷大样地——全球意义的森林研究范本
- 样地现场有标记的树
- 测量胸径所需要的工具，如卷尺等

拓展活动

- 植物样方调查练习
- 测量一棵树的「胸径」

拓展知识点

「胸径的测量」

- 树木的主干自下而上由粗变细，为了便于统计调查，一般采用人的胸部高度位置的树干直径，即“胸径”来作为测量标准。测量时无需弯腰或爬梯，十分方便。在中国，我们使用国际通用标准胸部高度，即为离地面 1.3m。

注：不是所有的树木都长得笔直标准，实际胸径测量会根据树形以及树木生长环境的不同等因素而有差异。

其次，你要认识它们是谁，并为每棵树佩戴一个独一无二的身份证——一个带弹簧的金属“生长环”的物种标识牌；

接下来，你就要开展一项长期坚持的工作：你需要定期“拜访”每一棵树，观察它们的生长变化，并记录下来。

听起来好像也不是很难，不过，不要忘了我们所在的这片样地有 $24hm^2$，这种调查研究和记录需要涵盖样地内所有胸径 1cm 以上的树，需要被定位、编号、挂牌与监测的树木总数超过 14.7 万棵。

我们假设每棵树的定位工作需要 10min，按照每人每天工作 8h，每个月 22 个工作日估算，完成整个样地的研究记录工作需要一个专业工作人员大约 140 个月，合 11 年半的时间。

当然，这是最传统的方法。现代科技的进步，尤其是专业图鉴、移动互联网的应用，大大提高了野外工作的效率。同时，很多志愿者也可以借助公民科学家的方式参与到这些日常监测工作中来，这使得大样地的系统、持续、深入、动态的研究成为可能，也为今天我们更为科学、有效地评估、管理整个大样地，以及建设、管理国家公园的重要科学和技术支撑。

古田山的科研监测工作已经坚持了十几年，取得了丰硕的成果，也成为国家公园体制试点单位中科研工作的标兵。

TD01010110

Suzhuang Town | Observatory Tower | Scientific Research Nature Experience Route

Interpretation No. SZ-LWT-03

Stop 03: 24 hectares plot

● Interpretation theme

Research on dynamic monitoring of 24-hectares of large plot

● Key points

1. The concept of the plot and the introduction of the 24-hectare plot
2. Scientific research work on a large plot of 24 hectares

● Site overview

A site that leads to the woods at first

"The plot is a representative area of scientific research"

Due to the influence of the East Asian monsoon, the evergreen broad-leaved forest in East Asia has formed its unique feature, which is called the "typical evergreen broad-leaved forest". The research here has representative significance for not only China, but also East Asia, and the entire world.

However, the situation in the wild is complex and diverse. If a scientist wants to study a forest, it is impossible to investigate every tree and every piece of land. In this case, some representative areas are selected as samples. This is the plot.

What is a plot? In Chinese, it literally translates into a square sample area. As the name implies, the plot is a square area containing various research samples.

The size of the earliest plot is determined by the research object. Generally, the plot size of the herbaceous plant is the smallest which equals approximately the area where four people stretch out holding hands, followed by shrubs. The plot of the arbor is relatively larger, sometimes over 100－200 square meter, which is equivalent to 1－2 times the size of a three-bedroom apartment .

"Introduction to 24-hectare plot"

With the continuous development of scientific research, our demand for understanding the integrity and dynamic changes of ecosystems has become stronger and stronger. At the same time, with the in-depth research and the improvement of research technology methods, the accuracy of the sample can no longer fulfill the needs, so the larger plots were created. The larger the plot, the more accurate data we get.

The forest we are facing is a large plot with a total area of 24 hectares built in 2004 by the Institute of Botany of the Chinese Academy of Sciences and Zhejiang University, which is equivalent to the area of nearly 34 standard football fields (105m × 68m). A plot in such big size is quite rare in the world. As a result, this is one of the first five plots established by China's forest biodiversity monitoring network, and it is also an important part of the monitoring network of the World Tropical Forest Research Center.

"Scientific research work on a 24-hectare plot"

If we compare this plot with a village, the trees are the permanent residents here. If you want to do scientific research, you need to know every resident and conduct regular surveys and visits. So as a scientist, what needs to be done?

First, you need to know where each tree "lives" in the plot and document their location.

Secondly, you have to know who they are, and wear a unique ID card for each tree: a species identification plate with a metal spring "growth ring".

Next, you need to conduct a long-term persistent work: you need to "visit" each tree regularly, observe their growth changes, and record them.

Attentions

- Try to combine the actual situation and choose a suitable angle in order to get good viewing points.

Recommended auxiliary facilities and toolkits

- Interpretation signage: 24-hectare plot, a global forest research sample.
- Selected trees with marks on site.
- Tape needed to measure the diameter of the trees at breast height.

Recommended activities

- Plant sample survey
- Measurement of the tree diameter of breast height

Extra information

"Measurement of diameter of breast height"

- The trunk of the tree becomes thinner from bottom to top. In order to facilitate statistical investigation, the trunk diameter at the height of the plant's breast is generally used as the measurement standard. There is no need to bend down or climb stairs when measuring, which is very convenient. In China, we use the international standard breast height, which is 1.3 meters above the ground.

Notes: Not all trees are upright and straight. Actual diameter of breast height measurement will vary depending on the shape of the tree and the growth environment.

It doesn't sound like a difficult job, but don't forget that we have 24 hectares in this plot. This kind of research and records need to cover all trees with a diameter of breast height of more than 1cm, and the total number of trees that need to be located, numbered, listed and monitored are more than 147,000.

We assume that the positioning work of each tree takes 10 minutes, when one person works 8 hours per day, 22 working days per month, it will take 140 months for a professional staff to complete the research and record work of the entire plot. That's 11 and a half years time.

Of course, this is the most traditional method. Advances in modern technology, especially the application of professional pictorials and mobile internet, have greatly improved the efficiency of field work. At the same time, many volunteers can also participate in these daily monitoring work as citizen scientists, which makes systematic, continuous, in-depth, and dynamic research of large plots possible. It also became an important scientific and technical support for scientific and effective assessment and management of the entire plot as well as the construction and management of national parks.

The scientific research and monitoring work has been carried out in Gutian Mountain for more than ten years with fruitful results. What kind of stories will be staged in the future? Please wait and see.

古田山大样地内的树木胸径标注，挂牌编号与带弹簧的“生长环”

苏庄镇 | 瞭望台 | 国家公园科研体验路线　　解说编号：SZ-LWT-04

解说点 04 步道观察点 #1

29° 14′ 45.56″ N，118° 6′ 59.29″ E
海拔：510m

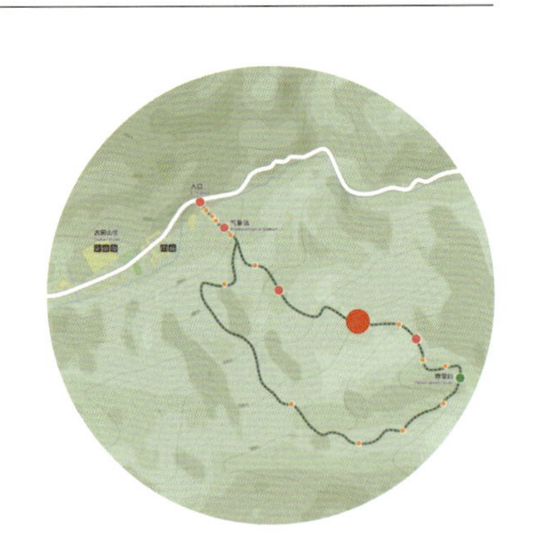

● 解说主题

森林碳汇与气候变化研究

● 解说要点

1. 森林碳汇的概念，及其气候变化的重要关系
2. 如何进行碳汇测量工作

● 场地概述

可以测量树木胸径的地方

解说方式 Interpretation Style

- ☑ 人员解说 Staff Interpretation
- ☐ 教育活动 Educational Activity
- ☑ 解说牌 Interpretation Signage
- ☐ 场馆解说 Hall Interpretation

解说季节 Interpretation Season

- ☑ 春季 Spring
- ☑ 夏季 Summer
- ☑ 秋季 Autumn
- ☑ 冬季 Winter

气候变化是当今世界最受关注的环境问题之一。有数据表明，工业革命以来，全球气候变暖，地球的平均温度升高了近1℃，其中的重要原因是大气中CO_2等温室气体浓度的升高。怎么解决这个问题？一是减少排放，二是增加吸收。其中吸收指想办法把这些多余的CO_2通过各种天然的或人为的途径收集起来，这个过程叫作“固碳”。

「森林碳汇与气候变化」

森林本身就是一个巨大的碳汇，它可以通过光合作用吸收CO_2，释放氧气，这样CO_2就会被固定在植物体内或者变成凋落物储存在土壤当中，从而降低大气中CO_2的浓度，对稳定全球气候具有重要作用。

研究表明，森林生物量越大，生态系统越稳定，其固碳效果越好。与此同时，森林也是对气候变化最为敏感的生态系统之一，伐木、烧荒、开发利用等人类活动或者温度的升降、降水的多少、极端气候等自然变化都会影响森林生态系统的健康，从而影响其固碳功能的发挥。

「如何进行碳汇测量工作」

为了对森林的碳汇能力进行监测与评估，我们需要计算森林的固碳量。怎么去测算一片森林的固碳量呢？让我们从一棵树开始。

根据研究，每立方米体积的木材平均约可吸存250kg的CO_2。想知道树木的体积大小，就必须先测量出树木的树高和胸径，算出材积，进而得到这一棵树的固碳量。

不过，要提醒大家的是，每棵树的固碳量受到树种、生长环境、气候条件、生长率等诸多因素的影响而表现出不同。同时，除了计算地上植被，我们还要考虑凋落物等生物残体的固碳量。

听起来是不是很简单？事实上，钱江源国家公园通过公民科学家项目，招募了像大家一样的来自全球各地的志愿者。这些志愿者通过碳汇测量等工作，以实际行动参与到应对气候变化的科学研究中。将来如有机会，也欢迎大家一起来参与。

注意事项

- 从二氧化碳、温室效应等熟悉的概念引入，介绍森林碳汇的概念与影响因素。
- 通过参与碳汇测量倡导公民科学家项目与自我实践。

辅助道具推荐

- 标识牌：森林碳汇——气候变化的重要启示
- 一棵树5年固碳量的测量卡片：

1. 测量胸径等指标计算当前树木的固碳量基数。
2. 记录并核算5年内树下落叶、残枝、掉落树种、树皮及虫粪等残体的含碳总量。
3. 5年后再次测量胸径，计算活树固碳量的增加值，加上残体含碳量的总和。

注：可以结合实际，算出一棵5年树木的固碳量可以抵消我们平时某种活动（如开车）的碳排放量。

拓展活动推荐

- 测量一棵树的“固碳量”

Suzhuang Town | Observatory Tower | Scientific Research Nature Experience Route

Interpretation No. SZ-LWT-04

Stop 04: Trail observation point # 1

● Interpretation theme

Research on forest carbon sink and climate change

● Key points

1. The concept of forest carbon sink and the important relationship between carbon sinks and climate change

2. How to conduct carbon sink measurement

● Site overview

A place where the diameter of trees at breast height can be measured

Climate change is one of the most concerned environmental issues in the world today. An important cause of global warming is the increase in the concentration of greenhouse gas such as CO_2. How to solve this problem? One is to reduce emissions, and the other is to increase absorption. Absorption refers to finding ways to collect these excess CO_2 through various natural or artificial means. This is what we call "carbon sinks".

"Forest carbon sinks and climate change"
The forest itself is a huge carbon sink, which can absorb carbon dioxide and release oxygen through photosynthesis. In this way, CO_2 will be fixed in the plant or stored in the soil as litter, so that the concentration of CO_2 in the atmosphere will be reduced, which plays an important role in stabilizing the global climate.

Studies have shown that the greater the forest biomass is, the more stable the ecosystem is and the better its carbon sequestration effect is. At the same time, forests are also one of the most sensitive ecosystems to climate change. Human activities such as logging, burning waste, development and utilization, or temperature, the amount of precipitation, extreme climate and other natural changes will affect the health of forest ecosystems and thus the capability of its carbon sink.

"How to conduct carbon sink measurement"
In order to monitor and evaluate the forest's carbon sink capacity, we need to calculate the forest's carbon sequestration. How do we measure the carbon sequestration of a forest? Take one single tree for example.

According to research, one cubic meter of wood can absorb about 250 kilograms of CO_2 on average. To know the size of a tree, you must first measure the tree's "height and diameter at breast height", calculate the volume, and then multiply it by 250 kilograms to get the tree's carbon sequestration.

However, it should be reminded that the carbon sequestration of each tree is affected by many factors such as tree species, growth environment, climatic conditions, and growth rate. At the same time, in addition to calculating the vegetation on the ground, we also need to consider the amount of carbon fixed by biological residues such as litter.

Sounds simple, isn't it? In fact, Qianjiangyuan National Park has recruited volunteers from all over the world through the citizen scientist project. Through carbon sink measurement and other works, volunteers have taken part in scientific research to combat climate change through practical actions. If you have an opportunity in the future, you are welcome to join us.

Attentions

- Introduce familiar concepts such as carbon dioxide and greenhouse effect to illustrate the concept and influencing factors of forest carbon sinks.
- Advocate the citizen scientist project and self-practice through participating in carbon sink measurement.

Recommended auxiliary facilities and toolkits

- Interpretation signage: forest carbon sinks
- "Little card": a method for measuring the carbon sequestration of a tree in 5 years

1. Calculate the diameter at breast height and other indicators to calculate the current tree carbon sequestration base.
2. Record and account the total carbon content of the fallen leaves, stumps, fallen tree species, bark and insect dung of the tree within 5 years.
3. After 5 years, the diameter at breast height is measured again to calculate the increase in the amount of carbon fixed by the live tree, plus the sum of the carbon content of the residual body.

Notes: It can be combined with reality to calculate the carbon sequestration of a five-year tree that can offset the carbon emissions of some kind of our usual activities (such as driving).

Recommended activities

- Measuring the "carbon sequestration" of a tree.

苏庄镇 | 瞭望台 | 国家公园科研体验路线　　解说编号：SZ-LWT-05

解说点 05 步道观察点 #2

29°14′45.56″ N，118°6′59.29″ E
海拔：543m

● **解说主题**

种子雨收集器

● **解说要点**

1. 种子雨的概念及其对于森林的意义
2. 关于种子雨的收集方法与研究（种子雨收集器）

● **场地概述**

没有种子雨收集器的地方

解说方式 Interpretation Style

- ☑ 人员解说 Staff Interpretation
- ☐ 教育活动 Educational Activity
- ☑ 解说牌 Interpretation Signage
- ☐ 场馆解说 Hall Interpretation

解说季节 Interpretation Season

- ☑ 春季 Spring
- ☑ 夏季 Summer
- ☑ 秋季 Autumn
- ☑ 冬季 Winter

对于森林的研究，仅限于观察植物的现状是不够的，还需要了解其繁衍更替的过程。那么，对种子的研究就非常重要，不仅要了解种子的形态、结构，还要了解它们发生的时机，传播和繁衍的方式。

在古田山的大样地中，布设有很多这样的方形网状设施。大家知道它是做什么用的吗？其实，这是特别用来研究种子的装置，叫作种子雨收集器。

在一定时间和空间内，种子会大量地从母树脱落，就像下雨一样，这种现象被生动地描述为“种子雨”。就好像不同的人有不同的习惯一样，不同植物的种子雨发生的时间、强度及方式都有很大的差异。此外，受外部环境和气候等因素的影响，在不同地点，不同时间，甚至不同年份，种子雨的情况都有很大的变化。 因此，研究不同植物的种子雨什么时候发生，量有多大，强度如何以及种子雨是如何扩散的，对于我们认识森林生态有很重要的指导意义。

大家仔细观察一下这个种子雨收集器（尽可能结合现场讲解，也可准备相关的简易模型或图片）。这是一种非常简单有效的工具。它的边长和高度都可以调节，用来满足不同物种，不同研究项目的需要。

每年 5 月到 8 月的幼苗生长旺盛期，科研人员每周都会将每个收集器上采集的植物种子、落叶等凋落物进行回收、分类，并进行物种、生物量等指标的记录、监测、分析和研究，探究不同树种存活和繁衍的机理，进而理解森林生态系统的存续方式等问题。

注意事项

- 种子雨和种子的意义稍有不同，种子一般是对于一棵树或者一个物种来说，而种子雨更偏向群落或者生态系统的尺度，对于群落更替具有重要意义。
- 种子雨的种子不一定来自收集器上空的树木，也可能来自森林其他地方或者其他森林。

辅助道具推荐

- 标识牌：种子雨记录的森林奥秘

拓展活动推荐

- 种子雨收集器 DIY 与运用 （P584：活动卡 23）

Suzhuang Town | Observatory Tower | Scientific Research Nature Experience Route

Interpretation No. SZ-LWT-05

Stop 05: Trail observation point # 2

Interpretation theme

Seed rain collector

Key points

1. The concept of seed rain and its significance to the forest
2. Research and methods on seed rain and its collecting methods (seed rain collector)

Site overview

A place were seed collector stands

For forest research, it is not enough to know the existing plants, but also need to understand the process of their reproduction and replacement. In this case, the study of seeds is very important. We need to not only know the morphology and structure of seeds, but also understand the timing of their occurrence, and the way of spreading and reproduction.

In the large plot of Gutian Mountain, there are many square mesh facilities like these. Do you know what it is used for? In fact, this is a special facility for seed research: seed rain collector.

In a certain time and area, the seeds will fall off from the mother tree in a large amount, just like rains. This phenomenon is called "seed rain". Just as different people have different personalities, the happening time, intensity and method of seed rain of different plants are very different. In addition, due to factors such as the external environment and climate, the situation of seed rain varies greatly at different locations, at different times, and even in different years.

So, when does the seed rain of different plants occur? How much is it? How strong is it? And how does the seed rain spreads? Answers to these questions are very important for us to understand forest ecology.

Let's take a closer look at this "seed rain collector" (It'll be the best if there is one on site, you can also prepare related simple models or pictures). This is a very simple and effective tool. The length and height of its sides can be adjusted to meet the needs of different research projects of different species.

Every year from May to August, when seedlings are growing vigorously, researchers will collect and classify the plant seeds, fallen leaves

and other litters collected on each collector every week, and record, monitor, analysis and research to explore the mechanism of survival and reproduction of different tree species, and then to understand the survival of forest ecosystems and other issues.

Attentions

- The significance of seed rain is slightly different from that of seeds. Seed is generally relative to a tree or a species, but the seed rain determines the scale of the community or ecosystem, and has great significance for community reproduction and replacement.
- Seeds of the seed rain do not necessarily come from the trees above the collector, they may come from other parts of the forest or other forest.

Recommended auxiliary facilities and toolkits

- Interpretation signage: Forest mysteries recorded by seed rain

Recommended activities

- DIY and application of seed rain collectors（P584；No. 23）

苏庄镇 | 瞭望台 | 国家公园科研体验路线　　解说编号：SZ-LWT-06

解说点 06 瞭望台

29° 14′ 45.56″ N，118° 6′ 59.29″ E
海拔：630m

解说主题

全域覆盖的网格化监测（红外相机）

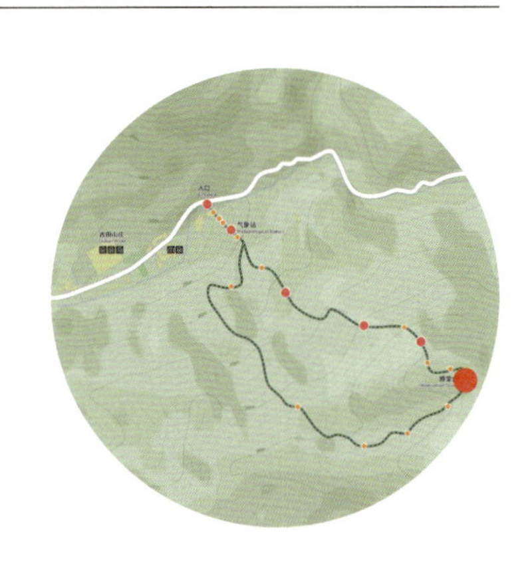

解说要点

1. 全域覆盖的网格化红外监测体系
2. 案例阐释：黑麂、白颈长尾雉、黑熊

场地概述

放置红外相机的场地

解说方式 Interpretation Style

- ☑ 人员解说 Staff Interpretation
- ☑ 教育活动 Educational Activity
- ☑ 解说牌 Interpretation Signage
- ☐ 场馆解说 Hall Interpretation

解说季节 Interpretation Season

- ☑ 春季 Spring
- ☑ 夏季 Summer
- ☑ 秋季 Autumn
- ☑ 冬季 Winter

「全域覆盖的网格化监测体系」

刚刚我们介绍了各种研究植物的方法，例如，如何收集样本、标记物种，甚至借助塔吊等设施，监测并记录必要的数据和信息以开展研究。 但是，对于行踪变化莫测的野生动物，这些研究方法就未必有效了。

不过不用担心，红外摄像结合全域化布点的方法，可以帮我们全方位、无死角地了解国家公园的资源分布——这就是全域网格化监测系统。具体而言，就是把整个国家公园划分为若干个监测网格，在每个网格内都布设监测设施，确保科研监测的无死角、无漏洞。我们熟悉的国家重点保护野生动物黑麂、白颈长尾雉、黑熊等在国家公园内的分布数据，就是通过这样的方法监测得来的。

大家一起看看我手中这三张不同物种的分布图：先找找看白颈长尾雉、黑熊和黑麂这三种动物谁的分布范围最小？没错，是黑熊，仅在北部临近安徽省界有出现。那哪种动物的分布比较广呢？是白颈长尾雉，它们分布集中在缓冲区和实验区，核心区反而较少，这是因为它们偏好低海拔的次生林、人工林和油茶林。相对来说，生性胆小需要安全感的黑麂则集中分布在国家公园的核心区域，很少进入人类活动较多的试验区。

野生动物大多生性警觉，一些保护物种的种群数量本来就很少，因此很难在野外被直接观测到。红外相机网格化监测系统的设置为人类在不施加过多干扰的前提下研究和认识大自然提供了非常有效的工具。

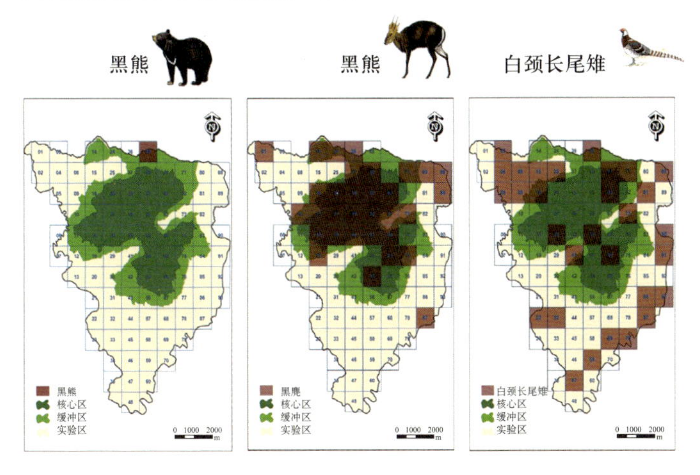

注意事项

- 从植物到动物，引出网格化监测的必要性；
- 结合明星物种的监测数据，说明监测的价值；
- 此处山路台阶较滑，需注意。

辅助道具推荐

- 标识牌：野外红眼，全域覆盖的网格化监测
- 红外相机拍摄的白颈长尾雉、黑麂、黑熊等的照片
- 古田山地区野生动物分布监测记录（附图片）

拓展活动推荐

「如何在野外布设一台红外相机」

步骤 1：根据预先的地图分析划分监测网络，确定监测区域。

步骤 2：红外相机理论上应布设在网格的中心位置附近。

步骤 3：红外相机的布设点应考虑以下要素。

(1) 附近有动物出没痕迹，有较高概率记录到动物活动踪迹；

(2) 高度为主要记录物种身高的一半左右，距离适宜，且镜头范围没有遮挡物，确保能有效触发拍摄并有效记录现场。

(3) 安装点稳固、隐蔽、安全。

步骤 4：安装成功后，应检查并测试以确保红外相机可正常使用。

步骤 5：定期检查、回收数据并维护设备。可灵活根据收集数据质量调整布点方案。

Suzhuang Town | Observatory Tower | Scientific Research Nature Experience Route

Interpretation No. SZ-LWT-06

Stop 06: Observatory tower

Interpretation theme

Grid infrared monitoring system with full coverage (infrared cameras in the wild)

Key points

1. Grid monitoring system with full coverage
2. Case explanation: the black deer, white-necked pheasant, black bear

Site overview

a place where you can see infrared cameras

"Grid infrared monitoring system with full coverage"

We just introduced various methods of studying plants, for example how to collect samples, mark species, and even use facilities such as tower cranes to monitor and record the necessary data and information to carry out research. But for wild animals with unpredictable traces, these research methods may not be effective.

Don't worry, infrared cameras combined with the method of globalized distribution can help us to understand the resource distribution of national parks in all directions without dead ends—this is the grid monitoring system with full coverage. Specifically, the entire national park is divided into several monitoring grids, and monitoring facilities are arranged in each grid to ensure that there are no dead ends and no loopholes in scientific research monitoring. We are familiar with the distribution data of protected species animals such as the black deer, white-necked pheasant, and black bear in the national park, which is monitored by this method.

Let's take a look at the monitoring distribution maps of these three different species in my hands: first look for the three animals, the white-necked pheasant, the black bear and the black deer, whose distribution range is the smallest? Yes, it is a black bear, which only appears in the north near the border of Anhui Province; what kind of animal is widely distributed? It is a white-necked pheasant. They are concentrated in the buffer zone and the experimental area, but not much in the core area. This is because they prefer low-altitude secondary forests, artificial forests and camellia forests. Relatively, the distribution of black deer that requires a sense of security is concentrated in the core area and buffer zone of the national park, and rarely enters the test

area with more human activities.

Wild animals are mostly vigilant and the population of some protected species is already very small, so it is difficult to be directly observed in the wild. The setting of the grid monitoring system of infrared cameras provides a very effective tool for humans to study and understand nature without too much interference.

Notes: When an infrared camera is placed, the following elements should be considered.
(1) There are traces of animals nearby, and there is a high probability to record traces of animal activities.
(2) The height is about half of the height of the main recorded species, the distance is appropriate, and there is no obstruction in the lens range, which can effectively trigger the shooting and record the scene effectively.
(3) The installation point is stable, concealed and safe.

Attentions

- From plants to animals, draw forth the necessity of grid monitoring.
- Combine the monitoring data of star species to explain the value of monitoring.
- The mountain road steps are slippery here, please caution.

Recommended auxiliary facilities and toolkits

- Interpretation signage: Gridded monitoring with universe coverability
- Photos of white-necked long-tailed pheasants, black bears and black muntjacs;
- Infrared monitoring records of wildlife distribution in Gutian Mountain.

Extra activities

"How to set up an infrared camera in the field"

Step 1: According to the pre-map analysis, divide the monitoring network and determine the monitoring area.
Step 2: The infrared camera should theoretically be placed near the center of the grid.
Step 3: After the installation is successful, it should be checked and tested to ensure normal use.
Step 4: Regularly check, retrieve data and maintain equipment.
The setup plan can be adjusted according to the quality of the collected data.

苏庄镇 | 瞭望台 | 国家公园科研体验路线　　解说编号：SZ-LWT-07

解说点 07 结束集散点

29° 14′ 45.56″ N，118° 6′ 59.29″ E
海拔：330m

● 解说主题

本条游线总结与其他游线推荐

● 解说要点

1. 总结本条游线
2. 对国家公园其他参访资源的介绍

● 场地概述

步道入口

解说方式 Interpretation Style

- ☑ 人员解说 Staff Interpretation
- ☐ 教育活动 Educational Activity
- ☑ 解说牌 Interpretation Signage
- ☐ 场馆解说 Hall Interpretation

解说季节 Interpretation Season

- ☑ 春季 Spring
- ☑ 夏季 Summer
- ☑ 秋季 Autumn
- ☑ 冬季 Winter

「游线总结」

各位朋友，本次国家公园科研体验路线的参访就到这里了。不知道大家在这一个多小时的时间里，有没有对国家公园的科学研究和科学家的工作有更深一步的认识呢？接下来，我们简单回顾下本次行程。

基于 24hm² 大样地，科学家认认真真对每一棵树进行户口普查与定期追踪。科学家们的研究工作深入且细致，从对每一棵树长短粗细的测量，林冠层的动态研究，到种子雨的收集，再到对于行踪不定的野生动物的研究，从一个小小的尺子、网框，到红外相机，再到塔吊这样的大型设备的应用，借助各种手段，从不同角度为我们一一揭开了钱江源的神秘面纱。

「其他游线推荐」

在人类文明的不断冲击下，钱江源国家公园仍能保存这么一片大面积连续的原真区域离不开当地山民的守候与科研工作者的保护。除了我们走的这条游线之外，钱江源还有很多地方值得大家去探索。比如，在古田山的森林王国，接受森林的洗礼；去苏庄镇的古村落走走，体验一下当地的风土民情。当然，既然来到了钱江源，你一定不能错过位于莲花塘的钱塘江寻源之旅。如果你想在最短的时间内体验最壮阔的风景，大峡谷会是不错的选择。

「游线分享」

在结束今天的行程之前，我想通过三个小问题跟大家一起聊聊此行的感受。

1. 令你印象最深的一个研究方法或工具是什么？

2. 如果你回去同家人或朋友分享此次自然体验的话，你会选择分享什么？

3. 如果以后你有机会以志愿者的身份参与到国家公园的守护工作中来，你最希望参与的工作是什么？

「行为倡议」

感谢大家的分享！钱江源国家公园的保护离不开当地人的守护和科研工作人员长期的坚持，当然也离不开大家的珍惜与爱护。谢谢你们一路的支持，希望大家未来的旅途愉快，返程平安！

注意事项

- 召集大家集合，做 15 分钟的小回顾和体力上的休整。
- 邀请伙伴分享，并给予小礼物。
- 通过大家的反馈，延伸启发与思考，提出回去后的行为倡议。
- 感谢大家的到访，感谢大家聆听自己的解说（请大家回去后向身边人宣传钱江源国家公园）。

Suzhuang Town | Observatory Tower | Scientific Research Nature Experience Route

Interpretation No. SZ-LWT-07

Stop 07: Assembly ending point

Interpretation theme

Summary and recommendation for other rates

Key points

1. Summarize this tour
2. Introduce other visiting resources of the national park

Site overview

Entrance to the trail

"Travel summary"

Dear friends, our visit to the Research Nature Experience Route is done. I wonder if you have a deeper understanding of the national park's scientific research and scientists' work in the past hour? Next, let's briefly review this trip.

Based on a large plot of 24 hectares, scientists diligently conduct census and regular tracking of each tree. The research work of the scientists is in-depth and meticulous, from the measurement of the length of each tree, the dynamic study of the forest canopy, to the collection of seed rain, the study of wild animals with uncertain whereabouts, from a small ruler, grid system, to infrared cameras, to the application of large equipment such as tower cranes. We use various means to help us unveiled the mystery of Qian Jiangyuan from different angles.

At the same time, under the constant impact of human civilization, Qianjiangyuan National Park can't still preserve such a large and continuous original area without the protection of local people and the protection of scientific researchers.

"Recommended other routes"

In addition to the tour we took, Qian Jiangyuan has many places worth exploring. For example, accept the baptism of the forest, in the forest kingdom of Gutian Mountain, and walk around the ancient village of Suzhuang Town and experience the local customs. Of course, since you came to Qianjiangyuan, you had better not miss the source-seeking journey of Qianjiangyuan in Lianhua Pond. If you want to experience the most magnificent scenery in the shortest time, the Grand Canyon is a good choice.

"Travel sharing"
I want to know about your feeling of this circle by asking you three questions before the circle is over.

The first question: Which research method or tool did impress you the most?

Second question: If you go back and share this walk with family or friends, what would you choose to share?

The third question: If you have the opportunity in the future, would you like to participate in the protection of the national park as a volunteer?

"Behavior initiative"
Thank you for your sharing. The protection of Qianjiangyuan National Park is inseparable from the protection of local people and the long-term adherence of scientific research staff. Thank you for your support along the way, and wish you all a happy future journey and a safe return journey!

Attentions

- Gather everybody, organize a 15-minute quick review, and take a break.
- Invite Tourists to share feedback and give away small gifts.
- Lighten inspiration and propose environmental protection strategy.
- Thank them all for visiting and listening to your interpretation.

2.4 Q01 莲花塘：钱塘江寻源自然体验路线

Qiantang River source exploration route

鸣界亭
Mingjie Lounge
问茶亭
Wencha Lounge
禅庐
Chanlu
经轩
Jingxuan
源头第一泉
The First Spring at the Source
莲花塘
Lotus Pond
抱松亭
Baosong Lounge
探源亭
Tanyuan Lounge
思夔亭
Sikui Lounge
源头碑
Headstone
岔口平台
Platform with Forked Road
越涧桥
Yuejian Bridge
三叠瀑
Sandie Waterfall
涵煦亭
Hanxu Lounge
彩虹飞瀑
Waterfall and Rainbow
润心亭
Runxin Lounge
停车场
Parking Lot
入口
Entrance
图例 LEGEND
钱江源国家公园
Qianjiangyuan National Park
水域
Water
等高线
Contours
步行道
Trail
车行道
Road
服务中心
Visitor Center
停车场
Parking Lot
商店
Shop
卫生间
Toilet
休憩亭
Rest Lounge
观景处
Recreation Spot
瀑布
Waterfall
设解说牌的点位
Point with Signage
设解说牌、人员解说的点位
Point with Signage and Interpretation
设解说牌、人员解说、课程的点位
Point with Signage, Interpretation and Course
0 200m 400m 600m
N

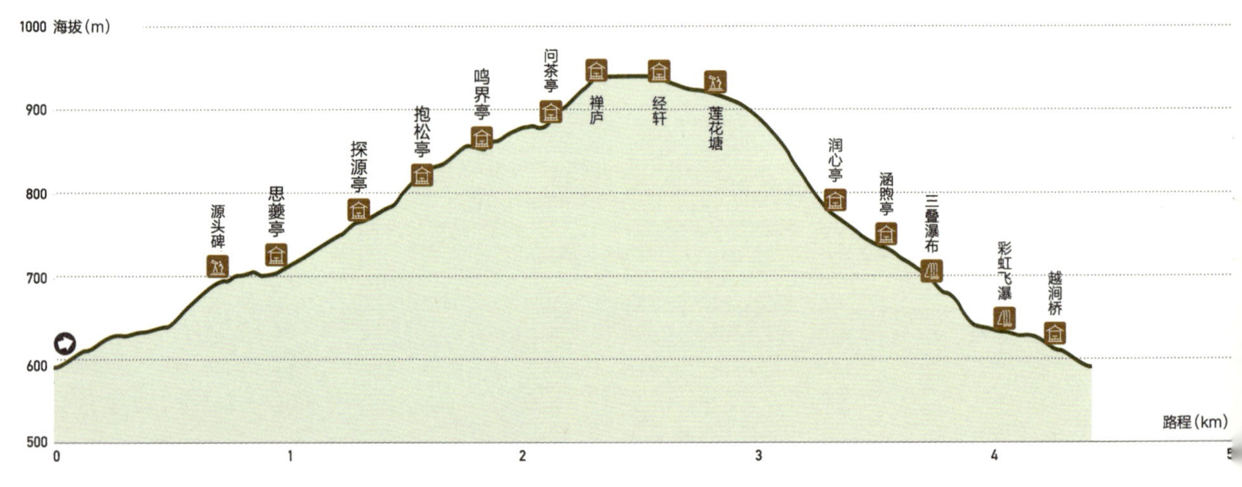

海拔(m)
1000
900
800
700
600
500
源头碑
思夔亭
探源亭
抱松亭
鸣界亭
问茶亭
禅庐
经轩
莲花塘
润心亭
涵煦亭
三叠瀑布
彩虹飞瀑
越涧桥
路程(km)
0
1
2
3
4
5

Q01 莲花塘：钱塘江寻源自然体验路线

解说主题 主题 2：润泽富庶的河源追溯
主题 3：丛林佑护的珍贵原真
主题 1：点亮北纬三十度的生机绿洲

步道里程 往返环线约 4.5km
体验时长 往返 3～3.5h
最低海拔 548m
最高海拔 930m
步道等级 大众型
基础设施 休憩亭 / 解说牌 / 活动场地 / 卫生间 / 无餐饮

作为钱塘江水源地所提供的水源涵养与供给功能是钱江源国家公园的一个重要特色资源。关于钱塘江之源，自古以来就有很多讨论，直到 1999 年才最终确定下来钱塘江正源是位于衢州市开化县齐溪镇的莲花尖。

本条路线位于钱江源国家公园北部齐溪镇，溯源而上，访客一路不仅可以感受到钱塘江源头水系的清秀灵动和沿途的自然风光，还可以更进一步了解塑就水源地的整个地质演化过程，以及森林生态系统对于水源地孕育的作用，直接地感受水源地的珍贵和脆弱，同时思考河流与健康、河流与人类社会发展之间的紧密联系。

本条自然体验路线全长约 4.5km，在整条探索路线上，我们一共为解说员和访客设计了 16 个推荐解说点，有从古至今人们对钱塘江正源求索历程的介绍，也有对河流源头形成和维续机理的介绍，还有因地制宜的针对本地动植物及生态系统多样性的解说内容。

同时，根据本书提供的通用型人员解说方案（非定点解说方案）资料，解说员也可以在合适的动植物资源停留点进行相应的引导性观察和解说。

注：本条路线可延伸至莲花尖，但需要提前预约并经专人引导前往。

Q01 Qiantang River source exploration route

Theme
- 02 The river source that breeds the fertile land
- 03 Mysterious wildlife in the forest
- 01 The 30° N miracle of green oasis

Trail Length About 4.5 kilometers round-trip
Time 3 – 3.5 hours round trip
Lowest elevation 548 meters
Highest elevation 930 meters
Level of trail Public
Infrastructures Lounge / interpretation board / event site / restroom / no catering

As the water source of the Qiantang River, the Qianjiangyuan National Park has another special contribution of water conservation and supply function. There has been much discussion about the source of the Qiantang River since ancient times. It was not until 1999 that the Lotus Peak, located in Qixi Town, Kaihua County, Quzhou City, was finally determined as the source of the Qiantang River.

This trail is located in Qixi Town in the north of Qianjiangyuan National Park. Tracing upwards, not only can visitors feel the beautiful and clear water system of the source of the Qiantang River and the natural scenery along the way, but also learn more about the entire geological evolution process of the water source, as well as the role of forest ecosystems in nurturing water sources, intuitively feel the preciousness and fragility of water sources, and think about the close connection between healthy water sources, rivers and the development of human society.

This nature experience route is about 4.5 kilometers long. We have arranged a total of 16 recommended interpretation sites during the entire experience trip.

There are reviews of people's search for where the Qiantang River Source is located from ancient times to the present, the introduction of formation and maintaining mechanism of the river sources, as well as interpretation of the

local flora and fauna and ecosystem diversity based on local conditions.

At the same time, according to the information provided in general interpretation (non-route-oriented interpretation) of this staff interpretation manual, interpreters can also find suitable animal and plant resources to conduct corresponding guided observation and interpretation.

Notes: This route can be extended to the Lotus Peak, but it needs reservation in advance and guided by a dedicated person.

齐溪镇 | 莲花塘 | 钱塘江寻源自然体验路线　　解说编号：QX-LHT-01

解说点 01 集合点

29°23′24.81″ N，118°12′39.57″ E
海拔：562m

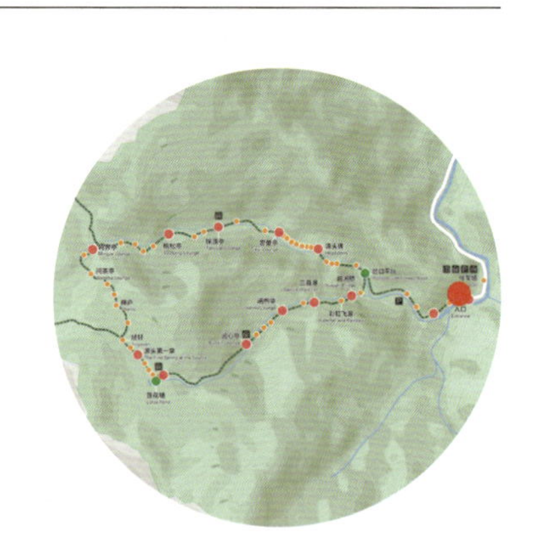

● **解说主题**

本条自然体验路线的总体介绍

● **解说要点**

1. 整条游线的总体介绍
2. 徒步注意事项

● **场地概述**

停车场或入口山门内的集合点，有步道总体导览牌

解说方式 Interpretation Style

☑ 人员解说 Staff Interpretation　☐ 教育活动 Educational Activity
☑ 解说牌 Interpretation Signage　☐ 场馆解说 Hall Interpretation

解说季节 Interpretation Season

☑ 春季 Spring　☑ 夏季 Summer
☑ 秋季 Autumn　☑ 冬季 Winter

解说词

各位朋友，欢迎来到钱江源国家公园，我是公园的解说员 XXX。今天将由我陪伴大家一起进行接下来的钱塘江溯源之旅。

本次路线全程大约有 4.5km，全程参访加上解说平均需要约 3h，其中有较多徒步难度较高的登山路段，对大家的体力会有一定的挑战。在我们出发前，请大家做好充分的准备工作，带足饮用水，并检查一下个人的装备。

“水”是我们此行的主角，探访钱塘江的源头是我们此行的主题。我们会顺着莲花溪溯流而上，见识溪流、瀑布、山泉等不同形式的水，欣赏沿途的景观变化，同时，我们也会从地形、气候、水文、植被和历史文化等不同的角度去认识钱塘江的源头。自古以来，人们进入山林最简单的方法就是循着溪流前进。因为水往低处流，一般情况下，沿着水流一定能走到山下。同理，溯流而上，我们也能找到河流的源头。当然了，溪流两岸的地形并不一定都适合修筑步道，大家也可以沿途感受一下我们如何利用桥、阶梯等形式，最合理地在山谷间和溪流旁布设这条自然体验路线的。

当然了，除了有莲花溪为我们指路以外，在行程中我们可能还会收到很多额外的礼物：沉醉在中国最优质的天然氧吧，观察一下亿万年形成的岩石，抑或与这里的野生生物们不期而遇，直到我们最终找到母亲河的源头。钱塘江的源头到底在哪里呢？我看大家已经迫不及待想要出发了。让我们边走边慢慢为大家解说。

在出发之前，我还要提醒大家，为了个人安全，更为了保护国家公园的宝贵资源，一定要跟随我的指引，不要走出游线范围，更不要去做任何危险的动作。有任何问题都可以随时与我沟通。让我们像拜访朋友一样，安静地到来，满足地离开，除了照片和脚印，不留下任何痕迹。

接下来，让我们把时间和期待交给大自然！我们出发吧！

注意事项

记住，请注意集合点的开场。当你出现在听众们的面前，解说员的工作已经开始了。如何吸引听众的注意力，树立自身的解说员身份，一个简短、内容明确、带有互动的开场白对解说员来说非常重要。以下几点提示可以帮你达到这一目标。

- 选择一个合适的、可以帮助大家聚焦这次步道体验的集合地点，比如，步道导览牌边上。
- 体现你身份的制服、解说员的标志和便携的解说装备包。
- 你的介绍中一定有独特的、导览牌上未曾涉及的内容。
- 根据团队成员的构成，针对性地提供步道过程中的注意事项建议。

辅助道具推荐

- 游线总体导览牌

拓展活动推荐

- 钱江源国家公园打卡地图（P562：活动卡 02）

Qixi Town | Lotus Pond | Qiantang River Source Exploration Route

Interpretation No. QX-LHT-01

Stop 01: Assembly meeting point

Interpretation theme

General introduction to this nature experience route

Key points

1. An overall introduction of this route
2. Suggestions and precautions of the hiking

Site overview

The parking lot or the square beside the main entrance. Where the overall introduction board is installed.

Hello everyone, welcome to Qianjiangyuan National Park. I am the interpreter of the park XXX. Today, I will accompany you on the next Qiantang River source tracing tour.

Today's route is about 4.5 kilometers long. The entire visit and interpretation will take about 3 hours on average. There are quite a few sections that are relatively difficult to climb, which may pose some challenge to everyone's physical strength. Before leaving, please get yourself ready, bring enough drinking water, and check your personal equipment.

"Water" is the star of our trip, and our theme today is visiting the source of the Qiantang River. We will follow the Lotus Creek upward to see the different forms of streams, waterfalls, mountain springs and other waters, and appreciate the changing landscape along the way. At the same time, we will also understand the source of Qiantang River from different perspectives of terrain, climate, hydrology, vegetation, and historical culture, *etc*.. Since ancient times, the easiest way for people to enter the forest is to follow the streams. Because the water flows downwards, under normal circumstances, the water must be able to go down the mountain. In the same way, we can also find the source of the river when we go upstream. Of course, the terrain on both sides of the stream is not necessarily suitable for building trails. You can also feel how we use bridges, ladders and other forms along the way to build this nature experience route between valleys and the streams.

Of course, in addition to the Lotus Creek showing us the way, we may have many discoveries along the way, such as enjoying the best quality

natural oxygen bar in China, observing the rocks formed hundreds of millions of years ago, or encountering the wildlife, before we finally found the source of the mother river. Where is the source of the Qiantang River? I see everybody can't wait to leave. Let me tell you all about it while we walk.

Before we leave, I would also like to remind everyone that for the sake of your own safety and protection of the precious resources of the national park, please follow my lead, do not leave the group, and do not take any dangerous actions.

You can always communicate with me if you have any questions. Let us come and visit quietly and leave happily like visiting friends, leaving no trace except photos and footprints.

Next, let us hand over our time and expectations to nature! Let's go!

Attentions

Remember, pay attention to your debut at the assembly meeting point. When you show up to the audience, your job as an interpreter has already begun. As an interpreter, it is very important to attract the attention of the audience and establish your own personality, such as using a short, clear and interactive opening statement. The following tips can help you achieve this goal.

- A proper assembly place that can help to gather the crowd and start this natural experience, such as the open space next to the general guide sign at the start of the trail.
- Uniforms and costumes that reflect your identity: the logo of the Interpreter and supporting materials tool kit.
- There must be something unique in your introduction that is not covered on the guide board.
- Raise targeted safety requirements and precaution based on specific team members.

Recommended auxiliary facilities and toolkits

- Overall guide board of the route

Recommended activities

- Task list map of Qianjiangyuan National Park（P562：No. 02）

齐溪镇 | 莲花塘 | 钱塘江寻源自然体验路线　　解说编号：QX-LHT-02

解说点 02 莲花溪

29°23′24.81″ N，118°12′39.57″ E
海拔：582m

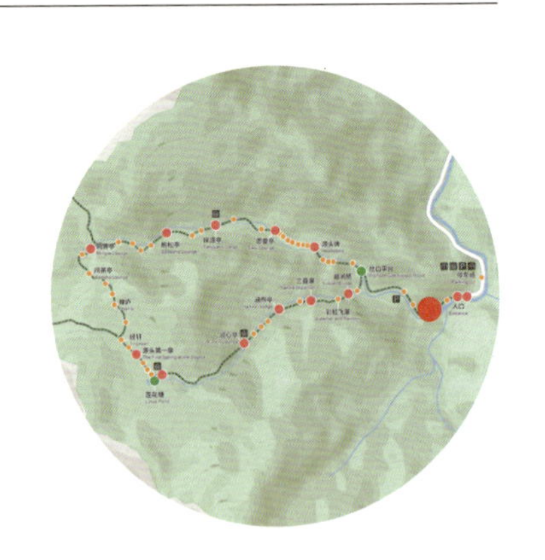

● **解说主题**

通过莲花溪认识钱塘江的历史

● **解说要点**

1. 莲花溪与钱塘江系的关系
2. 钱塘江得名的历史

● **场地概述**

步道开始段落的山道，沿线左侧有溪流

解说方式 Interpretation Style

☑ 人员解说 Staff Interpretation　☐ 教育活动 Educational Activity
☑ 解说牌 Interpretation Signage　☐ 场馆解说 Hall Interpretation

解说季节 Interpretation Season

☑ 春季 Spring　☑ 夏季 Summer
☑ 秋季 Autumn　☑ 冬季 Winter

「身边的莲花溪是钱塘江的源头」

各位朋友，说起钱塘江，大家脑袋里会浮现出哪些文字或者画面？钱塘江大潮！如果是浙江人的话，可能还会多一层涵义："母亲河"。

的确，钱塘江是浙江省第一大河，全长近700km，自西向东横穿浙江，在杭州湾注入东海，并形成举世闻名的钱塘江观潮盛景。我们身旁清澈的小溪流就是钱塘江的源头之水：莲花溪。莲花溪因为水质清澈，水流湍急，遇到石头溅起的水花犹如朵朵白莲而得此名。接下来的旅途，它会一直陪伴我们左右，指引我们走向钱塘江的源头所在。

「钱塘江古名浙江」

钱塘江自古以来就备受关注。关于钱塘江的最早文字记录出现在《山海经》中，因其流经古钱塘县，也就是今天的杭州而得名。不过，当时钱塘江仅指下游流经钱塘县的一部分河段，民国时期才作为全江的统称。

在钱塘江之前，它还有一个古称，是大家非常熟悉的。大家可以猜猜看是什么。提醒一下：我们现在在哪个省？浙江。这就是钱塘江的古称。中国汉字的魅力之一就在其音形兼具的丰富内涵，我们把"浙"字拆开来看，水和折，不正如钱塘江"蜿蜒曲折"的形态一样吗？是不是很形象？

「莲花溪与钱塘江水系的关系」

中国有句古话叫作："不积小流无以成江海。"作为浙江省流域面积最大的河流，也是浙江人的母亲河，钱塘江的浩荡水量是兼容并蓄的结果。不过，身处水系发达、水网密布的长三角地区，钱塘江的水源地只有我们眼前这么一条吗？如果不是的，我们又凭什么判定这里是钱塘江的源头呢？答案就在我们前方。

注意事项

- 从熟悉的钱塘江大潮出发，引出大家熟悉的钱塘江，同时通过点出"浙江"，指出钱塘江对于浙江人民的重要意义。
- 在与访客互动时，应结合实际情况灵活应对，不应拘泥于文本。

辅助道具推荐

- 标识牌：你好，这里是莲花溪
- 标识牌：千年寻源的探究之旅

拓展知识点

「钱江源对浙江和长江三角洲的生态服务价值」

- 一方山水养一方人。作为优质水源与充沛水量的涵养地，钱江源国家公园持续为流域内的衢州、杭州、绍兴等26个县（市）提供高质量的用水保障。同时，这里也是长江三角洲地区重要的生态屏障。

Qixi Town | Lotus Pond | Qiantang River Source Exploration Route

Interpretation No. QX-LHT-02

Stop 02: Lotus Creek

- **Interpretation theme**

  Introduce the history of Qiantang River from Lotus Creek

- **Key points**

  1. Relationship between the Lotus Creek and the Qiantang River
  2. The naming road of Qiantang River

- **Site overview**

  The mountain road at the beginning of the trail, where you can see the stream on the left

"The Lotus Creek besides us is the source of Qiantang River"

Dear friends, when talking about Qiantang River, what words or pictures will appear in your head? The tide of the Qiantang River! If you are from Zhejiang, there may be an additional meaning: "Mother River".

Indeed, the Qiantang River is the largest river in Zhejiang Province, with a total length of nearly 700 kilometers. It crosses Zhejiang from west to east and injects into the East China Sea in Hangzhou Bay, and has formed a world-famous scene, the Qianjiang tide. The clear stream beside us is the source of the Qiantang River: Lotus Creek. It is named because of the clear water and rapid flow. The following journey, Lotus Creek will be with us and lead us to the source of the Qiantang River.

"Qiantang River was once called Zhejiang"

The Qiantang River has attracted tons of attention since ancient times. The earliest written record of the Qiantang River appears in the *Shan Hai Jing*. It got its name because it flows through the ancient Qiantang County, which is today's Hangzhou. However, at that time, it only referred to a part of the river flowing through Qiantang County in the downstream. Not until the Republic of China time was it recognized as a general term for the whole river.

Before it was called the Qiantang River, it also had an old name, which is very familiar to everyone. Take a guess? A little tip, which province are we at now? Zhejiang. That's right. This is the ancient name of Qiantang River. One of the charms of Chinese characters

lies in the rich connotation of both sound and shape. When we take the Chinese character "浙" apart, it's made up of "氵" (meaning water) and "浙" (meaning curve) just like the shape of Qiantang River's "twisting".

"Relationship between the Lotus Creek and the Qiantang River"

There is an old Chinese saying: "No Ocean can be formed without the accumulation of streams." As the largest river in Zhejiang Province, it is also the mother river of the people of Zhejiang. The massive water volume of the Qiantang River is a result of accumulation. But in the south of the Yangtze River, where the water system is dense and well-developed, is the lotus creek in front of us the only water source of the Qiantang River? If not, how can we determine that this is the source of the Qiantang River? The answer is right in front of us.

Attentions

- Starting from the familiar Qianjiang tide, point out the familiar Qiantang River, lead to "Zhejiang" and demonstrate the importance of the Qiantang River to the people of Zhejiang.
- When interacting with visitors, you should respond flexibly according to the actual situation, and you don't have to stick to the text.

Recommended auxiliary facilities and toolkits

- Interpretation signage: Welcome to Lotus Creek
- Interpretation signage: A thousand years' journey tracing the river source

Extra information

"The ecological service value of Qianjiangyuan to Zhejiang and Yangtze River Delta"

The unique features of a local environment always give special characteristics to its inhabitants. As a conservation place for high-quality and abundant water sources, Qianjiangyuan National Park continuously provides high-quality water supply for 26 counties (cities), including Quzhou, Hangzhou, Shaoxing, and other regions within the basin. At the same time, as an important ecological barrier in the Yangtze River Delta region, it is like a backyard garden for central cities such as Shanghai, Suzhou, Nanjing, and Hangzhou.

齐溪镇 | 莲花塘 | 钱塘江寻源自然体验路线　　解说编号：QX-LHT-03

解说点 03 岔口平台

29°23′24.81″ N，118°12′39.57″ E
海拔：597m

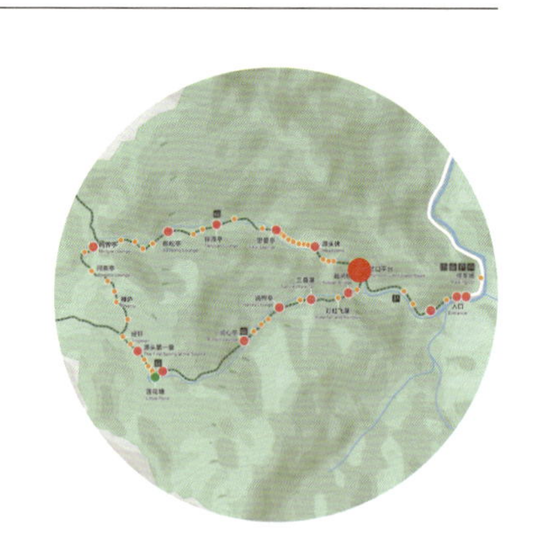

● 解说主题

丰富的生物多样性与孑遗物种资源

● 解说要点

1. 长江三角洲最后的原始森林的珍贵价值
2. 认识孑遗植物

● 场地概述

道路分叉口的小型平台，有青钱柳等多棵孑遗植物

青钱柳（果实）　杜仲（叶）　香果树

解说方式 Interpretation Style

- ☑ 人员解说 Staff Interpretation
- ☑ 教育活动 Educational Activity
- ☑ 解说牌 Interpretation Signage
- ☐ 场馆解说 Hall Interpretation

解说季节 Interpretation Season

- ☑ 春季 Spring
- ☑ 夏季 Summer
- ☑ 秋季 Autumn
- ☐ 冬季 Winter

解说词

行走在这样的环境中，大家会不会突然产生一些联想？儿时背诵过的很多诗句，好像在这里突然跃入眼中。比如，我就想到了“明月松间照，清泉石上流”“蝉噪林愈静，鸟鸣山更幽”“返景入深林，复照青苔上”。不知道此时此刻大家心里有没有涌现出哪些诗句。

「长江三角洲最后的原真森林与丰富的植物多样性」

中国陆地 960 万 km² 的土地上，森林所占的面积高达 1/5 以上，说起来并不算少，但其中大部分是人工林，只有不足 1% 的森林属于未受到人类活动明显干扰的原始森林。我们现在所在的钱江源国家公园便属于这极其珍贵的天然原始森林之列。

钱江源国家公园为 1500 多种种子植物提供了安居的家园，是已知 11 种模式物种的发现地，同时也是 32 种古老的孑遗植物的庇护之所。

「什么是孑遗植物」

什么是孑遗植物？这个名词大家听起来可能有些陌生。孑遗植物的祖先可以追溯到遥远的地质年代，它们比人类更早出现在地球上，曾经是地球上最繁盛的生物家族之一，其个体遍布全球。亿万年来，沧海桑田，很多生物都逐渐退出历史舞台，唯有孑遗植物虽经过了历史的大浪淘沙，但仍然坚强地活着。不过，它们如今仅零星分布在极小部分区域。它们的存在是千百万年的地球气候和地质历史变迁的见证。如今它们身上依然保留着祖先的一些原始性状，我们把它们形象地比喻为“阅尽千帆”的孤独旅人。多亏了钱江源这样的原始森林为它们提供的避难所，它们才得以熬过严酷的生存考验，让各位今天仍然有幸能看到它们。

在钱江源，我们能看到哪些孑遗植物呢？它们有号称“摇钱树”的青钱柳、“叶断丝连”的杜仲、美丽迷人的长柄双花木、芳香馥郁的香果树等。

接下来，我给大家几分钟自由时间。大家可以在附近休息一下，拍拍照，有兴趣的朋友可以根据这几张图片去寻找我刚刚提到的青钱柳等孑遗植物。最后，我们来比一比谁最先找到“摇钱树”以及谁认识的植物种类最多。请大家注意安全，我们 5 分钟后回到这里集合。

注意事项

- 观察身边典型常绿阔叶林、针叶林代表树种，引出这里植物的多样性。
- 解说孑遗植物从侧面印证了钱江源守护与被守护的价值与意义。
- 讲解具体植物（特别是青钱柳）可配合照片或者通用解说方案的内容介绍。

辅助道具推荐

- 标识牌：阅尽千帆的孤独旅人
- 捡拾的青钱柳果实
- 相关孑遗植物介绍卡片

拓展活动推荐

- 孑遗植物连连看（P576：活动卡 15）
- 寻找青钱柳：结合青钱柳的物种介绍卡片，让访客组队寻找青钱柳

Qixi Town | Lotus Pond | Qiantang River Source Exploration Route

Interpretation No. QX-LHT-03

Stop 03: Fork platform

● Interpretation theme

The precious value and rich biodiversity of the forest and relict plants in Qianjiangyuan National Park

● Key points

1. The precious value of the last virgin forest in Yangtze River Delta
2. Know relict plants in the forest

● Site overview

A small platform at the fork of the road with many relic plants such as wheel wingnuts

Walking in an environment like this, will everyone suddenly have some associations with the poetry we recited in childhood? For example, "a silvery moon is shining through the pines, the limpid brooks are gurgling over the stones" "The forest is more peaceful while cicadas are chirping, the mountain is more secluded while the birds are singing." I don't know if there is any poetry that comes into anybody's mind?

"The last primitive and indigenous forest and rich plant diversity in the Yangtze River Delta"

On the land of 9.6 million square kilometers in China, forest occupies more than one-fifth of it, which is not small, but most of it is artificial forest, and less than 1% of it is virgin forest that hasn't been significantly interfered by human activities. The Qianjiangyuan National Park where we are at now belongs to this extremely precious natural virgin forest. Qianjiangyuan National Park is home to more than 1,500 seed plants, the discovery site of 11 known species, and is also a sanctuary for 32 ancient relic plants.

"What is a relict plant"

What is a relict plant? This term may sound strange to everyone. But if we talk about living fossils, everyone must be familiar. Relic plants are a group of ancient relics that have survived since a geological age long ago. They appeared on the earth earlier than humans and were widely distributed in many places, but now only scattered in very small areas. They have silently witnessed millions of years of changes in the earth's climate and geological history, and they still retain some of their ancestors' primitive traits. We call them solitary travelers who have experienced everything. The sanctuary provided by the

Qianjiangyuan National Park allows them to survive the harsh survival test so that you can still see them today.

What relict plants can we see in Qianjiangyuan? They include the "money tree" wheel wingnut, *Eucommia* whose fiber is still connected when its leaves are broken, beautiful and charming *Disanthus* trees, and fragrant *Emmenopterys henryi*. Their distribution is also different.

Next, I will give everyone a few minutes. You can take a break and take photos nearby. You can use these pictures to find the relic plants such as the wheel wingnut that I just mentioned if you are interested in it. Let's see who is the first to find the "money tree" and who knows the most plant species?

Attentions

- Observe the representative tree species in typical evergreen broad-leaved forest and coniferous forest and lead to the diversity of plants here.
- Confirm the value and significance of protecting Qianjiangyuan through relic plant interpretation.
- Interpretation specific plants (especially wheel wingnut) can be combined with photos or the general interpretation.

Recommended auxiliary facilities and toolkits

- Interpretation signage: living ancestors
- Forest litter of *Cyclocarya* and cards of other relic plants

Recommended activities

- Relic plants link game (P576：No. 15)
- Find the cyclocarya: Ask visitors to team up to find the *Cyclocarya*.

齐溪镇 | 莲花塘 | 钱塘江寻源自然体验路线　　解说编号：QX-LHT-04

解说点 04 源头碑

29°23′31.54″ N，118°12′15.59″ E
海拔：636m

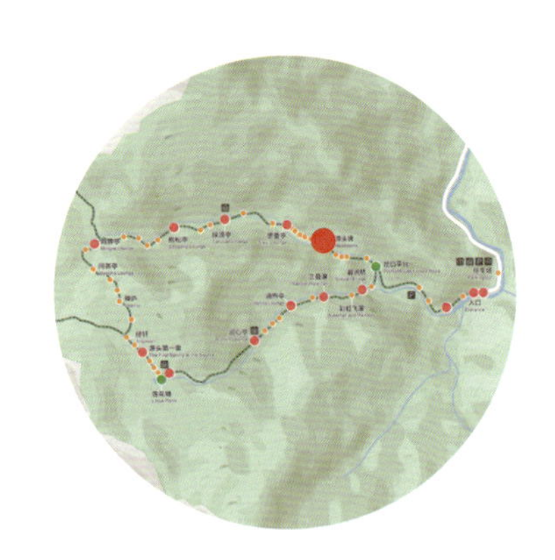

解说主题

从古至今钱塘江的寻源历程

解说要点

1. 历史上的三个主要寻源认知
2. 近现代对于钱塘江正源的综合判定方法

场地概述

石碑前，有较大的场地空间可向游客解说寻源的概念

解说方式 Interpretation Style

- ☑ 人员解说 Staff Interpretation
- ☑ 教育活动 Educational Activity
- ☑ 解说牌 Interpretation Signage
- ☐ 场馆解说 Hall Interpretation

解说季节 Interpretation Season

- ☑ 春季 Spring
- ☑ 夏季 Summer
- ☑ 秋季 Autumn
- ☑ 冬季 Winter

任何一条大江大河都是由众多涓涓细流汇聚而成的，钱塘江当然也不例外。大家来看下这张钱塘江的水系图，会发现这里的水系纷繁复杂，也因此有个问题历史上却迟迟得不到答案：钱塘江上游有这么多条河流，究竟哪一条才是钱塘江的正源呢？

「历史上的三个主要寻源认知」

很久很久以前，生活在钱塘江沿岸的古人和我们就有着相同的疑问。为了探究河流源头，他们只有一个办法：走！说起来也不难，逆流而行，一直走到尽头不就是河流的源头了吗？但是这一走却走了2000多年。随着科技的进步与社会的发展，人们对于江河源头的探寻和结论也被一一记录。古代关于钱塘江发源地最主要的观点有三种：北部安徽省的新安江源头论、浙江西部开化县的衢江源头论和浙江东部的东阳江源头论。

其中，北源新安江因为是三条河流中最长的一条，符合“河源唯远”的原则，是历史上较被广泛认可的钱塘江源头。《山海经》《汉书》《水经注》均持同样的观点。但是，北宋的《九域志》和清初的《读史方舆纪要》却分别提出衢江和东阳江为钱塘江正源的说法。直至20世纪末，钱塘江的源头也一直未能真正确定下来。

「近代钱塘江正源的科学研判」

随着近代科学的发展，我们可以用更客观、科学的方式以及统一的标准去寻找并界定江河的源头。“河源唯远”不再是唯一的原则，界定江河的源头一般要综合考量三个标准：干流长度、流域面积和水量贡献，同时，流向是否与干流主方向一致，是否符合地方文化符号的意义也会一并考虑。1999年，专家通过实地考察，对上述指标进行了综合论证，结果发现：衢江虽然长度略短于新安江，但流域面积和径流量均大大超过新安江，而且干流与钱塘江流域的中轴方向一致，从而使整个水系呈完美的羽状对称分布，符合国际国内通例，最终被认定为钱塘江的正源。

人类的对环境的认识就是在从古至今对自然的探究中得到逐步扩展的，寻源中的每一次探索都是人类文明留下的印迹。追随古人的寻源之路，我们发现的不仅是钱塘江的源头，也是一段河流灿烂的文明史。

注意事项

- 古人关于河流的认识全凭经验，是用脚步一点点丈量出来的，每走一步，就积累一分经验，通过文字传递给后人。
- 通过水系图可以清楚地看出钱塘江源头有三，没有哪一个一定是绝对正确的，只是源于不同的参考标准。
- 综合标准下，最终于1999年确定衢江为钱塘江正源。

辅助道具推荐

- 标识牌：千年寻源的探究之旅
- 标识牌：源出何处的科学研判

Qixi Town | Lotus Pond | Qiantang River Source Exploration Route

Interpretation No. QX-LHT-04

Stop 04: Headstone

Interpretation theme

The history of Qiantang River' s sourcing from ancient times

Key points

1. Three main source-finding cognitions in history
2. The method of determining the source of the Qiantang River in modern times

Site overview

In front of the stele, where there is a large space to explain the concept of sourcing to tourists

Walk in an environment like this. Well, all the big rivers are made up of many small streams, and Qiantang River is of course no exception. If you look at this water system diagram of the Qiantang River, you will find that the water system here is complicated, and there is a question that has not been answered for a long time: the main stream of the lower reaches of the Qiantang River is abundant and well established, but there are so many rivers in the upstream, which one is the source of the Qiantang River?

"Three main cognitions of source in history"
A long time ago, the ancients living along the Qiantang River had the same question as you. In order to explore the source of the river, they had only one way: walk. It's not hard, going against the flow, and when you reach the end, it is the source of the river. But this walk has lasted for more than 2000 years. With the development of human technology and science, people's explorations and conclusions about the source of rivers have also been recorded. There are three main opinions about the origin of the Qiantang River in ancient times: the source theory of the Xin'an River in the northern Anhui Province, the source theory of the Qujiang River in the west of Zhejiang, and the source theory of the Dongyang River in the east of Zhejiang.

Among them, the Xin'an River in the north is the more widely recognized source of the Qiantang River in history, because the Xin'an River is the longest of the three rivers, which is in line with the principle that "the longest river is the source". *Shan Hai Jing*, *Han Shu*, *Shui Jing Zhu* all hold the same view. However, the *Nine Territory Records* in the Northern Song Dynasty and the *Essentials of Geography for Reading History* in the early Qing Dynasty put forward the claims that Qujiang River and Dongyang River were the source of Qiantang River respectively. Until the end of the 20th

century, the source of the Qiantang River could not be truly determined.

"Modern scientific research judgment on the real source of Qiantang River"

With the development of modern science, we can use more objective and scientific methods and uniform standards to find and define the source of rivers. "The longest river is the source" is no longer the only principle. Generally, three criteria should be considered comprehensively: the length of the mainstream, the area of the basin and the contribution of water volume, and whether the flow direction is consistent with the main direction of the mainstream. In addition, for rivers with particularly important resources and cultural values, it is generally required to comply with the meaning of local cultural symbols. In 1999, experts conducted a comprehensive demonstration of the above indicators through field investigations, and found that although the length of Qujiang River is slightly shorter than that of Xin'an River, the area and runoff of the river basin are much larger than that of Xin'an River, and the main stream is consistent with the central axis of Qiantang River Basin so that the entire water system is perfectly feather-like and symmetrically distributed. In line with international and domestic general regulations, Qujiang River was finally recognized as the official source of the Qiantang River.

Human's understanding of the environment has been gradually expanded in the exploration of nature since ancient times. Every exploration in the source trace is a footprint left by human civilization. Following the ancient people's path of finding the source, what we discovered is not only the source of the Qiantang River but also a splendid history of civilization.

Attentions

- The understanding of ancient people about rivers was based on experience. It was measured step by step, and every step taken accumulated a tiny experience and was passed it on to future generations through words.
- It can be clearly seen from the water system diagram that there are three sources of the Qiantang River, and none of them can be absolutely correct, depending on different standards.
- Under the comprehensive standards, Qujiang was finally identified as the main source of the Qiantang River in 1999.

Recommended auxiliary facilities and toolkits

- Interpretation signage: A thousand years' journey tracing the river source
- Interpretation signage: Discovery of the river source location

齐溪镇 | 莲花塘 | 钱塘江寻源自然体验路线　　解说编号：QX-LHT-05

解说点 05 思夔（音 kuí）亭

29°23′34.04″ N，118°12′6.50″ E
海拔：678m

解说主题

杉木的遥远旅途

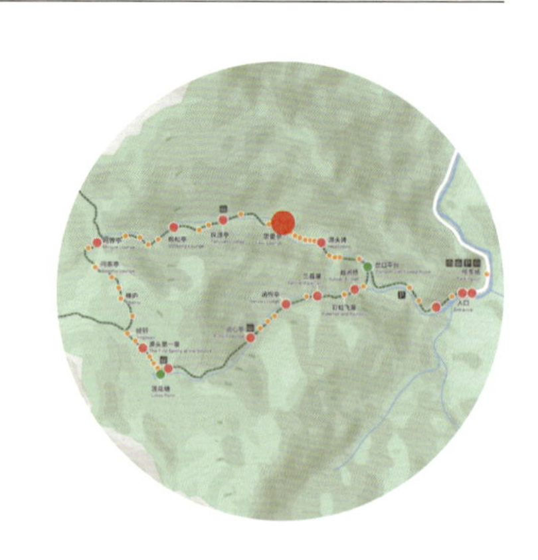

解说要点

1. 杉木林背后的人地关系的演变历程
2. 识别杉木的主要特征

场地概述

杉木林

解说方式 Interpretation Style

☑ 人员解说 Staff Interpretation　☐ 教育活动 Educational Activity
☑ 解说牌 Interpretation Signage　☐ 场馆解说 Hall Interpretation

解说季节 Interpretation Season

☑ 春季 Spring　☑ 夏季 Summer
☑ 秋季 Autumn　☑ 冬季 Winter

解说词

一路走来，我们遇到的树木大多数生长得非常自由随意，唯独眼前的这片杉（音 shā）木林像一批士兵一样整整齐齐地排列着。它们虽然长势很好，但是和周围充满野性的常绿阔叶林相比，似乎多了几分规整。

其实，这片林子并不是该地天然生长的树种，而是人工种植的次生林。号称“九山半水半分田”的开化，由于群山环抱，可耕地比较少，但山林资源丰富，自古以来木材经营就是其主要的经济来源之一。

杉木因树干通直，生长迅速，且耐虫蛀，一直以来都是深受人们喜爱的一种木材。明清时期，以“盐客木商”闻名的徽商几乎垄断了整个江南地区的木材贸易，而开化就是徽商木材生意的主要供货地。开化杉木曾随着徽商的足迹走遍大江南北，北京紫禁城修建太和殿所用的巨杉“皇木”就来自这里。

不要看眼前的杉木似乎很瘦弱，但古时杉木一般生长在深山里，长大成材之后才会被利用。不过，这么大块的木材怎么运出深山呢？聪明的古人想到了借助水力。人们会在固定时节将杉木伐倒，待到雨季河水上涨时，利用水力将其运载出山。一根根浑圆笔直的杉木就这样顺着钱塘江水系的河流一路向东，最终被送达当时繁华的商贸口岸杭州，然后再销往全国各地。

林场转型为保护地后，伐木活动被禁止，这片杉木林也幸运地逃脱了被砍伐的命运，从而成为见证这片山林历史的印迹之一。

接下来，大家可以捡拾地上杉木的落叶和果实，观察下它和我们之前介绍过的针叶树有什么不同。

注意事项

- 注意杉（shā）木的发音。
- 杉木的经济价值与过去的用途。
- 通过古代杉木的运输，指出钱塘江水系所具有的连接与航运的意义。
- 可以结合通用解说方案中的生物多样性部分，带领访客观察与识别杉木。
- 莲花塘附近也有一片杉木林，解说员可以在那里停留，带领访客重温杉木的故事或者进行更加深入的解读。

辅助道具推荐

- 标识牌：杉木的遥远旅途
- 标识牌：从林场到国家公园——岁月见证的改变
- 提前捡拾的杉木凋落物

拓展知识点

- 宋代匠人曾这样评价杉木：“杉木直干似松叶芒心实似松蓬而细，可为栋梁、棺淳、器用，才美诸木之最。”
- 和马尾松相比，杉木整体给人感觉比较硬朗且锐刺明显，而马尾松相对来说则柔软得多。

Qixi Town | Lotus Pond | Qiantang River Source Exploration Route

Interpretation No. QX-LHT-05

Stop 05: Sikui Lounge

Interpretation theme

The long journey of the Chinese fir

Key points

1. The evolution of the relationship between man and land behind the Chinese fir forest
2. Identify the main characteristics of Chinese fir

Site overview

Chinese fir forest

Along the way, most of the trees we met grow very casually, but the Chinese fir trees in front of us are arranged neatly like a group of soldiers. However, although they look good, they seem to be a bit more rigid than the surrounding wild evergreen broad-leaved forest.

In fact, this forest is not a tree species that grows naturally in this area, but a secondary forest planted manually. Known as the feature of "90% mountains, 5% water, 5% farm", because of the mountains, there is few arable land but rich forest resources in Kaihua. Since ancient times, timber business has been one of its main economic sources.

Chinese fir has been loved by people because of its straight wood, rapid growth, and resistance to insects. During the Ming and Qing Dynasties, Huizhou merchants, known as "salt and wood merchants", almost monopolized the timber trade in the entire Jiangnan region, and Kaihua was the main supplier of Huizhou merchants' timber business. Kaihua's Chinese fir has followed the footprints of Huizhou merchants from north to south, and the giant cedar "imperial wood" used in the construction of the Taihe Hall in the Forbidden City in Beijing comes from here.

Though the Chinese fir in front of us seems to be very thin, in ancient times, they generally grew deep in the mountains and will only be used when they mature, but how can such large pieces of wood be transported out of the mountains? The clever ancient thought of using the power of water. People will cut down the firs at a fixed time, and when the rainy season comes, the river will rise and use water to carry it out of the mountains. The round and straight firs, along the river at the source of the Qiantang River, went eastward, and were finally delivered to the bustling commercial

and trading port of Hangzhou at that time, and then sold to all parts of the country.

After the timber mill was transformed into a protected area, logging activities were banned, and this fir forest was lucky enough to escape the fate of being cut down.

Next, you can pick up the fallen leaves and fruits of the Chinese fir on the ground and observe how it differs from the conifers we met such as Masson pines.

Attentions

- Note the pronunciation of fir.
- The economic value of Chinese fir and its past use history.
- Through the transportation of ancient Chinese fir, the significance of the connection and shipping of the Qiantang River water system is pointed out.
- It can be combined with the biodiversity section of general interpretation to lead visitors to observe and identify Chinese fir.
- There is also a fir forest near the Lotus Pond, where the interpreter can stay and lead visitors to relive the story of the fir or make a more in-depth interpretation.

Recommended auxiliary facilities and toolkits

- Interpretation signage: China fir, the traditional building material
- Interpretation signage: From forest farms to national park
- Litters of China fir.

Extra information

- The Song Dynasty artisans once commented on the Chinese fir: “The straight stem of the Chinese fir is like a pine leaf and the heart is like a pine, and can be used for beams, coffins, and utensils.
- The Chinese fir as a whole feels tough and has sharp spines, while the masson pine is relatively soft.

齐溪镇 | 莲花塘 | 钱塘江寻源自然体验路线　　解说编号：QX-LHT-06

解说点 06 探源亭

29° 23′ 34.30 N，118° 11′ 59.74″ E
海拔 751m

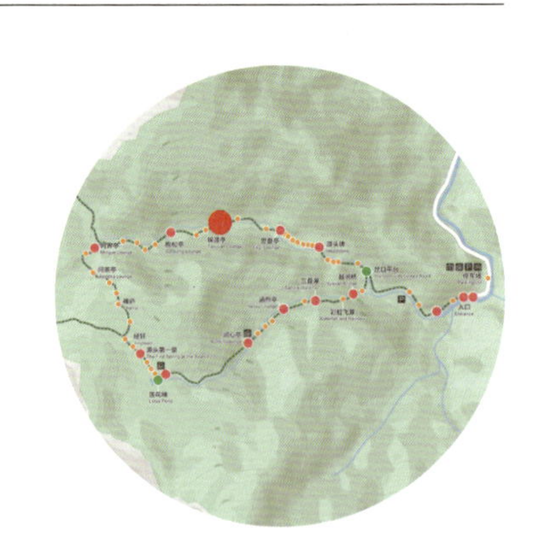

解说主题

从海拔变化的角度解说植物对环境的适应性

解说要点

1. 森林植被的垂直地带性分布规律
2. 影响垂直方向上植被分布的不同环境因素

场地概述

解说场地条件较好，有解说牌和休憩亭配套

解说方式 Interpretation Style

- ☑ 人员解说 Staff Interpretation
- ☐ 教育活动 Educational Activity
- ☑ 解说牌 Interpretation Signage
- ☐ 场馆解说 Hall Interpretation

解说季节 Interpretation Season

- ☑ 春季 Spring
- ☑ 夏季 Summer
- ☑ 秋季 Autumn
- ☑ 冬季 Winter

钱江源国家公园的森林十分独特。从全球尺度来看，这里位于北纬 30°，在世界上同纬度其他地区几乎均为荒漠，唯独中国到日本的这片区域被绿色的常绿阔叶林填充。从区域尺度来看，这里被经济发达、人口密集、城市化水平高的城市群所包围着，是长江三角洲地区最后一片大面积保存完整的原真森林。

此外，这里还拥有丰富的植物多样性，一方面是因为这里处于植被的过渡带上，兼有华东、华南、华北和华中等地的植被类型，另一方面，不同海拔的气温、降水、地形等的差异也是导致本地植物种类多样的重要原因。

「森林植被的垂直地带性分布」

这是一张本地植被的垂直分布图。图上非常清楚地展示了随着海拔的升高，植被呈现出由常绿阔叶林、常绿落叶阔叶混交林、针叶阔叶混交林再到针叶林逐步过渡的变化规律。大家可以观察下我们身边，根据我们现在所处的海拔高度，判断这里的植被应该属于哪种类型。我们现在属于常绿阔叶林带，海拔高度大约是……（可以借助手机指南针查看海拔）。

这里的常绿阔叶林的树种以四季常绿的甜槠、木荷、青冈等常绿树种为主。随着海拔的升高，温度逐渐降低，我们会观察到森林里会逐渐掺杂一些耐寒的落叶阔叶树种和针叶树种。到了海拔 1000m 以上，这里几乎就是黄山松、马尾松和杉木等耐寒的针叶树的天下了。上山的时候，大家可以留心观察森林中的物种构成，看看能否判断出它们属于哪一类植被。

当然了，这只是理论上的划分，实际植被分布随海拔的变化并非泾渭分明。我们可能在阔叶林中看到马尾松等针叶树种的身影，也有可能在海拔较高的针叶林中看到木荷、青冈等阔叶树种的存在。

植被在垂直方向上的分布与海拔带来的热量下降、坡度和坡位带来的光照和水肥条件等都相关。不过植物深谙生存的智慧，总能准确选择适合自己落脚的地方，并充分利用环境给予的条件，展现出“适者生存”的智慧。

注意事项

- 首先要让访客了解不同海拔植被类型的变化，可以从叶片的角度去认识其中的规律。
- 其次，要让访客了解随着海拔的升高，不同植被类型分布的规律背后的影响因素可能是什么。
- 注意，影响植被类型的往往是多种因素的组合。

辅助道具推荐

- 标识牌：一山一世界
- 标识牌：常绿阔叶林——顶极群落的生生不息
- 标识牌：常绿落叶阔叶混交林——过渡地带的森林变化
- 标识牌：针阔叶混交林——攀向高处的森林变化
- 标识牌：针叶林——适应高海拔的佼佼者

Qixi Town | Lotus Pond | Qiantang River Source Exploration Route

Interpretation No. QX-LHT-06

Stop 06: Tanyuan Lounge

Interpretation theme

Explain the adaptability of plants to environment from elevation

Key points

1. Vertical distribution of forest vegetation

2. Different environmental factors affecting the distribution of vegetation types in the vertical direction

Site overview

A well-equipped interpretation site with interpretation board and resting lounge

The forest in Qianjiangyuan National Park is very unique. In a horizontal direction, it is located at thirty degrees north latitude, and almost all other regions in the world at the same latitude are deserts. Only this area from China to Japan is filled with green evergreen broad-leaved forest; at the same time, this area that is surrounded by an economically developed, densely populated and highly-urbanized urban agglomeration, is the last large-scale preserved primitive and indigenous forest in the Yangtze River Delta region. In addition, it also has rich plant diversity. On the one hand because it is located in the vegetation transitional zone, there are also vegetation types in East China, South China, North China, and Central China; on the other hand, differences in temperature, precipitation, and terrain at different elevations are also important reasons for the variety of local plants.

"Vertical distribution of forest vegetation" This is a vertical distribution map of local vegetation. The figure clearly shows the regularity of the gradual transition of vegetation from evergreen broad-leaved forest, evergreen deciduous broad-leaved mixed forest, coniferous broad-leaved mixed forest to coniferous forest as the elevation increases. Everybody looks around us, based on the elevation we are at now, could you tell me what type of vegetation should be here, how many our current elevation is about... (you can check the elevation on your phone compass), which belongs to the distribution area of evergreen broad-leaved forests.

The gugertree and sweet oachestnut around us are typical representatives of evergreen broad-leaved forest. As the elevation increases, the temperature gradually decreases, and we

will observe that the vegetation gradually transitions from evergreen broad-leaved forest to evergreen deciduous broad-leaved mixed forest, which is also the layer with the most abundant color change in the four seasons; and to 1000 meters above sea level, the above is almost the world of coniferous trees such as Huangshan pines, Masson pines and Chinese firs. When you go up the mountain, you can observe the species composition in the forest and see if you can tell which type of vegetation they belong to.

Of course, this is only a theoretical division, and the actual distribution of mountain vegetation is not clear. We may see the presence of coniferous trees such as Masson pines in the broad-leaved forests, or the presence of broad-leaved tree species such as ring-cupped oak in coniferous forests at higher elevations.

The vertical distribution of vegetation is related to the heat drop caused by elevation, the light and water and fertilizer conditions caused by slope and slope position, *etc*.. However, plants are well equipped with the wisdom of survival, and they can always choose the place that suits themselves, and make full use of the conditions given by the environment to show the wisdom of "survival of the fittest".

Attentions

- First of all, let visitors understand the changes in vegetation types at different elevations, and you can understand the rules from the perspective of the leaves.
- Secondly, let visitors understand what may be the influencing factors behind the law of the distribution of different vegetation types as the elevation increases.
- Note that it is often a combination of factors that affect vegetation types.

Recommended auxiliary facilities and toolkits

- Interpretation signage: diverse vegetation varied with elevation
- Interpretation signage: evergreen broadleaved forest climax community — circle of life
- Interpretation signage: deciduous and evergreen broadleaved forest — transition zone
- Interpretation signage: coniferous and broadleaved forest —forest succession
- Interpretation signage: coniferous forest— mountaintop dominants

齐溪镇 | 莲花塘 | 钱塘江寻源自然体验路线　　解说编号：QX-LHT-07

解说点 07 抱松亭

29°23′26.27″ N，118°11′28.58″ E
海拔：710m

● 解说主题

公园多样的植物区系

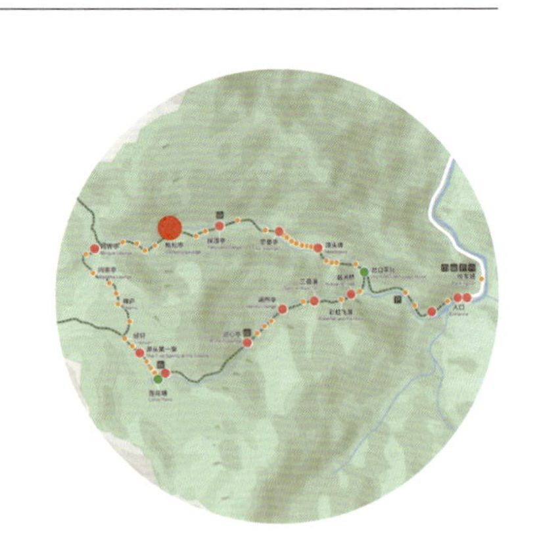

● 解说要点

1. 认识本地的植物区系
2. 本地植物区系多样的原因

● 场地概述

解说场地条件较好，有解说牌和休憩亭配套

解说方式 Interpretation Style

- ☑ 人员解说 Staff Interpretation
- ☐ 教育活动 Educational Activity
- ☑ 解说牌 Interpretation Signage
- ☐ 场馆解说 Hall Interpretation

解说季节 Interpretation Season

- ☑ 春季 Spring
- ☑ 夏季 Summer
- ☑ 秋季 Autumn
- ☑ 冬季 Winter

「公园汇聚了多样的植物区系」

我想问问大家的家乡在哪里。（做一个简单的问答互动）

钱江源国家公园每年都迎来来自全国各地的访客，其实公园里的植物也和你们一样，来自全国各地。

中国国土面积广阔、地质历史悠久、地理环境复杂多样，为多种植物的生长提供了不同的生活条件，因此成为世界上植物区系最为丰富和复杂的国家之一。而在钱江源国家公园，我们可以看到华东、华北、华中和华南等地区的典型植被，堪称“中国植物多样性的天然博物馆”。

这里除了华东地区代表性的甜槠、木荷等常绿阔叶树种之外，也为青钱柳、杜仲、香果树、八角莲等来自华中与西南区系的典型物种，桑树、构树、连香等来自华北植物区系的落叶阔叶树种和木莲、含笑、蚊母树、猴欢喜等来自华南植物区系的典型树种提供了栖息的家园。

「造就公园多样植物区系的自然地理因素」

就像人类在长期的地理环境的影响下会产生不同肤色和文化一样，不同地理环境下生长的植物经过漫长时间的演化也会产生外形、性状等的差异。植被类型是地形、气候、水文、地质等自然环境长期演变结果最直观的呈现，钱江源国家公园的植物多样性背后其实反映的是地理环境的多样性。

从地理位置上来看，钱江源国家公园地处我国华北和华南、华东与华中交界，多个不同植物区系分界线的边缘地带，加之浙江省自西南向东北倾斜的地势，使得不同地区的代表性植物都能进入并在当地繁衍生息，形成这里多元、交融的植物种类与系统结构，让我们一次游览就能看到华夏大地的各种植物的风采。从地质演变的历史来看，这里的山体由于多次地壳运动而形成的山系格局，为南来北往的植物提供了生存的庇护所。

接下来，我将带领大家简单认识下我们身边的植物。（这部分可以根据现场情况，结合通用型解说方案中的生物多样性部分进行解说。）

注意事项

- 钱江源国家公园丰富的植物多样性背后是地理环境的复杂与多样。
- 植物区系的概念比较抽象，现场如果有具体的物种，能结合具体物种解说效果更佳。

辅助道具推荐

- 标识牌：一园一世界

Qixi Town | Lotus Pond | Qiantang River Source Exploration Route

Interpretation No. QX-LHT-06

Stop 07: Baosong Lounge

- **Interpretation theme**

  Diverse floras in Qainjiangyuan National Park

- **Key points**

  1. Know the local flora
  2. Reasons for the diversity of local flora

- **Site overview**

  A well-equipped interpretation site with interpretation board and resting lounge

"Diverse floras"

I want to ask everyone where is your hometown?

Every year, our Qianjiangyuan National Park welcomes visitors from all over the country. In fact, the plants in the park, like you, come from all over the country.

China has a vast territory, a long geological history, and a complex and diverse geographical environment. It provides different living conditions for the growth of a variety of plants. Therefore, it has become one of the countries with the most abundant and complex flora in the world. In Qianjiangyuan National Park, we can see the typical vegetation in East China, North China, Central China, and South China, which can be called the "Natural Museum of Plant Diversity in China".

Therefore, in addition to the evergreen broad-leaved tree species such as sweet oachestnut and gugertree, which are representative of East China, they are also typical species from Central China and Southwestern flora such as wheel wingnut, hardy rubber trees, *Emmenopterys henryi*, and *Dysosma*. Deciduous broad-leaved tree species such as mulberry, paper mulberry, *Katsura*, *etc*. come from the plants of the North China flora and typical tree species from the South China flora such as Vang Tam, *Magnolia figo*, *Distylium*, and *Sloanea*.

"Nature factors that lead to diverse floras"

Just as humans will produce different skin colors and cultures under the influence of a long-term geographical environment, plants grown in different geographical environments will also produce distinctions in shape and characters after a long period of evolution. The type of plant is the most intuitive presentation of the long-term evolution of the natural environment such as terrain,

climate, hydrology, and geology. The diversity of plants in Qianjiangyuan National park actually reflects the diversity of geographical environment.

From the geographical point of view, Qianjiangyuan National Park is just located in the transitional zone of North China and South China, the border between East China and Central China, the boundary edge zone of different flora, addition to the mountain topography that slopes from southwest to northeast of Zhejiang Province, creating a convenient passage for the gathering of various plants; from the perspective of the history of geological evolution, as a result of multiple crustal movements, the mountain pattern formed has provided a refuge for plants from south to north. All these allow us to see various plant around China within one tour.

Now, I will lead you to have a brief understanding of the plants around us. (This part can be interpreted according to the real-time site conditions. The General Interpretation of Biodiversity is a very useful reference here.)

Attentions

- Behind the rich plant diversity in Qianjiangyuan National Park is the complexity and diversity of the geographical environment.
- The concept of flora is relatively abstract, and if there are specific species on site, it can be better combined with specific species to interpret.

Recommended auxiliary facilities and toolkits

- Interpretation signage: a collective display of floras throughout the country

齐溪镇 | 莲花塘 | 钱塘江寻源自然体验路线　　解说编号：QX-LHT-08

解说点 08 鸣界亭

29°23′28.17″ N，118°11′39.58″ E
海拔：816m

● **解说主题**

森林的四季景观变化

● **解说要点**

钱江源国家公园四季分明的特色

● **场地概述**

视野开阔，可以远眺森林景观的停留点

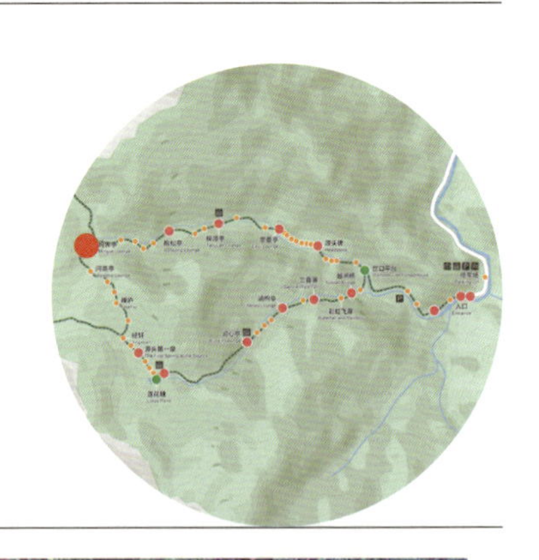

解说方式 Interpretation Style

- ☑ 人员解说 Staff Interpretation
- ☐ 教育活动 Educational Activity
- ☑ 解说牌 Interpretation Signage
- ☐ 场馆解说 Hall Interpretation

解说季节 Interpretation Season

- ☑ 春季 Spring
- ☑ 夏季 Summer
- ☑ 秋季 Autumn
- ☑ 冬季 Winter

解说词

各位朋友，如果说多样的植物区系、地形与海拔高度的变化从空间上为钱江源丰富的植物多样性定下了错落有致的基调，那么这里四季各异的景观，就像森林上演的变装秀。

「钱江源国家公园四季分明的特色」

随着季节的交替，森林会呈现出不同的样子。俗话说“春有百花，秋有月，夏有凉风，冬有雪，若无闲事挂心头，便是人间好时节。”如果你能在不同时节造访，一定会发现钱江源的一年四季都会有独特的风景等着你。

春天，沉睡了一个冬天的山林从寂静中苏醒，生命喷薄而出，丛林青翠，云海萦绕，春花盛开，莺歌燕舞，最是游览的好时候。

这里夏日的平均气温只有 24℃，和动辄温度就逼近 40℃的城市相比，绝对是不可多得的避暑胜地。宁静的湖泊，潺潺的溪涧，阴凉的林下，是不是想象一下都感觉有一股清凉袭来?

秋意正浓时，最适合登高望远，看万山红遍、层林尽染。此时步入林间，脚踩着刚刚铺就的落叶，一路寻访各种挂梢野果，偶尔还可以捡拾几片落叶、几颗种子，感受大自然的丰收馈赠。

冬天，大雪有时会不期而至。此时来到银装素裹的国家公园，恍若进入了冰雪女王的世界。若从高处眺望，无论是密生的林冠层，还是位于半山腰的古村落，都被瑞雪装点出一番别样的景致，并酝酿着来年的新生。

注意事项

- 注意结合参观季节适当修改解说词。

辅助道具推荐

- 标识牌：季相轮回的四季舞台

拓展活动推荐

- 四季的钱江源，就像大自然这位画家的画室，单是绿就不下数十种：嫩芽的黄绿、新叶的鲜绿、老叶的深绿、革质叶的油绿、毛茸茸叶片的糙绿、苔藓的水绿、地衣的粉绿、落叶的黄绿、流水的清绿、草的绿、石头的绿、森林的绿……
- 除此之外，这里还有哪些颜色呢？我们不妨来玩一个颜色接龙的游戏，看看我们能找到多少种颜色。

Qixi Town | Lotus Pond | Qiantang River Source Exploration Route

Interpretation No. QX-LHT-08

Stop 08: Mingjie Lounge

- **Interpretation theme**

  Landscape changes in the forest of four seasons

- **Key points**

  Distinctive features of four seasons of Qianjiangyuan National Park

- **Site overview**

  A place with a open view that can overlook the forest landscape

Dear friends, since the diverse flora, changes in terrain and elevation has established a colorful background for rich plant diversity of Qiangjiangyuan National Park, the different seasons here are just like a drag race performed by the forest.

"The distinctive features of four seasons of Qianjiangyuan National Park"

As the seasons change, the forest will show different appearances. As the saying goes, "There are flowers in spring, moon in autumn, cool breeze in summer, and snow in winter. If you have nothing to worry about, it is a good time." If you can visit again in another season, you will find that Qianjiangyuan National Park has unique landscapes waiting for you each time.

In the spring, after sleeping for a winter, the mountain forest awakens from the silence, life spurts out, flowers bloom, chirping birds dance in the sky, which is the best time to visit.

The average summer temperature here is only 24 ℃, which is definitely a rare summer resort compared with cities where the temperature is constantly approaching 40℃. There are tranquil lakes, gurgling streams, and shady forests here. You can feel the cool air just by imagining it.

When it's mid autumn, it is most suitable for hiking and overlooking the mountains covered in red. If you step into the forest at this time, you can stand on the fallen leaves, looking for all kinds of wild hanging fruits along the way. Occasionally, you can pick up a few fallen leaves and seeds to feel the harvest of nature.

In winter, heavy snow sometimes arrives unexpectedly. At this time, the national park is wrapped in silver, just like the world of the *Frozen*. Overlooking the National Park, both the dense forest canopy and the ancient village in the middle of the mountain are decorated by snow, creating a unique scenery and brewing for the next year.

Attentions

- Pay attention to the appropriate modification of the interpretation based on the visiting season.

Recommended auxiliary facilities and toolkits

- Interpretation signage: the stage of four seasons

Recommended activities

- The four seasons of Qianjiangyuan National Park are like the painter's studio of nature. There are more than dozens of types of green alone: the yellow green of the bud, the fresh green of the new leaf, the dark green of the old leaf, the oily green of the leathery leaf, the rough green of the furry leaves, the water green of the moss, the pink green of the lichen, the yellow green of the fallen leaves, the fresh green of the flowing water, the green of the grass, the green of the stones, the green of the forest…
- Besides, what other colors are there? We might as well play a color solitaire game and see how many colors we can find.

齐溪镇 | 莲花塘 | 钱塘江寻源自然体验路线　　解说编号：QX-LHT-09

解说点 09 源头第一泉

29°23′14.11″ N，118°11′49.74″ E
海拔 918m

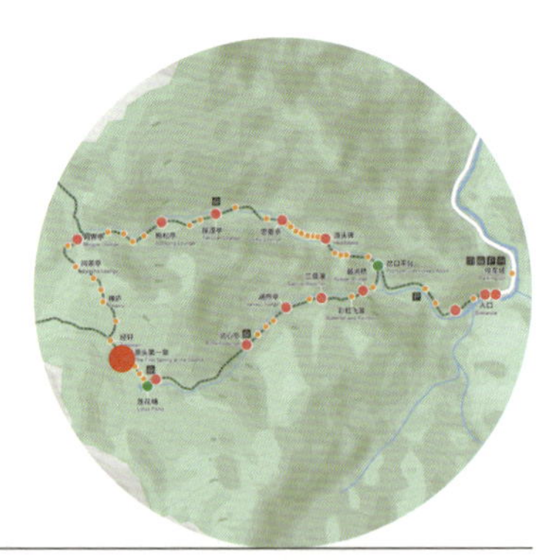

● **解说主题**

孕育河源的山川群像

● **解说要点**

钱江源国家公园的整体山势格局对水源汇聚的影响

● **场地概述**

树立有源头第一泉标示牌处

解说方式 Interpretation Style

☑ 人员解说 Staff Interpretation　☐ 教育活动 Educational Activity
☑ 解说牌 Interpretation Signage　☐ 场馆解说 Hall Interpretation

解说季节 Interpretation Season

☑ 春季 Spring　☑ 夏季 Summer
☑ 秋季 Autumn　☑ 冬季 Winter

解说词

各位朋友，眼前的“源头第一泉”就是我们一路追随着莲花溪所要寻找的钱塘江上游。如果不是我们一路追踪，是不是很难将这样一条小小的集水沟与钱塘江联系起来？不过，这可不是钱塘江唯一的水源，钱塘江是由无数条这样的涓流汇聚而成的。

水源的形成离不开山势地形的引导。大家知道，因为我国地势西高东低，所以一些较大的河流，比如，长江、黄河，大都是发源于西部高山，自西向东流入海洋。钱塘江流域整体上处于我国地势的第三级阶梯上，相对低平；但从流域尺度来看，钱江源所在的浙江西部地区是以中山丘陵为主的地形，地势较高，也更为复杂，因此河流也是自西向东流的。

钱江源国家公园地处江苏、浙江、江西三省交界处，与黄山、婺源等地的山峰同出一族，都属于白际山脉，岩石成分以花岗岩为主，并不容易形成蓄水层。但这里群山绵延、沟谷纵横，有莲花尖、高楼尖、外溪岗等众多山峰矗立，而在山峰之间的低洼处，水源汇聚，为河流的孕育提供了可能。

莲花塘所在的这片区域刚好被群峰环绕，形成一片天然的凹陷，周边山泉的水便在此汇聚，成为钱塘江源头最重要的水源涵养地。被众峰环绕的莲花塘就像莲花的花心，而莲花尖等山峰则像是莲花的花瓣，自地面向天空伸展开，共同组成了一朵绽放的“莲花”，它像是被群山环绕的莲花塘空中俯瞰的缩影，也是钱塘江源头的一幅画卷。

注意事项

- 钱塘江的源头之水是由众多小溪流共同汇聚而成的。
- 钱塘江的水源形成受到山系格局的制约。

辅助道具推荐

- 标识牌：河源于此
- 标识牌：孕育河源的山川群像

拓展知识点

- 莲花塘的山体岩质以质地坚硬且强度高、耐久抗压的花岗岩为主，并不利于在地下形成蓄水层，但被这几座山峰所围绕的自然凹陷的莲花塘区域却形成了一块容易蓄积水资源的天然湿地；同时，这里丰富的森林植被也为涵养水源提供了有利条件，森林和湿地的共同作用才让这里成为浙江省母亲河钱塘江的发源地。

Qixi Town | Lotus Pond | Qiantang River Source Exploration Route

Interpretation No. QX-LHT-09

Stop 09: The first spring at the source

Interpretation theme

Mountains and rivers that nurture the river source

Key points

The influence of the overall mountain pattern of Qianjiangyuan National Park on water concentration

Site overview

A stream with Riverhead No.1 spring board

Dear friends, the "first spring at the source" in front of us is the upper reaches of the Qiantang River we are looking for. If we didn't walk along the lotus creek, would it be difficult to connect such a small spring with the Qiantang River? However, this is not the only water source of the Qiantang River. The Qiantang River is formed by countless streams like that.

The formation of water sources is inseparable from the guidance of mountain terrain. As we all know, because our country is high in the west and low in the east, some of the larger rivers, such as the Yangtze River and the Yellow River, mostly originate from the high mountains in the west and flow into the ocean from west to east. The Qiantang River Basin as a whole is on the third step of the terrain in China, which is relatively low and flat; however, from the perspective of the watershed scale, the western Zhejiang area where the Qiantang River source is located is mainly composed of mountainous hills, with higher and more complex terrain, and in this case the river flows from west to east.

Qianjiangyuan National Park is located at the junction of the three provinces of Jiangsu, Zhejiang and Jiangxi. It belongs to the same family as the mountains of Huangshan and Wuyuan. They all belong to the Baiji Mountains. The rock composition is mainly granite, and it is not easy to form an aquifer. But there are rolling hills and valleys, many mountain peaks such as Lotus Peak, Gaolou Peak, Waixigang, *etc*.. So the low-lying areas between the peaks, where the water gathers, make the birth of rivers possible.

The area where the Lotus Pond is located is surrounded by mountains, forming a natural depression, where the water flows from the high mountains will converge here. Therefore it becomes the most important water source

conservation area at the source of the Qiantang River. The Lotus Pond is like the heart of a lotus flower, while the mountains such as the Lotus Peak are like petals of the lotus, stretching from the ground to the sky, together forming a blooming "lotus". It is like a miniature of the mountain peaks as well as a beautiful portrait of the Qiantang River.

Attentions

- The water at the source of the Qiantang River is made up of many small streams.
- The water source formation of the Qiantang River is constrained by the mountain pattern.

Recommended auxiliary facilities and toolkits

- Interpretation signage: the source of the Qiantang River
- Interpretation signage: mountains at the river source

Extra information

- The mountains around Lotus Pond are mainly composed of hard granite that is durable and resistant to compression. It is not conducive to the formation of an aquifer underground. However, due to the natural depression by the mountains, there forms a natural wetland and that is easy to accumulate water resources; at the same time, the rich forest vegetation here also provides favorable conditions for conserving water. The co-effect of forests and wetlands makes it the birthplace of the Qiantang River, the mother river of Zhejiang Province.

齐溪镇 | 莲花塘 | 钱塘江寻源自然体验路线　　解说编号：QX-LHT-10

解说点 10 莲花塘 #1

29° 23′ 14.11″ N，118° 11′ 49.74″ E
海拔 912m

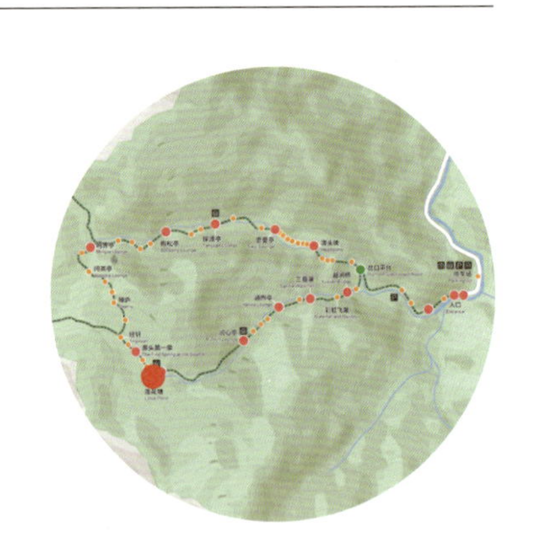

● **解说主题**

钱江源地区的地质形成

● **解说要点**

1. 简单介绍钱塘江的水源地——莲花塘
2. 莲花塘湿地的形成原理

● **场地概述**

莲花塘区域适合停留，有遮阴的林间空地

解说方式 Interpretation Style

☑ 人员解说 Staff Interpretation　☑ 教育活动 Educational Activity
☑ 解说牌 Interpretation Signage　☐ 场馆解说 Hall Interpretation

解说季节 Interpretation Season

☑ 春季 Spring　☑ 夏季 Summer
☑ 秋季 Autumn　☑ 冬季 Winter

「钱塘江的水源地——莲花塘」

各位朋友，现在位于我们眼前的这片水域，就是大家此行的目的地莲花塘了。和大家期待中的莲花塘有什么不同吗？有朋友说，在遇见它之前的期待是“百闻不如一见”，见到之后觉得还是“相见不如怀念”的好。

的确，这片小小的水域，可能很难和闻名遐迩的浙江第一大河钱塘江产生联想。但你知道吗？其实，真正的河流源头就是这样在自然中若隐若现的湿地。它们积蓄着水资源，为孕育钱塘江的大小溪流常年不停地输送水源，即使在枯水期，这里的水也从未完全干涸过。

「莲花塘湿地的形成原理」

那么莲花塘的水从哪里来呢？这还得从莲花塘的形成开始说起。8 亿年前，这里曾是一片汪洋，后来由于多次地壳运动才形成如今群山起伏的地貌格局，莲花塘即是在地壳运动过程中形成的一片被四周山峰包围的塌陷洼地，是一个天然形成的集水区域，日久年深，枯落物和腐殖质堆积形成沼泽湿地，积水越聚越多，最终形成了一片片这样的小型湖泊。

湖泊的出现为水草等水生动植物的生长提供了条件，但与此同时，泥沙及腐殖质也不断积累，湖泊水位下降，这里逐步演变成松软的草甸。但可别小瞧这些草甸，它们就像海绵一样，具有很强的水源存储与涵养能力。《开化水利志》曾经这样描述过这种湿地草甸：“水草茂密，望之水不过膝，人若走进，立即下陷，深不可测”。

可惜的是，由于人为的毁林开荒和自然环境的变化，如今在莲花塘区域已经没有这样的丰茂景象了。不过仔细观察你的周围，这些密林其实就是一片巨大的隐形水库，是钱塘江的真正“源头”。

注意事项

- 说明湿地变迁的几个关键过程：地质历史时期的塌陷—水源的汇聚—湖泊形成—腐殖质积累—亚高山草甸。

辅助道具推荐

- 标识牌：莲花塘下的“隐形”水库

Qixi Town | Lotus Pond | Qiantang River Source Exploration Route

Interpretation No. QX-LHT-10

Stop 10: Lotus Pond #1

Interpretation theme

Geological formation in Qianjiangyuan area

Key points

1. Source of the Qiantang River — Lotus Pond
2. The formation of the Lotus Pond wetland

Site overview

Forest glade next to the Lotus Pond area

"Source of the Qiantang River — Lotus Pond"

Dear friends, the water in front of us is the legendary Lotus Pond. Is it different from the Lotus Pond that everyone expects? A friend said that the expectation before meeting it was "seeing is believing", and after seeing it, it was better to "Better to miss than to meet". Indeed, this small body of water may be difficult to associate with the famous Qiantang River, the largest river in Zhejiang. But, do you know, in fact, the real source of rivers is such wetlands that are looming in nature? They accumulate water resources and continuously transport water for the large and small streams that breed the Qiantang River. Even in the dry season, the water here has never completely dried up.

"The formation of Lotus Pond wetland"

So where does the water from the Lotus Pond come from? This has to start from the formation of the Lotus Pond. 800 million years ago, there was a vast ocean. Later, due to multiple crustal movements, the current mountainous landscape pattern was formed. The Lotus Pond is a collapsed depression surrounded by the mountains formed during the crustal movement, a naturally formed water catchment area. Year after year, litters and humus has accumulated, forming this marsh wetlands. When water is gathered more and more, it eventually forms this small lake.

The emergence of lakes provides conditions for the growth of aquatic plants, at the same time, sediment and humus are also constantly accumulating in the lake as the water level drops, and gradually evolves into a soft meadow. But don't underestimate these meadows, they are like sponges and have strong water storage and conservation capabilities.

Kaihua hydraulic documentary once described

this wetland meadow as follows: "The aquatic plants are dense, and look like the water level is lower than the knees. If people walk in, they will immediately sink, which is unfathomable."

Unfortunately, due to human deforestation and changes in the natural environment, there is no such lush scene in the Lotus Pond area. However, if you look closely around you, these dense forests are actually a huge hidden reservoir, which is the real "source" of the Qiantang River.

Attentions

- Explain several key processes of wetland changes: collapse during geological history – water gathering – lake formation – humus accumulation – subalpine meadow.

Recommended auxiliary facilities and toolkits

- Interpretation signage: invisible reservoir nurtures the river source

齐溪镇 | 莲花塘 | 钱塘江寻源自然体验路线 解说编号：QX-LHT-11

解说点 11 莲花塘 #2

29° 23′ 14.11″ N, 118° 11′ 49.74″ E
莲花塘区域 海拔 907m

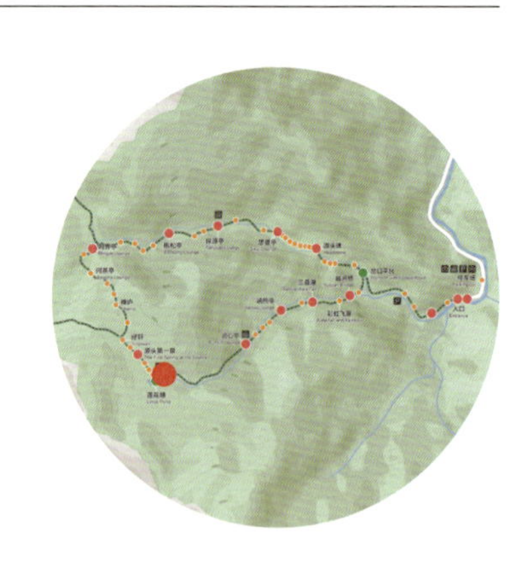

解说主题

森林涵养水源的功能

解说要点

1. 森林是如何涵养水源的
2. 保护森林的重要性

场地概述

有森林、有休憩点的地方

解说方式 Interpretation Style

- ☑ 人员解说 Staff Interpretation
- ☐ 教育活动 Educational Activity
- ☑ 解说牌 Interpretation Signage
- ☐ 场馆解说 Hall Interpretation

解说季节 Interpretation Season

- ☑ 春季 Spring
- ☑ 夏季 Summer
- ☑ 秋季 Autumn
- ☑ 冬季 Winter

解说词

各位朋友，虽然莲花塘看起来和普通的小池塘无二，但是它却内藏玄机。据专家估算，莲花塘的集水面积约 $1km^2$，水深也不过 2m 左右，但是每年从这里流出的水量超过 $4000000m^3$。即使把这里所有的降水量加起来，也只有从莲花塘流出的水流量的一半，那么另一半水源来自哪里呢？答案就在我们四周的这片森林里。

「森林是如何涵养水源的」

我们不妨想象这样一个画面：一场雨落在森林上，然后，放慢雨的脚步……它们会去往哪里呢？首先，应该是落在森林的帽子——林冠层上，然后，一部分雨水会透过枝叶的缝隙或者是顺着树干往下流，最终到达地面。

到达地面的降水会浸湿地面的枯落物，或者被底层的苔藓、地衣吸收，然后再慢慢渗进土壤，沿着土壤空隙继续扩散，其中还有一部分会顺着地势低洼处由高向低流入小溪、河流，最终汇入大海。

多亏了林冠的遮挡、枯枝落叶层的保护和根部的吸收，只有少部分的水会白白蒸发掉，这也是森林能够涵养水源的原因。

「保护森林的重要性」

如果我们把刚才的场景换成寸草不生的被破坏的生境(如森林被破坏以后的场景)，降落的雨水又会去哪里呢？渗入地下、白白流走或者直接蒸发掉。如果这是一场暴雨呢？水分可能来不及渗透进土壤就会流走，水流越聚越多，最终就会有水土流失甚至山洪暴发的威胁，土壤也会因为被冲刷而越来越贫瘠。

注意事项

- 从森林涵养水源到如何涵养水源的进一步提问。
- 进一步讲大自然里的雨水收集和水源涵养。
- 说明森林对水源涵养的意义。
- 引申城市中的实践，如海绵城市的概念。

辅助道具推荐

- 标识牌：下雨时，森林在……
- 标识牌：水系绵延的杰出贡献

拓展活动推荐

- 小雨滴的旅行（P568：活动卡 08）

拓展知识点

- 我们不妨再切换一个场景，假设降水落在了城市上空，然后会发生什么？大家可以畅所欲言。

注：了解“城市热岛效应”“城市雨岛效应”“海绵城市”“雨水花园”等名词，以便与访客交流时灵活调用。

Qixi Town | Lotus Pond | Qiantang River Source Exploration Route

Interpretation No. QX-LHT-11

Stop 11: Lotus Pond #2

Interpretation theme

The function of forest on water conservation

Key points

1. How forests conserve water

2. The importance of protecting forests

Site overview

Where there are forests and rest areas

According to expert estimates, the water catchment area of the Lotus Pond is about 1 square kilometer and the water depth is only about 2 meters, but the amount of water flowing out from here is as high as more than 4 million cubic meters per year. Even if all the precipitation here is added up, it is only half of the water flow from the Lotus Pond, so where does the other half of the water come from? The answer lies in the forest around us.

"How do forests conserve water"

We might as well imagine this picture: rain falls on the forest, and then, where will they go after they have been slowed down? First, it should fall on the hat of the forest, the canopy, and then part of the rainwater will pass through the gap between the branches and leaves, or go down the tree trunk, and finally reach the ground.

The rainfall that reaches the ground will soak the litter on the ground, or be absorbed by the bottom moss and lichen, and then slowly penetrate into the soil and continue to spread along the gaps in the soil. Some of them will flow into the creeks, rivers, and finally into the sea along the low-lying terrain. Thanks to the cover of the canopy , the protection of the litter layer and the absorption of roots, only a small part of the water will evaporate in vain, which is why the forest can conserve water.

"The importance of protecting the forest"

If we replace the scene before with a ruined habitat where there is no grass (imagine when the forest has been destroyed), what will happen after the rain falls, seep into the ground, run away in vain or evaporate directly? What will happen if this is a heavy rain? The water may run away before it penetrates into the soil, and the more the water flows, the more it will lead to soil erosion and even the flash flood. As a consequence, the soil will become more and more barren as it is washed away.

Attentions

- Further question from forest water conservation to how to overall conserve water.
- Further talk about rainwater collection and water conservation in nature.
- Explain the significance of forests for water conservation.

Recommended auxiliary facilities and toolkits

- Interpretation signage: how do forest capture the water.
- Interpretation signage: river source safeguards the basin.

Recommended activities

- The journey of little raindrops（P568：No. 08）.

Extra information

- Let's imagine another scenario, assuming that precipitation falls over the city, what will happen then? Everyone can say what you want.

Notes: Understand the terms "urban heat island effect" "urban rain island effect" "sponge city" "rainwater garden" *etc.* in order to use it flexibly when communicating with visitors.

齐溪镇 | 莲花塘 | 钱塘江寻源自然体验路线　　　　解说编号：QX-LHT-12

解说点 12 润心亭

29°23′17″ N，118°12′4″ E
海拔：798m

● 解说主题

钱江源国家公园的动物多样性

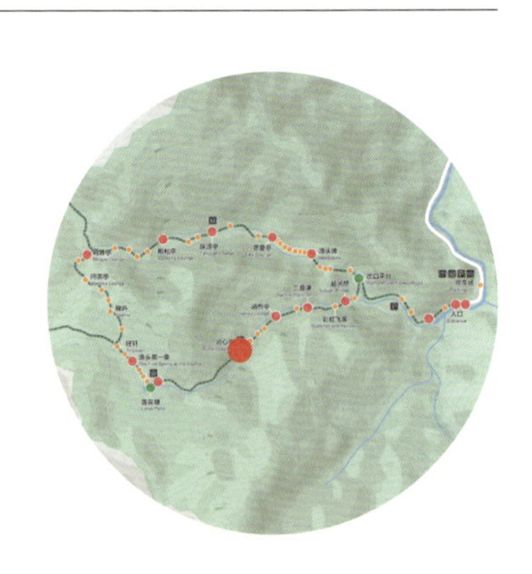

● 解说要点

1. 游线的重要资源与动物多样性介绍
2. 说明物种共存的重要性

● 场地概述

润心亭，有解说牌配套

解说方式 Interpretation Style

☑ 人员解说 Staff Interpretation　☐ 教育活动 Educational Activity
☑ 解说牌 Interpretation Signage　☐ 场馆解说 Hall Interpretation

解说季节 Interpretation Season

☑ 春季 Spring　☑ 夏季 Summer
☑ 秋季 Autumn　☑ 冬季 Winter

「游线的重要资源与生物多样性介绍」

钱江源国家公园茂密的丛林、纵横交错的山涧溪流、温暖的气候等因素综合起来，使得这里成为众多野生动物栖居的天堂。这里不仅是数千种野生动植物的家园，同时也是国家一级重点保护野生动物黑麂、白颈长尾雉和国家二级重点保护野生动物白鹇、黑熊等的庇护所。

要想在野外看到黑麂、小麂、黑熊等野生兽类是一件很困难的事。一方面是由于受栖息地和食物等因素的限制，它们的野外种群数量有限；另一个原因是很多兽类生性胆小，行踪隐蔽，不愿意和人接触，若发现人类闯入，可能早就悄悄离开，因此我们一般只能在红外相机中一睹它们的样子。

和野生兽类相比，鸟类的数量就丰富得多。沿途我们不仅可以在溪流边见到活泼好动的燕尾、红尾水鸲和白鹡鸰等傍水而生的鸟类，还可以在开阔地带看到天空中盘旋着的猛禽，而密林之中，经常有朋友偶遇悠闲散步的白鹇夫妇，或者听到啄木鸟“笃—笃—笃—”的啄木声。夜晚，这里还会有猫头鹰出来活动。

两栖和爬行类动物也比较适合被观察。比如，青蛙和蟾蜍在夜间或者雨后比较容易观察到，而天气好的时候，在路边还有可能见到一种名为蓝尾石龙子的小蜥蜴。

当然，最容易看到的非昆虫和蜘蛛莫属。无论是树干、叶片、树枝、枯木上还是草丛中，随处都会见到它们的身影。不管你喜不喜欢，它们可都是自然界中重要的一员。

「物种共存的重要性」

多种多样的野生动物生活在这里，经过世世代代的繁衍生息，互相适应，早已联结成环环相扣的一个有机整体，缺失其中的任何一环，都有可能导致整个生态系统的变化甚至崩溃瓦解。首当其冲的，就是位于食物链顶端的肉食动物。

20 世纪 90 年代，这里记录到我国四大猫科动物之一的云豹捕食家畜，但此后再无记录。云豹是否还存在于这片山林？没有人说得清楚。但是由于人类活动的影响，作为云豹食物的小型哺乳动物的种群数量已经明显减少，同时栖息地也大面积消失，这些问题都是不争的事实。但实际上人类才是这片山林的外来者。所以，大家来到这里，一定要心存敬畏之心，记住：我们是访客，它们才是这片土地的原住民。

注意事项

- 结合沿途遇到的一些动物进行解说，效果会更好。
- 解说过程中尽可能涵盖不同生境下的物种。
- 在介绍鸟类的时候，结合鸟类声音的模仿是一种很吸引人的方式。
- 大多数野生动物会自然地躲避人，因此在访客参观过程中是很难被直接发现的，作为解说员需要提醒访客适当降低预期，同时通过展示照片等方式丰富你的解说。
- 对于需要解说的具体物种，可以结合通用物种解说的内容进行介绍。

辅助道具推荐

- 小麂、黑麂、黑熊、野猪、黑熊、云豹等野生动物卡片
- 白冠燕尾、小燕尾、灰背燕尾、红尾水鸲、白鹇、白颈长尾雉等鸟类介绍卡片

拓展活动推荐

「昆虫记录」

- 如果大家感兴趣的话，不妨将你遇到的每种昆虫都拍下来，最后看看我们一路会遇到多少种昆虫。

「鸟类观察」

- 如果你想在钱江源的密林中观鸟，找准时机很重要。一般鸟儿在早晨和傍晚会相对活跃一些。所以，如果你能更进一步了解鸟类的生活习性和作息规律，找到它们会更容易一些。

Qixi Town | Lotus Pond | Qiantang River Source Exploration Route

Interpretation No. QX-LHT-12

Stop 12: Runxin Lounge

Interpretation theme

Animal diversity in Qianjiangyuan National Park

Key points

1. Introduction of important resources and animal diversity
2. Explain the importance of coexistence of species

Site overview

At Runxin Lounge, where there is a supporting interpretation board

"Introduction to important resources and biodiversity of the route"

The combination of dense jungles, criss-cross mountain streams and warm climate in Qianjiangyuan National Park makes this a paradise for many wildlife. This is not only the home of thousands of wild animals and plants, but also the sanctuary of the national first-level protected animals such as black muntjacs, Elliot's pheasants, snow leopards and the national second-level protected animals such as silver pheasants, black bears, giant salamanders.

It is very difficult if you want to see animals in the wild, such as black muntjacs, Chinese muntjacs, black bears. On one hand, due to the restrictions on habitat and food, their wild populations are limited; on the other hand, many animals are timid, hidden, or unwilling to come to interact with humans. If they find us intruding, they may be long left. In this case, we are only able to see them in infrared cameras.

Compared with wild land animals, the number of birds is much richer. Along the way, we can not only see various lively birds such as swallowtails, water redstarts, and white wagtails along the stream, but also see raptors circling in the sky. You may also encounter the silver pheasants couple wandering in the forest, or hear the sound from woodpeckers "Tuk—Tuk—Tuk—" . At night, there will be owls coming out.

Amphibians and reptiles are also relatively easy to be observed. For example, frogs and toads are easier to be observed at night or after rain, and when the weather is nice, there is a chance to see a small lizard called Shanghai elegant skink on the roadside.

Of course, the most easily seen creatures are insects and spiders. They can be seen everywhere, whether on trunks, leaves, branches, dead wood or in grass. However, whether you like it or not, they are very important members of nature.

"The importance of coexistence of species"

A variety of wild animals live here. After generations and generations, they have adapted to each other and have long been linked into an organic circle that interlocks. If any one of them is missing, it may lead to changes in the entire ecosystem or even collapse and disintegrate. The first to bear the brunt is the carnivore at the top of the food chain. In the 1990s, it was documented that one of the country's four major cats, snow leopards, preyed on domestic animals, but no records since then. No one can say for sure whether snow leopards still exist in this mountain forest. However, due to the impact of human activities, the population of small mammals that have been considered as food for snow leopards has decreased significantly along with the largely disappeared habitats. These problems are indisputable facts.

But in fact human beings are outsiders in this mountain forest. Therefore, everyone here must be aware, remember: we are visitors, they are the indigenous people of this land.

Attentions

- To make the interpretation more vivid, you can integrate some introduction of animals encountered along the trail.
- Your interpretation should try to integrate more diverse biodiversity, such as mammals, birds, amphibians, reptiles, insects and other species in different habitats such as streams, jungles and marshes.
- When interpreting birds, it is an attractive way to display some audio of bird sounds, such as the song of a redstart.
- Please always remember that most of the wildlife will hide to avoid encounter with people. As an interpreter, you should remind your visitors to lower their expectations, as well as presenting some picture to enrich your interpretation.
- To interpret specific species, you can refer to the General Interpretation of Biodiversity.

Recommended auxiliary facilities and toolkits

- Interpretation cards for small muntjacs, black muntjacs, black bears, wild boars, black bears, clouded leopards and other wildlife.
- Interpretation cards for birds such as white-crowned swallowtails, small swallowtails, grey-backed swallowtails, plumbeous water-redstart, silver pheasants, and white-necked pheasants.

Recommended activities

"Insect observation"

- If you are looking for birds in the dense forest of Qianjiangyuan National Park, find a right timing is an important principle. Birds tend to be relatively active in the morning and evening. So if you know more about their habits and routines, it' s easier to find them.
- If you are interested in birding, you can also find a birding organization in your city and participate their birding activities. Your will find many amazing wildlife are actually just close to you.

"Birds observation"

- If you are interested in birding, you can also find a birding organization in your city and participate their birding activities. Your will find many amazing wildlife are actually just close to you.

齐溪镇 | 莲花塘 | 钱塘江寻源自然体验路线　　解说编号：QX-LHT-13

解说点 13 涵煦亭

29°23′23″ N，118°12′12″ E
海拔 710m

● 解说主题

森林的垂直分层

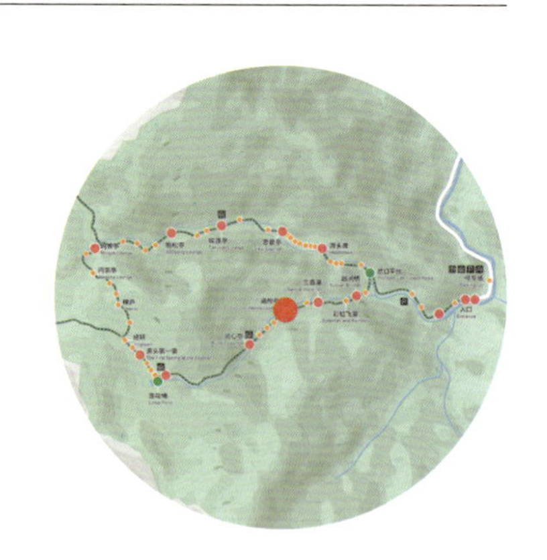

● 解说要点

1. 森林的垂直分层
2. 森林垂直分层的意义

● 场地概述

下山路上涵煦亭附近，便于停留及讲解的林下空地

解说方式 Interpretation Style

☑ 人员解说 Staff Interpretation　☐ 教育活动 Educational Activity
☑ 解说牌 Interpretation Signage　☐ 场馆解说 Hall Interpretation

解说季节 Interpretation Season

☑ 春季 Spring　☑ 夏季 Summer
☑ 秋季 Autumn　☑ 冬季 Winter

「森林的垂直分层」

各位朋友，你会怎么认识一棵树呢？通过根、茎、叶、花、果实，或者从观察一颗种子的萌发开始？你会怎么认识一片森林呢？我们可以从高处俯瞰森林的整体面貌、也可以从时间演替的历史来认识一片森林的养成，又或者就像我们现在一样，可以深入丛林，了解森林的内部结构。

简单来说，森林的植物种类按其生长的高度可以分为四大类：最高的是乔木层，乔木层之下是灌木层，最下面则是草本层和地被层。

乔木层无疑是一片森林的主宰，它们体量庞大，同时占据着森林最顶层的阳光和最底层的养分以及绝大部分的森林空间。由于太过高大，我们又可以把乔木层分为林冠层和下木层两层。林冠层是森林顶部枝叶最繁茂的部分，也是森林与外界接触最密切的界面之一，无论是阳光雨露还是风雨雷电，它们都是最直接的接收者。远远看去，钱江源国家公园的林冠层就像一颗颗花菜一样起起伏伏，又像是森林中撑起的一把把绿色的小伞。

林冠下方的下木层是树木彼此交叠最多的地方，由于受到林冠层的保护，气候条件相对比较稳定。许多附生植物，如兰花、蕨类喜欢生活在这一层，一方面可以得到较为充足的光线，另一方面又可以受到一定的遮挡，不至于因曝晒而失水过多。

灌木层位于乔木层之下，环境较为阴暗、潮湿。灌木层的植株看起来不如乔木那样棵棵分明，甚至有些杂乱，但却是森林生物多样性最高的一层，除了常见的灌木之外，有很多乔木的幼苗和幼树也生长在这里。

经过了乔木层与灌木层的一层层过滤，位于森林最下层的草本层和地被层的环境有些昏暗。草本层是最早感知森林季节变化的那一层，地被层主要由苔藓和地衣构成，常因体形矮小容易被忽略。地被层是林下植物养分和水分的重要来源，也是昆虫等小动物的庇护所，在整个森林生态系统中起着不可或缺的作用。

「森林垂直分层的意义」

不是所有的森林都有明显的垂直分层。大家有没有关注过城市公园中的植物？有明显的分层结构吗？可能是不常见的。而在钱江源国家公园的森林中，这种垂直分层却分外显著。一般来说，只有保存时间较长、未被破坏的森林才会拥有这样清晰完整的垂直分层。

所以你看到的钱江源国家公园这样丰富的、复杂的、有明显分层的结构，实际上不仅是这里环境条件好的反映，也是经历漫长时间守护而较少被干扰到的结果。这片区域也因具有较高的物种丰富度和独特而复杂的植物分层特征，成为植物研究者的天堂。

注：乔木有一根主干，树冠上有许多枝，而灌木一般从地面上长出许多枝干。对于专家来说，一株植物要被称作乔木的话，通常高度最少达到 3m。

Qixi Town | Lotus Pond | Qiantang River Source Exploration Route

Interpretation No. QX-LHT-13

Stop 13: Hanxu Lounge

● Interpretation theme

Vertical stratification of the forest

● Key points

1. Vertical stratification of the forest
2. The significance of vertical forest stratification

● Site overview

The forest on the downhill way, near Hanxu Lounge

"Vertical stratification of the vegetation"

How would you know a tree? By roots, stems, leaves, flowers, fruits, or by observing the germination of a seed? How would you know the forest? We can overlook the overall appearance of the forest, and understand the formation of a forest from the history of time succession, or, as we do now, we can go deep into the jungle and understand the internal structure of the forest.

In short, the types of forest plants can be divided into four categories according to their growth height: the highest is the tree layer, the shrub layer is below the tree layer, followed by the herb layer and the ground layer.

The arbor layer is undoubtedly the master of a forest. Relying on its huge sizes, it occupies the sun on the top of the forest, the nutrients at the bottom as well as most of the forest space. Because it is too tall, the tree layer can be divided into two layers: the canopy layer and the underlayer. The canopy layer is the most lush part of the top of the forest and one of the most intimate interfaces between the forest and the outside world. Whether it is sunny or rainy, windy or thundering, the canopy layer is the most direct recipient. Overlooking the forest, the canopy of Qianjiangyuan National Park has ups and downs like cauliflowers, or small lushing umbrellas propped up in the forest.

The underlayer layer under the canopy is where the trees overlap the most. Due to the protection of the canopy, the climate conditions are relatively stable. Many epiphytes, such as orchids and ferns, like to live in this layer. On one hand, they can get more sunlight; on the other hand, they can be protected from excessive water loss.

The shrub layer is underneath the arbor layer, and the environment is dark and humid. The plants in the shrub layer do not look as clear as trees, even a little messy, but the shrub layer has the highest forest biodiversity. In addition to common shrubs, there are many seedlings and trees of trees growing here.

After filtering through the layers of arbors and shrubs, the environment of the herb layer and the ground layer at the bottom of the forest is a bit dim. The herb layer is the layer that first senses the seasonal changes of the forest. The ground layer is mainly composed of moss and lichens which are often overlooked because of their small sizes. The ground layer is an important source of nutrients and water for understory plants, as well as a refuge for insects and other small animals, thus playing an indispensable role in the entire forest ecosystem.

"The meaning of vertical forest stratification"

Not all forests have obvious vertical stratification. Have you paid attention to the plants in the city park? Do they have a clear layered structure? Probably not. However, in the forests of Qianjiangyuan National Park, this vertical stratification is particularly remarkable. In general, only forests that have been preserved for a long time without being destroyed can have such clear and complete vertical stratification.

So the rich, complex and clearly layered structure of Qianjiangyuan National Park in front of us is actually not only a good reflection of the water and heat combination, but also the result of a long time preservation with little interference. In this case, this area has become a paradise for plant researchers due to its high species richness as well as unique and complex characteristics of plant stratification.

齐溪镇 | 莲花塘 | 钱塘江寻源自然体验路线　　解说编号：QX-LHT-14

解说点 14 彩虹飞瀑

29°23′24.28″ N，118°12′16.52″ E
海拔：680m

解说主题

钱江源国家公园的瀑布和彩虹

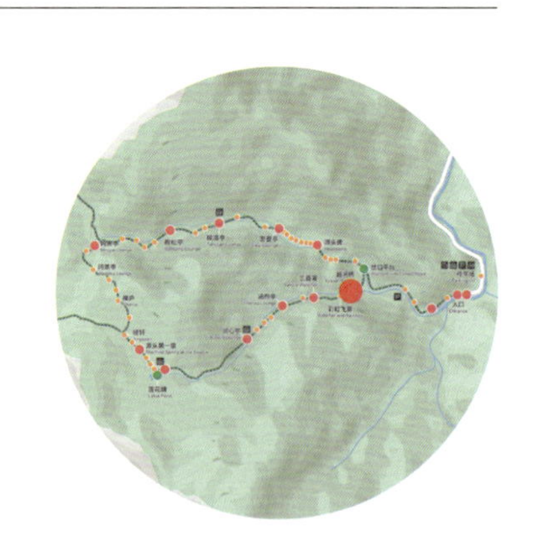

解说要点

1. 瀑布见证的大地运动
2. 彩虹飞瀑的景观与光学现象

场地概述

瀑布周边方便停留及讲解的空间

解说方式 Interpretation Style

- ☑ 人员解说 Staff Interpretation
- ☐ 教育活动 Educational Activity
- ☑ 解说牌 Interpretation Signage
- ☐ 场馆解说 Hall Interpretation

解说季节 Interpretation Season

- ☑ 春季 Spring
- ☑ 夏季 Summer
- ☑ 秋季 Autumn
- ☑ 冬季 Winter

「瀑布见证的大地运动」

各位朋友，我们知道罗马不是一日建成的，高山峡谷的形成也不是一蹴而就的，而是经过了漫长的地壳运动和地质变化。那么在国家公园中，我们如何才能见证大地的运动呢？眼前的瀑布或许就能告诉我们答案。

之前我们说过，因为亿万年前的造山运动使得这里地壳抬升，形成了一座座连绵的山峰，但是如今钱江源的山谷地形的形成，除了地质运动之外，还离不开水流的精心雕琢。

钱江源的水流顺着复杂的地势一路飞奔向下，像一把温柔却锋利的刀刃，不断切割着岩石，在流水落差较大的地方，形成了我们看到的这种瀑布。

这里的瀑布数量多，层叠不断，可达几十米，像一串串珍珠一样，镶嵌在我们徒步的路线上。同时，由于山区降水充足，水流量大，加上地形的落差，使得这些瀑布的声势十分浩大，往往“未见其形，先闻其声”。

「彩虹飞瀑的景观与光学现象 」

经年累月的冲刷，雕刻了峡谷独特的地貌景观。当阳光透过峡谷的缝隙照进山谷，经常会与瀑布的小水滴共舞出一条条缤纷的七彩虹链。

平时大家可能觉得很稀奇，但实际上只要空气中有足够的小水滴，阳光又充足的情况下是很容易在瀑布附近看到彩虹的。

在钱江源国家公园的大峡谷、古田山、枫楼坑等其他游线上，也可以体会瀑布带给我们的变幻景色。

注意事项

- 结合飞瀑景观进行解说，下山时路窄且陡，注意找到相对安全的一处平台进行解说。
- 沿途溪流有可能会遇到其他野生动植物，可以结合通用解说方案中的生物多样性部分的内容进行介绍。

辅助道具推荐

- 标识牌：峡谷间的彩虹光影

拓展活动推荐

- 入口到瀑布这一段路线，道路宽阔，基础设施较好，可以用于开展相关自然活动课程。

Qixi Town | Lotus Pond | Qiantang River Source Exploration Route

Interpretation No. QX-LHT-14

Stop 14: Waterfall with Rainbow

Interpretation theme

Rainbow Kingdom in Qianjiangyuan National Park

Key points

1. The earth movement witnessed by the waterfall
2. The landscape and optical phenomena of waterfall with rainbow

Site overview

A relatively large resting platform where tourists can sit and stay, along the waterfall

"Earth movement witnessed by waterfall"

Dear friends, we know that Rome was not built in a day, and the formation of mountains and valleys was not done overnight, but after a long crustal movement. So how can we witness the feat of the movement of earth here? The waterfall in front of us may be able to tell us the answer.

As we mentioned before because of the mountain-building movement from hundreds of millions of years ago, the earth's crust was raised here, forming a series of peaks. However, in addition to the geological movement, the valley of Qianjiangyuan is inseparable from the careful sculpting of water flow.

Qianjiangyuan's water flowing down the complicated terrain, like a gentle but sharp blade, constantly cut the rock, and in the place where the water drop is large, formed the waterfall we are seeing right now.

Waterfalls here are so numerous that they fold up to tens of meters, like a string of pearls, embedded on our hiking route. At the same time, due to the abundant rainfall in the mountain areas, the large water flow and the drop in the terrain, the sound of these waterfalls is very large. So, people often "hear the sound before seeing it."

"Extraordinary landscape and optical phenomena of waterfall with rainbow"

The unique landscape of the canyon was carved after years of erosion by the waterfall. When the sun shines through the gaps of the canyon and sheds into the valley, it often dances with the tiny droplets of the waterfall knitting a colorful rainbow.

Some may find it fascinating, but in fact, as long as there are enough tiny water droplets in the air and the sunlight is sufficient, you can

easily see the rainbow near the waterfall.

We can also experience the changing scenery brought to us by the waterfall in the Grand Canyon, Gutian Mountain, Fenglou Pit and other tour routes in Qianjiangyuan National Park.

Attentions

- Combine the waterfall landscape to conduct interpretation. The road is narrow and steep when going down, please find a relatively safer platform.
- Walk along the streams you may encounter other wildlife, which can be interpreted in combinational with the biodiversity section of general interpretation of this manual.

Recommended auxiliary facilities and toolkits

- Interpretation signage: rainbow beside the waterfall

Recommended for extra activities

- The route from the entrance to the waterfall has wide roads and good infrastructure, and can be used to carry out courses related to natural activities.

齐溪镇 | 莲花塘 | 钱塘江寻源自然体验路线　　　　解说编号：QX-LHT-15

解说点 15 越涧桥

29°23′26.27″ N，118°12′28.58″ E
海拔：630m

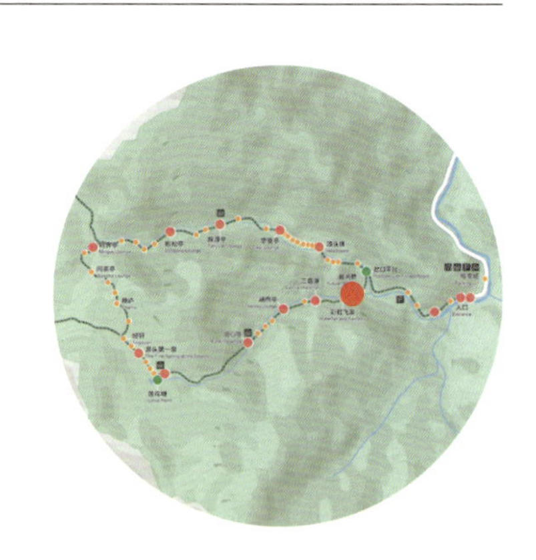

解说主题

森林氧吧

解说要点

1. 中国天然氧吧的评价标准
2. 钱江源国家公园负氧离子高的原因

场地概述

环境清幽，有长椅可以休息

解说方式 Interpretation Style

- [x] 人员解说 Staff Interpretation
- [ ] 教育活动 Educational Activity
- [x] 解说牌 Interpretation Signage
- [ ] 场馆解说 Hall Interpretation

解说季节 Interpretation Season

- [x] 春季 Spring
- [x] 夏季 Summer
- [x] 秋季 Autumn
- [x] 冬季 Winter

解说词

置身山林之中，并且有潺潺的溪水为伴，大家的各种感官是不是都被充分调动起来了？眼睛里看到的是巍巍青山，耳朵里听到的是潺潺水声，用手触摸岩石粗糙的肌理，感受清凉的水温，就连平时怠惰的嗅觉都变得灵敏，除了植物的芬芳之外，我们闻到的空气也是清新的。

「森林氧吧与负氧离子」

崇尚养生的中国人喜欢把这种地方叫作“森林氧吧”。“氧吧”这个词大家并不陌生。不过，什么是氧吧呢？是这里的氧气含量更高吗？恭喜你，猜对了一半。氧吧实际是指空气中的负氧离子浓度很高并达到特定标准。

负氧离子对人体健康十分有益，又被称为“长寿素”或“空气维生素”。据说，中国很多长寿村负氧离子浓度都非常高。负氧离子可以消除烟雾粉尘，净化空气。空气中负氧离子浓度越高，反映出的空气质量也就越好。

一般城市住宅的负氧离子含量约为 100 个 /cm^3，世界卫生组织规定，空气中负氧离子浓度不低于 1000～1500 个 /cm^3 时可被视为清新空气。钱江源国家公园的负氧离子含量有多少，大家可以猜一猜么？平均值相当于世界卫生组织给定的清新空气标准的 3 倍还多。开化县也因此早在 2016 年就被评为首批“中国天然氧吧”第一名。

「钱江源负氧离子高的原因」

为什么开化县会获此殊荣呢？这还要从负氧离子的产生机理开始讲起。

负氧离子的形成需要一些特定的条件。自然界的闪电、光合作用、瀑布、喷泉以及一些植物分泌的芳香类物质都能促进负氧离子的产生。

看看我们身边吧，这里不仅有奔腾不息的溪涧、声势浩大的瀑布，还有连绵茂密的原始森林，都为负氧离子的生成创造了优越的条件。同时，森林的滞尘作用、溅起的水花等还可以降低负氧离子的损耗。在产出较多，损失较少的情况下，这里的负氧离子含量自然也更高，空气也更清新，更有利于人类的健康。

注意事项

- 在引入森林氧吧的概念之前，先请大家做一次深呼吸，感受一下森林里清新的空气。
- 注意对负氧离子的介绍可以结合不同的受众群体选择解说内容。
- 比较一下自己生活的地方的空气质量指标与本地的差异（与游客的真实生活经验建立关联）。
- 负氧离子浓度高与国家公园纯净而未被干扰的自然资源条件是密不可分的。

辅助道具推荐

- 标识牌：天然氧吧的最佳范本

拓展知识点

「负氧离子与人体健康」

- 负氧离子在人体生命活动中发挥着中重要作用。人体本身就是一个复杂的奇妙的系统。细胞就是一个微型电池，通过细胞电池不断充放电来维持大脑指挥各部器官的正常生理活动。而细胞的充放电又必须由负氧离子来维持。因而负氧离子是维持人体器官正常活动所必需的生命食粮。因此通过增加空气中的负氧离子，从理论上的确有利于人们的身心健康，在临床研究中也证实了其对人体的功效。这也就能解释在古代，缺少现代科学知识的人们在身体不舒服时，去拜祭深山古树，或者在古树下静卧即可改善身体状况的现象。

Qixi Town | Lotus Pond | Qiantang River Source Exploration Route

Interpretation No. QX-LHT-15

Stop 15: Lounge Bridge

● Interpretation theme

Forest oxygen bar

● Key points

1. Evaluation standard of natural oxygen bar in China
2. Reasons for the high negative oxygen ion in Qianjiangyuan National Park

● Site overview

The lounge with bench, where the enviroment is quiet and beautiful

"Evaluation criteria of natural oxygen bar in China"

When in the park, do you feel that the air in Qianjiangyuan National Park is extra fresh? We are now bathing in a natural "oxygen bar". But what is an oxygen bar? Is the oxygen content here higher? Congratulations, you Are half correct. The oxygen bar actually means that the concentration of negative oxygen ions in the air is very high and reaches a certain standard.

Negative oxygen ions are very beneficial to human health, and are known as "longevity element" or "air vitamin". It is said that the concentration of negative oxygen ions in many longevity villages in China is very high. Negative oxygen ions can eliminate smoke and dust, purifying the air. The higher the concentration of negative oxygen ions in the air, the better the air quality is.

The negative oxygen ion content of general urban residences is about 100 / cm^3. The World Health Organization (WHO) stipulates that it can be regarded as fresh air when the concentration of negative oxygen ions in the air is not less than 1000 – 1500 per cubic centimeter. Can you guess the amount of negative oxygen ions in Qianjiangyuan National Park? It is about three times the clean air standard given by WHO, or 39 times that of urban dwellings. As a result, Kaihua County was named the first place in the first batch of "China Natural Oxygen Bar" as early as 2016.

"Why is the negative oxygen ion high in Qianjiangyuan"

Why did Kaihua County get this honor? This should start from the generation mechanism of negative oxygen ions.

The formation of negative oxygen ions requires some specific conditions. Lightning, photosynthesis, waterfalls, fountains and aromatic substances secreted by some plants in nature can promote the production of negative oxygen ions. Let's take a look at the Grand Canyon around us. Not only are there endless streams and magnificent waterfalls, but also the dense virgin forests have created excellent conditions for the generation of negative oxygen ions. At the same time, the dust-holding effect of the forest and splashing water can also reduce the loss of negative oxygen ions.

In the case of more output and less loss, naturally the content of negative oxygen ions here is also higher, and the air is cleaner, which is more conducive to human health.

Attentions

- Invite all to take a deep breath and feel the clean air in the forest first, before introducing the concept of a forest oxygen bar.
- Remember to use customized interpretation skills to introduce negative oxygen ions according to different types of your visitors.
- Compare the air quality indicators where you live with those of a national park (to make a connection with the visitor's real-life experience).
- The high concentration of negative oxygen ion is closely related to the superior pure and undisturbed natural resources of the national park.

Recommended auxiliary facilities and toolkits

- Interpretation signage: best model of natural oxygen bar

Extra information

"Negative oxygen ions and human health"

- The human body itself is a complex and wonderful system. A cell is a miniature battery, through which the battery is continuously charged and discharged to maintain the normal physiological activities of the brain to direct various organs. The charge and discharge of cells must be maintained by negative oxygen ions. Therefore, negative oxygen ions are the necessary food for maintaining the normal activities of human organs.

齐溪镇 | 莲花塘 | 钱塘江寻源自然体验路线　　解说编号：QX-LHT-16

解说点 16 集散点

29°23′24.81″ N，118°12′39.57″ E
海拔：560m

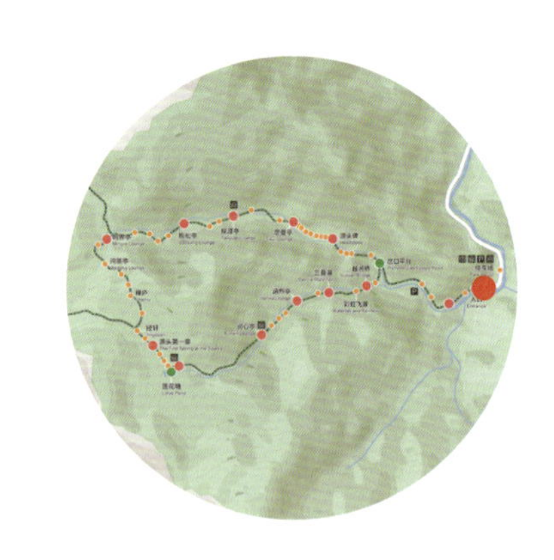

解说主题

游线总结与分享

解说要点

1. 回顾分享
2. 游线总结
3. 行为倡议

场地概述

莲花塘景区入口处，便于集合、停留的空地

解说方式 Interpretation Style

- ☑ 人员解说 Staff Interpretation
- ☐ 教育活动 Educational Activity
- ☑ 解说牌 Interpretation Signage
- ☐ 场馆解说 Hall Interpretation

解说季节 Interpretation Season

- ☑ 春季 Spring
- ☑ 夏季 Summer
- ☑ 秋季 Autumn
- ☑ 冬季 Winter

「游线总结」

各位朋友，我们今天关于钱塘江的溯源之旅就要结束了，非常感谢大家一路上的相伴扶持，我们才能顺利地走完全程。

在莲花溪的指引下，我们一路认识了各种各样的植物、动物，同时也更加深入地了解到了钱江源国家公园作为钱塘江水源孕育之地的意义。有句古诗道："问渠那得清如许？为有源头活水来。"正是钱江源国家公园茂密的森林、复杂的地形和丰富的降水才成就了这一片清澈纯净、源源不断的活水。

「其他游线推荐」

除了我们今天体验的这条游线之外，钱江源国家公园还有很多地方值得大家去探索。比如，去探索古田山的森林王国，或者化身科学家，去钱江源的科研样地看看，来一次特别的科学考察之旅；也可以去古村落走走，体验一下当地的风土民情。如果你想在最短的时间内体验最精致的风景，大峡谷会是不错的选择，顺道还可以体验下枫楼坑的湿地自然体验路线。

「游线分享」

在今天溯源之旅的自然体验行程结束之前，我想邀请大家一起再次回顾一下我们沿途的所观所感，然后告诉我最打动你让你印象深刻的一个点是什么？你可以用一句话，一个故事或者一张照片告诉我（做一个简短的和游客的问答互动）。

「行为倡议」

感谢大家的分享，也谢谢你们一路的支持，和你们交流也给了我很多新的启发。

今天大家的到访，让我们与自然更加亲近。希望大家回去后，谈起今天的经历，可能除了讲述这里的秀水青山、清新空气以及爬山途中挥洒汗水之外，也能将这些有意思的小故事同身边的人分享。在忙碌的工作生活之余，能想起这里有条绵延千里的江河，它的源头在自然秘境之中悄悄守护着我们，能时时感恩大自然给予我们的馈赠，即使身在城市，也要多多留意身边的自然。

祝福大家未来的旅途愉快，返程平安！也欢迎大家有机会再回来看看！

注意事项

- 对本次自然体验过程进行回顾和总结，并为访客推荐其他可选择的游线。

Qixi Town | Lotus Pond | Qiantang River Source Exploration Route

Interpretation No. QX-LHT-16

Stop 16: Assembly ending point

● Interpretation theme

Summary and sharing

● Key points

1. Review and share
2. Tour summary
3. Behavioral initiative

● Site overview

The entrance of Lotus Pond Scenic Area

"Travel summary"

Dear friends, our journey of tracing the source of the Qiantang River today is coming to an end. Thank you very much for your support along the way so that we can successfully complete the journey.

Following the Lotus Creek, we get to know all kinds of plants and animals along the way, and at the same time, deeply understand more about the significance of Qianjiangyuan National Park as the breeding ground for the Qiantang River. There is a saying: "How can it be so clear as has the springhead." It is the dense forests, complex terrain, and abundant precipitation of the Qianjiangyuan National Park that have made this piece of clear, pure, and endless running water.

"Recommended other tours"

In addition to the tour we took, Qianjiangyuan National Park has many places worth exploring. For example, go to explore the forest kingdom of Gutian Mountain, or incarnate as a scientist to take a look at Qianjiangyuan's scientific research site, and take a special scientific investigation tour; you can also go to an ancient village to experience the local customs and culture. If you want to experience the most exquisite scenery in the shortest time, the Grand Canyon is a good choice. Along the way, you can also experience the wetland nature experience route of Fenglou Pit.

"Travel sharing"

Before ending the natural experience journey of the source tracing tour today, I would like to invite everyone to review again our

observations and feelings along the way. Then tell me what touches you the most? You can use a sentence, a story, or a photo.

"Behavior initiative"

Thank you for sharing, and thank you for your support along the way. Communicating with you also gave me a lot of new inspiration.

Today's visit has brought us closer to nature. I hope that everyone will talk about today's experience after going back. In addition to telling about the beautiful water, the fresh air and the sweat on the way, you can also share these interesting little stories with the people around you. In addition to the busy work life, I can think of a river that stretches for thousands of miles. Its source is quietly guarding us in the natural secret. I can always thank nature for giving us a gift. Even in the city, pay more attention to it. The nature is around you.

I wish you all a pleasant journey in the future and a safe return home! Also welcome everyone to come back often!

Attention

- Review and summarize the whole trip, and provide some recommendations of other nature experience routes for your visitors.

2.5 Q02 里秧田—大峡谷：遇见峡谷飞瀑自然体验路线

Canyon and waterfall experience route

500m
500m
550m
入口
Entrance
映翠亭
Yingcui Lounge
水车
Water Wheel
墨池
Mo Pond
思源亭
Siyuan Lounge
神龙飞瀑
Shenlong Waterfall
N
图例 LEGEND
钱江源国家公园
Qianjiangyuan National Park
水域
Water
等高线
Contours
步行道
Trail
车行道
Road
餐厅
Restaurant
商店
Shop
卫生间
Toilet
休憩亭
Rest Lounge
观景处
Recreation Spot
瀑布
Waterfall
设解说牌的点位
Point with Signage
设解说牌、人员解说的点位
Point with Signage and Interpretation
设解说牌、人员解说、课程的点位
Point with Signage, Interpretation and Course
0
100m
700m
700 海拔(m)
650
600
550
500
450
映翠亭
水车
墨池
思源亭
神龙飞瀑
(返程)
路程(km)
0
1

Q02 里秧田—大峡谷：遇见峡谷飞瀑自然体验路线

解说主题 主题3：丛林佑护的珍贵原真
主题2：润泽富庶的河源追溯

步道里程 单程约 0.9km
体验时长 往返 1～1.5h
最低海拔 505m（步道起点）
最高海拔 646m（瀑布观赏平台）
步道等级 大众型
基础设施 休憩亭/解说牌/活动场地/卫生间

钱江源国家公园山系起源古老，其中让访客最能直观地看到这里古老的地质运动留下痕迹的地方非大峡谷莫属。峡谷飞瀑自然体验路线位于公园北部齐溪镇，游线不长，单程不到 1km，步道纤巧但在瀑布区坡度很陡，提供了非常独特的自然体验。

一路上伴随着潺潺的流水，游客可以尽情地在这座天然的氧吧中畅快呼吸，观察与溪流相伴而生的游鱼飞鸟，欣赏阳光与跌水瀑布共同谱就的七彩光影与曼妙音韵，同时还可以顺手捡起一块石头，听它讲述亿万年前地球经历的海陆变迁的故事。

在整条自然体验路线上，我们一共为访客安排了 9 个推荐解说点，同时，根据本书提供的通用解说方案（非定点解说方案）资料，解说员也可以在合适的生物资源等分布点进行相应的引导性观察和解说。

Q02 Canyon and waterfall experience route

Theme
- 03 Mysterious Wildlifes in the forest
- 02 The river source that breeds the fertile land

Trail Length | About 0.9 km oneway trip
Time | 1 – 1.5 hours round trip
Lowest elevation | 505 meters (outset of the trail)
Highest elevation | 646 meters (waterfall landscape platform)
Level of trail | Public
Infrastructure | Lounge / interpretation board / event site / restroom

The mountains in the Qianjiangyuan National Park originated from ancient times, among them it is the Grand Canyon where visitors can see the traces left by ancient geological movements. The Canyon Falls Natural Experience Route is located in Qixi Town, north of the park. The tour is not long (1.4 kilometers). The trail is narrow and has a steep slope in the waterfall area, but provides a very unique natural experience.

Along with the gurgling water along the way, visitors can enjoy breathing in this natural oxygen bar, embrace the colorful sunshine and rainbow when the light and the waterfall shed together, in the meanwhile pick up a stone and listen to it talking about the story of the changes of land and sea on the earth tens of billions of years ago.

We have arranged a total of 9 recommended interpretation stops for interpreters and visitors during the entire experience route. At the same time, according to the information provided in general interpretation (non-route-oriented interpretation) of this interpretation manual, the interpreter can also find suitable resources to conduct corresponding guided observation and interpretation.

齐溪镇 | 里秧田—大峡谷 | 遇见峡谷飞瀑自然体验路线　　解说编号：QX-DXG-01

解说点 01 集合点

29°23′56.09″ N，118°13′2.24″ E
海拔：473m

● **解说主题**

本条自然体验路线的总体介绍

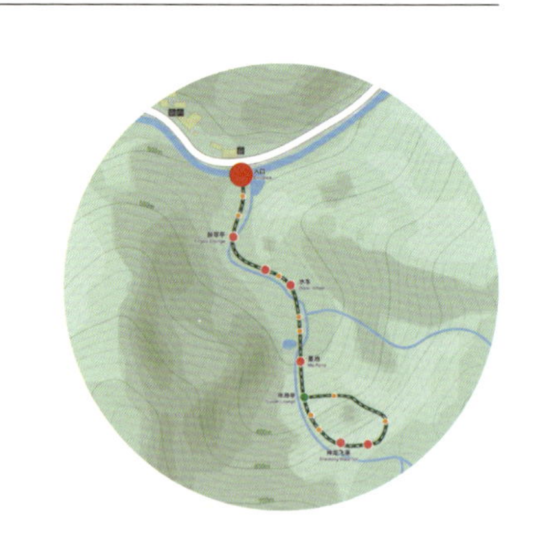

● **解说要点**

1. 开场白：对整条游线资源、设施进行总体介绍
2. 大峡谷的形成以及其中溪流与公园水系的关系

● **场地概述**

大峡谷入口集合点，门前有步道总体导览牌

解说方式 Interpretation Style

☑ 人员解说 Staff Interpretation　☑ 教育活动 Educational Activity
☑ 解说牌 Interpretation Signage　☐ 场馆解说 Hall Interpretation

解说季节 Interpretation Season

☑ 春季 Spring　☑ 夏季 Summer
☑ 秋季 Autumn　☑ 冬季 Winter

「开场白」

各位朋友，欢迎大家来到钱江源国家公园，我是各位今天的解说员 XXX，您可以叫我 ***(自然名)。欢迎大家来到大峡谷，很荣幸能在接下来的一个半小时左右，引导大家一起感受钱江源国家公园美丽的自然景观和珍贵的自然资源。接下来，我们的自然体验路线主题为遇见峡谷飞瀑。

峡谷瀑布在钱江源国家公园比较常见，但我们即将参访的大峡谷拥有最良好的自然体验基础设施，方便游客深入其中。在这里，不仅可以欣赏到美妙的峡谷景观，还能够在溪流边驻足停留，观察独特而丰富的野生动物。在峡谷最高处，也是这条步道的目的地，有一条落差超过 120m 的瀑布从山顶到山谷垂直落下，绝对会让你不虚此行。除此之外，沿途我还为大家准备了一些小的惊喜，所以大家一定紧紧跟着我，不要走散哦。

本路线全程不到 1km，大概 1 到 1.5 小时即可走完全程，沿途大部分路段都铺有石质路面，行走较为轻松。但是有一段爬山观赏瀑布的路线比较陡，行走会有一些吃力，且步道的宽度和容量有限，靠近瀑布顶端的路段不仅陡峭还比较湿滑。所以，大家途中一定注意看清路面，扶好护栏，不要拥挤，保持安全距离。

接下来，让我们一起开始峡谷飞瀑的体验之旅吧！

「大峡谷的形成」

（面向峡谷）首先我们来认识下峡谷。峡谷的“峡”字意思是两山之间夹着的狭长地带，我们面前就是非常典型的峡谷地貌：这条小溪位于两山之间峡谷的最低处，整个峡谷的横剖面呈“V”形（结合手势对着大峡谷的方向比画）。别看现在这里流水不断，曾几何时，这里可是一大片隆起的平地。后来因为地质运动，高耸的山体初现，然后是水流汇聚形成瀑布和溪流，并且不断切割着岩层，最终形成我们今天看到的景象。

顺着山势往下，溪流一路蜿蜒奔腾，最终会去向哪里呢？这还要从我们所在的公园的名字——钱江源开始说起。说到钱塘江的水源地，大家可能听到最多的就是莲花溪或者莲花塘，但实际上整个钱江源国家公园有众多溪流最终都是汇入钱塘江的，大峡谷中的这条溪流，就是这众多溪涧中的一条。这些溪流和国家公园的整个水系共同形成河流的源头，这条河流也就是被称为浙江母亲河的钱塘江。

注意事项

记住，请注意集合点的开场。当你出现在听众的面前，解说员的工作已经开始了。如何吸引听众的注意力，树立自身的解说员身份，一个简短、内容明确、带有互动的开场白对解说员来说非常重要。以下几点提示可以帮你达到这一目标。

- 选择一个合适的、可以帮助大家聚焦这次步道体验的集合地点，比如，步道导览牌边上。
- 体现你身份的制服、解说员的标志和便携的解说装备包。
- 你的介绍中一定有独特的、标识标牌上未曾涉及的内容。
- 根据团队成员的构成，针对性地提供步道过程中的注意事项建议。
- 大峡谷步道狭窄，尤其是最上面有一段路行走不方便，注意提前观察访客年龄和身体素质，以便提前安排。
- 需要控制每次活动的人数，建议在 10 人以内。

辅助道具推荐

- 游线总体导览牌

Qixi Town | Liyangtian–Grand Canyon | Canyon and Waterfall Experience Route

Interpretation No. QX-DXG-01

Stop 01: Assembly meeting point

Interpretation theme

General introduction to this nature experience route

Key points

1. Overall introduction of the route
2. Make sure the audience is listening and establish the identity as an interpreter

Site overview

The assembly meeting point outside the entrance of the mountain gate, where there is a general guide board in front of the gate

"Gathering visitors to the entrance of the Big Canyon"

Dear friends, welcome to Qianjiangyuan National Park. I am the interpreter today XXX. You can also call me ***. It's my pleasure to guide you all to experience the beautiful scene and treasurable natural resources in the coming one and a half hours. We are going to take a tour to the canyon and waterfall experience route together.

Canons and waterfalls are very common in the national park, but the one we are going to visit has the well-developed infrastructure thus visitors can conveniently go deep into the canyon and enjoy the amazing landscape, as well as stopping by the streams to observe its unique and diverse wildlife. At the highest point of the trail, the destination of this experience route, there is a waterfall that drops more than 120 meters from the mountain top to the deep valley, which will mark the most memorable moments for you. In addition, along the way I have also prepared some little surprises for you. Please follow me closely, do not get separated.

The whole route is less than 1 km and will take us about 1 to 1.5 hours to experience. Most of the trail along the route are paved with stone and easy to walk. However, there is a steep route to climb the mountain and take a close view to the waterfall, so it will be quite difficult to walk. Furthermore, the width and capacity of the trail are limited, especially the part near the top of the waterfall is steep, wet and slippery. So you must pay attention to the road condition on the way, hold the guardrails, do not be crowded, and keep a safe distance. Don't rush, safety comes first.

Now everyone please follow me and let us

start today's journey.

(Facing the canyon) First of all, let's talk about what is a canyon. In Chinese, the word "canyon" means a narrow strip of land between two mountains, and we had in front of us a very typical canyon: the waterfall cuts off the mountain into a V-shaped cross-section and the stream is formed at the lowest point of the canyon and flows to external river system (combined with the gesture to point out the direction of the Canyon). Due to geological movement, the towering mountains first appeared, and then the water came together to form waterfalls and streams, cutting through the rocks and the mountain, and finally formed the landscape we see today. Where does the stream wind its way down the hill? Yes, you must remember the name of the park: Qianjiangyuan. It will finally flow to the source of the Qiantang River. You are probably most familiar with the Lotus Creek or Lotus Pond, but in fact there are many streams throughout the National Park that eventually flow into the Qiantang River, and this stream in the Grand Canyon is just one of them. Together all the streams and water system of the national park forms the source, i.e. the Qiantang River, the so-called mother river of Zhejiang Province.

Attentions

Remember, pay attention to your debut at the assembly point. When you show up to the audience, your job as an interpreter has already begun. As an interpreter, it is very important to attract the audience's attention and establish your own personality, such as using a short, clear and interactive opening statement. The following tips can help you achieve this goal.

- A proper assembly place that can help gather the crowd and start this nature experience, such as the open space next to the general guide sign at the start of the trail.
- Give yourself an name that's easy to remember as well as a short but interesting self-introduction to make sure the tourists remember you from the beginning.
- Uniforms and costumes that reflect your identity: the logo of the interpreter and supporting materials tool kit.
- There must be something unique in your introduction that isn't mentioned in the ID card.
- Spoil some of the upcoming nature experience trip, including the theme, time, and particularly worthy interpretation, *etc.* in order to arouse everyone's curiosity.
- Based on specific team members, pose targeted safety requirements and precautions.
- Don't forget to remind everyone not to pick and take away any natural objects in the park, and don't throw garbage on the ground, *etc.*.

Recommended auxiliary facilities and toolkits

- Overall guide board of the route

齐溪镇 | 里秧田—大峡谷 | 遇见峡谷飞瀑自然体验路线　　解说编号：QX-DXG-02

解说点 02 映翠亭

29°23′53.38″ N，118°13′1.23″ E
海拔：484m

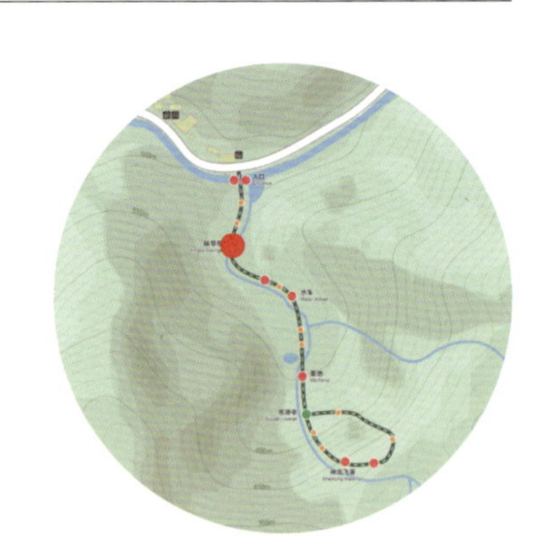

● 解说主题

溪流多样的生境下生活着的多种生物

● 解说要点

1. 适应湍急溪流环境而生活的鱼类和青蛳
2. 在溪流周边环境常出现的鸟类、昆虫等

● 场地概述

溪流边，可观察到与溪流相伴而生的动植物

解说方式 Interpretation Style

- ☑ 人员解说 Staff Interpretation
- ☐ 教育活动 Educational Activity
- ☑ 解说牌 Interpretation Signage
- ☐ 场馆解说 Hall Interpretation

解说季节 Interpretation Season

- ☑ 春季 Spring
- ☑ 夏季 Summer
- ☑ 秋季 Autumn
- ☐ 冬季 Winter

「溪流中生活的鱼类和青蛳」

大峡谷的溪流水质清澈，河底多砾石，且水有深有浅，速度有缓有急，多种多样的生境孕育了丰富的生物多样性。这里有喜欢生活在平静的浅水区的石斑鱼和青蛳，也有适应湍急水流的虾虎鱼、华南湍蛙等。水面之上，常见各种昆虫和小鸟儿的身影。如果你在夜晚来到大峡谷，舞台就被转交给两栖类和鸣虫，它们的合唱是大峡谷夜晚最美的旋律。

在溪流底部碎石较多且坡度稍缓的静水区，常有一些小鱼游来游去，仔细观察，你会发现它们黄褐色的身体之上有一些非常明显的横条纹均匀地排布着。这种小鱼就是当地人耳熟能详的石斑鱼（又叫光唇鱼）。这是一种非常“挑剔”的鱼类，它们只生活在水质清澈、水流湍急且河底多碎石的山涧溪流中，以石头上的苔藓和藻类为食。溪流里的野生石斑鱼体形较小，成年鱼类的个头也就 10cm 左右。

大峡谷流动的溪流浅水区中常有一些颜色青黑、体形修长的螺蛳贴附在水底的大石头上。开化青蛳肉质灰绿色，味道鲜美又带有一点清苦，是一道当地风味美食，也因为著名的纪录片《舌尖上的中国 2》而蜚声国内。青蛳主要生长在没有污染的活水中，以水生植物和有机碎屑物为食，因此它们天生就是水质优良与否的质量检测官。《舌尖上的中国 2》热播之后，各地食客纷至沓来，只为品尝这一美味。但因为它们只能生活在原生态的溪流冷水中，生长缓慢，数量有限，因此慢慢供不应求。据当地老人说，现在溪流中的青蛳不仅越来越少，而且很难见到个头大的青蛳了。

大峡谷的地势落差较大，水流速度快，溪流中有一种鱼类为了适应湍急的水流，干脆放弃了精进游泳的技能，将腹鳍进化成吸盘状，遇到急流，便运用吸盘攀附在石头上慢慢往上爬。这种鱼叫虾虎鱼，大多数时候它们会静静地平贴在溪底部的石头上，身上的保护色让人很不容易发现它们。

注意事项

- 可以选择有相应鱼类的溪流进行观察，但要注意安全。
- 解说具体物种的时候，也可以参考通用型生物多样性解说方案的内容。

辅助道具推荐

- 标识牌：生活在溪流中
- 标识牌：溪流中的生活
- 标识牌：与溪流相伴的飞羽精灵

拓展活动推荐

- 昆虫连连看（P573：活动卡 12）

「溪流边常见的鸟类」

沿途除了潺潺的水声之外，可能时不时还会有几声鸟叫传到大家的耳朵里，抬头看，你却很难发现它们的身影。溪流为蜻蜓、豆娘、石蝇、蜉蝣及其幼虫等水生昆虫提供了栖居的场所，这些昆虫则是依靠溪流生活的一些鸟类（如燕尾、红尾水鸲、白鹡鸰）的食物。溪流的开阔环境为我们近距离观察这些鸟类提供了很好的条件。

燕尾全身羽毛多黑白两色，尾巴像燕子的一样形似剪刀，常在水边走来走去觅食，双腿偏肉粉色，纤细瘦长，比较容易辨认出来。不过，燕尾是一个大家族，这边的溪流中常见的燕尾就有 3 种：那种头顶和背部都是灰色的，叫作灰背燕尾；额头或者头顶上有一丛白色羽毛的，叫作白冠燕尾；还有一种体形较小的，叫作小燕尾。燕尾生性好动，一般停歇在水边或者水边的石头上，以水中的昆虫或者昆虫幼虫为食，寻找食物的过程中尾巴会一张一合，煞是可爱。

和燕尾一样，红尾水鸲的尾巴也经常摆动，像一把扇子似的不停地打开合拢。不过红尾水鸲雄鸟的尾巴为非常醒目的砖红色，身体其他部位偏蓝色，个头较燕尾小得多，很容易辨认出来。相比之下，雌鸟就低调得多，上体为灰褐色，身上布满鳞状斑，要认出它们来还有一定的难度，但如果你在雄鸟的身边留心观察，说不定会在石块之间或者岸边植物丛下会发现雌鸟的身影。这种雌雄异色的现象多发生在一些雁形目或者鸡形目的鸟类身上，比如，大家熟悉的鸳鸯、家鸡，但是对于身材娇小的红尾水鸲来说，这实在是非常独特的外形特征。

白冠燕尾　小燕尾　灰背燕尾　红尾水鸲（雌幼）　红尾水鸲（雄幼）

「溪流的昆虫世界」

溪流也是一个弱肉强食的世界，各种生物之间通过捕食与被捕食的关系连接起来，形成不可分割的一个整体。其中，昆虫作为食物链的重要一环，是连接动物与动物，动物与植物的桥梁，在整个食物网络中起到不可取代的作用。但是因为它们个头较小，且善于隐蔽，一般很容易被忽略。今天，就让我们一起来努力寻找沿途的昆虫踪迹吧！

溪流边常见的蜻蜓，应该是大家最为熟悉的依水而生的昆虫之一了。它们体形修长，有一双圆圆的大眼睛，翅膀近似透明，非常容易辨认。但如果你仔细观察，会发现有的蜻蜓体形健硕，有的纤细瘦小，实际上后者并不是蜻蜓，而是蜻蜓的“孪生姐妹”——豆娘。它们无论从外形还是生活习性上都有几分相似之处。那我们如何区分它们呢？首先，蜻蜓体形比豆娘更粗壮一些；其次，蜻蜓停留时翅膀是张开的而豆娘是合拢的；另外，豆娘的两只眼睛中间有明显的界限，而蜻蜓的复眼则是连接在一起的。

赤基色蟌（雄幼）

萤火虫

蜻蜓

蜻蜓和豆娘一般会将卵产在靠近水面的植物上，比如，我们常说的蜻蜓点水就是蜻蜓产卵的现象。但是，由卵孵化成幼虫再到长成成虫，却是一个非常漫长的过程，一共要经历约 11 次蜕皮。蜻蜓的幼虫又叫水虿（chài），它们在水中的生活时间长达约 2 年，而且它是一个彻底的肉食者，喜欢各种小型的水生生物，包括水生昆虫，小虾、蝌蚪甚至小鱼等。当然，它们也是其他大型的鱼类或者鸟儿等生物的食物。

除了蜻蜓之外，水边常见的昆虫还有萤火虫。我国的萤火虫约有近 2000 种，但由于水体的污染、栖息地的丧失、人工光源的干扰等问题，如今我们在城市已经很难看到它们。但在钱江源国家公园，依然为很多萤火虫保留了生存的栖息地。每年的七八月，是萤火虫的繁殖季，夜晚，它们常常提着自己的小灯笼在大峡谷中飞来飞去，造就了夏夜里的灯光盛宴。

无论是蜻蜓、萤火虫还是其他生活在水边或水中的昆虫，都对它们生活的环境比较挑剔，多数喜欢在环境优美、植被茂密、没有外部干扰的洁净的山涧、溪流和河流中，以各种水生昆虫和小型水生生物为食，对水质的变化很敏感。它们的存在其实就是溪流水质的天然指标。

Qixi Town | Liyangtian–Grand Canyon | Canyon and Waterfall Experience Route

Interpretation No. QX-DXG-02

Stop 02: Yingcui Lounge

Interpretation theme

Mysterious wildlifes along stream

Key points

1. Aquatic wildlife in streams
2. Birds and insects living along the stream

Site overview

Along the stream, where you can see various creatures living along the stream

"Aquatic wildlife in streams"

The clean, free-flowing, gravel-bottomed streams provide rich biodiversity with its variety of habitats. You can find aquatic organisms preferring rapid currents, such as the groupers and South China turbulent frog, as well as those adapting to slow and shallow steams, such as goby and river snails. Beside the stream, you can easily find various insects and birds. If you have a night tour to the canyon, the amphibians and sound bugs will be the protagonists of the stage. Their chorus is the most beautiful melody of the Grand Canyon's night.

In the clean water of the stream, there are often some small fish swimming at the bottom. If you look closely, you will find that their yellow-brown bodies have some very distinct horizontal stripes evenly arranged. They are a kind of locally well-known groupers (also called light lipped fish). They live only in clear, free-flowing mountain streams with gravel-bottoms. Their thick lips allow them to graze moss and algae on rocks. Wild groupers in streams are relatively small, growing to around 10cm as adults.

There are also some small and blue-black river snails attached to the rocks in the steam. It is a local delicacy and made famous in China by the famous documentary *A Bite of China 2*. They can only live in unpolluted water and live on aquatic plants and organic debris, so they are natural inspectors of water quality. Since it became famous, people from all over the country came to enjoy the delicious and unique taste. However, since they can only live in the natural stream and grow slowly, their productivity is limited and is gradually short in supply. According to the local old people, they are getting fewer and smaller today.

In order to adapt to the turbulent streams,

some fishes simply gave up the skills of swimming, and evolved the ventral fins into a sucker shape, so as to firmly stick to the bottom of the stream, wait for the prey to come close and then capture them in one fell swoop. This kind of fish is called goby fish, and they only occasionally swim in a "snap-and-swim" way, and use "quiet as a virgin, as a rabbit as a move" to describe them. The native stream environment in Qianjiangyuan National Park provides a shelter for many native fishes, and it is also an important area for the study of goby and chub.

"Common birds by streams"

In addition to the gurgling water along the way, there may be some birds' chirping that reaches everyone's ears from time to time, but looking up, you will find it's not easy to spot their existence. Streams provide habitat for insects such as dragonflies, damselflies, mayflies and their larvae, which in turn feed on birds such as swallowtails, plumbeous water-redstarts and white wagtails. The open environment of the stream provides good conditions for us to observe birds at close range.

There are several small birds here that often appear in mountain streams and are nourished by the water. They are similar in size to sparrows or swallows. Some may think those birds look alike and really hard to distinguish, but in fact, if you pay attention to observe and master some bird watching skills, it is not difficult to distinguish them.

We often use scissors to describe the tail of a swallow. There is also a type of bird with a swallowed tail called forktail.

Forktails are a big family: the kind with gray head and back is called slaty-backed forktail; the one with white feathers on the forehead or top of the head is called white-crowned forktail; there is also a smaller body called the little forktail.

Forktails are active creatures that usually rest on the water's edge or stones near the water. They feed on insects or insect larvae in the water. During the process of searching for food, their tails will open and close from time to time, which is very cute.

Attentions

- You can choose streams with corresponding fishes for observation, but pay attention to safety;
- For the introduction of some species, please refer to the content in the "General Interpretation of Biodiversity".

Recommended auxiliary facilities and toolkits

- Interpretation signage: living in the creeks
- Interpretation signage: creek lifestyle
- Interpretation signage: creek-living birds

Recommended activities

- Insects link game (P573: No. 12)

There is also a dioecious small bird: the plumbeous water-redstart. Generally speaking, only large birds such as mandarin ducks and white pheasants will appear dioecious. For petite plumbeous water-redstarts, this is a very unique feeling.

The male bird of the plumbeous water-redstart is bright and dark blue, and its tail has a striking brick red. The two colors are strongly contrasted and are generally easier to find. The female bird is relatively low-key, with a grayish brown upper body and scaly spots on the body, which is generally not easy to identify, but if you pay close attention to the male bird, a figure of the female bird may be found between the stones or in the shore vegetation. Like the forktail, the tail of the plumbeous water-redstart often swings, opening and closing like a fan in the hand.

"Insect world of streams"

As the saying goes: "Big fish eat small fish, small fish eat shrimp, shrimp eat what? Shrimp eat mud (plankton and other aquatic organisms)." A variety of biological choices settled down in the stream, forming a network of wildlife with complex interrelationships. Insects, as an important component in the food chain, are the bridge connecting animals and animals, animals and plants, and play an irreplaceable role in the whole ecosystem network. But because they are small and good at concealment, they are generally easy to be neglected. Today, let us try to pay attention to whether we can find "insect traces" scattered everywhere.

Dragonflies, perhaps the most familiar insects for us who live near streams. They are easy to identify with their long and strong body, big round eyes and nearly transparent wings. But if you look closely, you'll see that some dragonflies are big and some are skinny, they are quite different in shapes. In fact, the thinner are not dragonflies at all, but their close relatives — damselflies. It is easy to get confused because they are somewhat similar both in shape and in habits. So how do we tell them apart? There are some tips. First, dragonflies are a bit bulkier than damselflies; second, when dragonflies stay, their wings are open horizontally while damselflies are closed vertically. In addition, damselflies have a distinct boundary between two eyes, while dragonflies have two compound eyes connected.

Dragonflies and damselflies always lay their eggs on plants near the surface of the water. But it is a very long process from the egg to the larva, then to the adult, which takes about 11 times molt cycles. Dragonfly larvae spend about two years in the water and they are a type of voracious meat-eaters, loving all kinds of small aquatic creatures including insects, shrimps, tadpoles and even small fish. Of course, they are also food for other large fish or birds.

In addition to dragonflies, we can also see fireflies at the water's edge in the evening. There are nearly 2000 species of fireflies in China, but due to water pollution, habitat loss, artificial light interference, and other problems, it is difficult to see them in cities today. But in Qianjiangyuan National Park, there is still good habitat for many fireflies. Every year in July and August as the breeding season of fireflies, we can enjoy the amazing night scene while they fly around near the water with their small lanterns, creating a fantastic light show in the summer night.

Dragonflies, fireflies or other insects living in or beside the water, are very critical to the quality of their living environment. They prefer clean mountain streams and rivers with beautiful environment, dense vegetation and no external disturbance. They eat a variety of aquatic insects and small aquatic creature and are very sensitive to changes in water quality. In fact, their presence is a natural indicator of stream water quality.

stick insect

齐溪镇 | 里秧田—大峡谷 | 遇见峡谷飞瀑自然体验路线　　解说编号：QX-DXG-03

解说点 03 长柄双花木

29° 23′ 52.63″ N，118° 13′ 2.31″ E
石壁刻有：碰头是福　海拔 490m

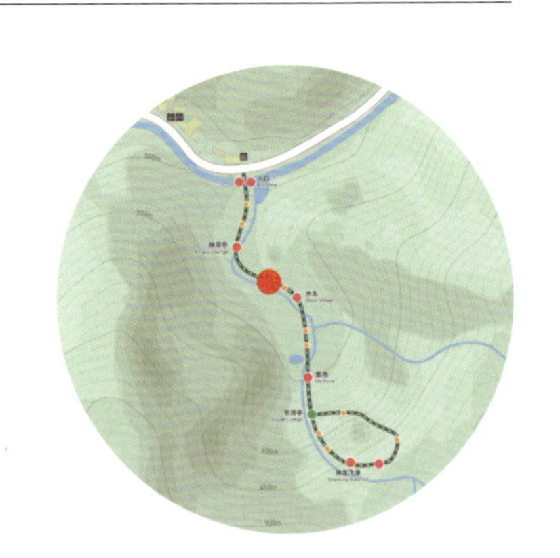

● 解说主题

原真自然庇佑的孑遗植物：长柄双花木

● 解说要点

1. 认识长柄双花木的外形特征
2. 长柄双花木的珍稀性与就地保护价值

● 场地概述

1. 溪流对岸，可以远观长柄双花木群落
2. 道路狭窄，注意团队安全

解说方式 Interpretation Style

☑ 人员解说 Staff Interpretation　☐ 教育活动 Educational Activity
☑ 解说牌 Interpretation Signage　☐ 场馆解说 Hall Interpretation

解说季节 Interpretation Season

☑ 春季 Spring　☑ 夏季 Summer
☑ 秋季 Autumn　☑ 冬季 Winter

「认识长柄双花木的外形特征」

钱江源国家公园不仅是亚热带常绿阔叶林的王国，千万年来相对稳定的环境也为许多珍稀和古老的孑遗植物提供了庇护所。我们在古田山可以看到杜仲、青钱柳等孑遗植物，而在大峡谷中，孑遗植物的明星非长柄双花木莫属。

（注意停留的地方路面较窄，水深流急，提醒访客注意安全。）

现在我们对岸就有一片长柄双花木群落。乍看之下，可能很难发现我们的目标。那么，就让我们先来借助图片了解下它的样子：首先，作为一种落叶灌木，长柄双花木整体树型不会太高大；其次，它的叶子近似心形，新叶颜色偏红紫色，可以和绿叶明显区分开来。根据这两点，大家可以找到溪流对岸哪棵树是长柄双花木吗？

没错，就是那里（手指长柄双花木方向）。别看它们现在毫不起眼，花开时却非常独特！作为金缕梅家族的一员，长柄双花木拥有家族标志性的丝丝缕缕的花瓣，花开时两朵并蒂，共生于一根长梗之上，因此得名长柄双花木。每年秋末冬初（10~12 月）是长柄双花木的花期。这时候不仅可以观花，还可以赏叶。和新叶一样，长柄双花木的花朵也是偏红紫色的，此时叶片也逐渐由绿色转红色，直至掉落。此时，枝头的花朵更加显眼，成为秋冬的一道独特的风景。

「钱江源国家公园的长柄双花木的珍稀性与保护价值」

长柄双花木区系起源古老、数量稀少，属于国家二级重点保护野生植物和濒危物种。从分类上来看，金缕梅科在全世界一共有 30 个属，其中，双花木属是金缕梅家族最原始、最古老的属，而长柄双花木是金缕梅科双花木属唯一的一个种。最早的种子化石表明，长柄双花木在白垩纪（和恐龙同一时期）曾繁盛一时。白垩纪末期的第五次生物大灭绝使得恐龙等大型动物从地球上消失，长柄双花木却幸运地逃过了一劫。不过它的生存区域狭窄，目前只零星分布在浙江、湖南和江西三省的部分区域。

请大家看，我们对面的这片林子，就集中生长有大量的长柄双花木。钱江源国家公园就是长柄双花木仅存的孤岛之一，连续分布面积逾 $333hm^2$，是全球长柄双花木天然分布连片面积最大的种群地之一。这里也是长柄双花木自然海拔分布的下限（480m）和最北的群落。

长柄双花木“佛系”的繁殖策略，决定了它们不可能广泛分布。它们的果实从初生到成熟，差不多经历了 3 个季节，而且成熟后的果实主要依靠弹力将种子传播出去，影响范围十分有限。此外，种子皮质较厚、休眠期较长也是它们繁殖后代的限制因素。最后，作为一种植物，它们不可避免地会受到病虫害、自然灾害等自然因素的影响和人为的干扰与破坏，这些因素综合在一起导致了它们孤岛式的分布状态。

长柄双花木虽美丽，也很脆弱，钱江源就这样静静守护了它们千万年，但今天的它们依然脆弱，经不起人类活动的任何干扰。这就是为什么我们只能这样隔着溪流观察，保持适当的距离。不去打扰，静静地观赏就十分美好。

（解说员需要了解长柄双花木的特征和位置，并根据合适的季节选择是否进行本点位的人员解说。）

Qixi Town | Liyangtian-Grand Canyon | Canyon and Waterfall Experience Route

Interpretation No. QX-DXG-03

Stop 03: Longstipe Disonthus (*Disanthus cercidifolius* var. *longipes*)

Interpretation theme

Mysterious wildlife in the forest: longstipe disanthus

Key points

1. Understand characteristics of longstipe disanthus
2. The rareness and protection value of longstipe disanthus in Qianjiangyuan National Park

Site overview

The other side of the stream, where longstipe disanthus can be observed from a distance and the road is narrow

"Understanding the characteristics of the longstipe disanthus"

Qianjiangyuan National Park is not only famous for its subtropical evergreen broad-leaved forests, but also a relatively stable natural environment for thousands of years that has provided a shelter as a sanctuary for many rare and ancient relic plants. We can see many of these rare and relic plants here. For example, in the Grand Canyon, you can see the star of the relic plants here, the longstipe disanthus *(Disanthus cercidifolius* var. *Longioes*).

Now there's a patch of it right over there on the opposite bank to us. It might be hard to spot our target at a first glance. Let's try to find its key characteristics from this picture. First of all, as a deciduous shrub, it is not too large in its shape. Second, its leaves are almost heart-shaped, and the new leaves are reddish and purple in color, which can be clearly distinguished from the other green leaves. Based on these two points, can you find out where is this rare plant on the other side of the stream?

Yes, it is over there. Maybe you will think it is just an ordinary-looking plant. But when they bloom, they are unique! As a member of the witch hazel family, it has the same signature strands of petals. When they bloom, two flowers are combined, which are born on a long stem, hence the name in Chinese is long-handle double flowers. Every year in late autumn and early winter (October to December) is the flowering time of it. At that time, you can not only watch the flowers, but also admire the leaves. Like the new leaves, the flowers are reddish purple, and the leaves gradually turn from green to red until they fall off. Then the flowers on the branches are

more conspicuous, making a unique scenery of autumn and winter in the forest.

This plant is anciently-originated, rare and endangered species, ranked the second-level national protected key wild plant. There are a total of 30 genera in the witch-hazels family in the world. It is the most primitive and oldest genus in the witch-hazels family, which is endemic to East Asia. There is only one species in the genus. Fossil studies show it lived alongside dinosaurs during the Cretaceous period. During the fifth mass extinction at the end of the Cretaceous period, which wiped out large animals like dinosaurs, it had luckily escaped the disaster and survived until today. However, their distribution is very narrow, currently only scattered in parts of Zhejiang, Hunan and Jiangxi provinces.

Look! A large number of the trees were concentrated in this forest right in front of us. Qianjiangyuan National Park is one of the only sanctuaries for it and many other relic species, with a continuous distribution area of more than 333 ha, and it is one of the largest populations of natural population in the world. This is also the northernmost and the lowest elevation (480m) of longstipe disanthus population's natural distribution.

In addition, the number of *Dianthus* is very rare because of their "Buddha-like" breeding strategy. They have more flowers but fewer fruits. The ripened fruits rely on the elastic force to spread the seeds out, and the range of influence is very limited. In addition, their thick seed cortex and long dormancy period are also the limiting factors for their reproduction. Finally, as a plant, they are inevitably affected by natural factors such as diseases and insect pests, natural disasters, and human interference and destruction. The combination of these factors leads to their island-like distribution.

This beautiful plant is fragile as well. Here in Qianjiangyuan national park, they have been well conserved for millions of years. However, they are still vulnerable to any interference of humans. That's why we should just stand here looking at them over the stream with a proper and safe distance. Don't disturb this ancient and precious plant, let's enjoy their beauty by just standing here quietly.

齐溪镇 | 里秋田—大峡谷 | 遇见峡谷飞瀑自然体验路线　　解说编号：QX-DXG-04

解说点 04 水车

29°23′48.92″ N，118°13′5.34″ E
海拔：503m

解说主题

与山水共生的生活智慧

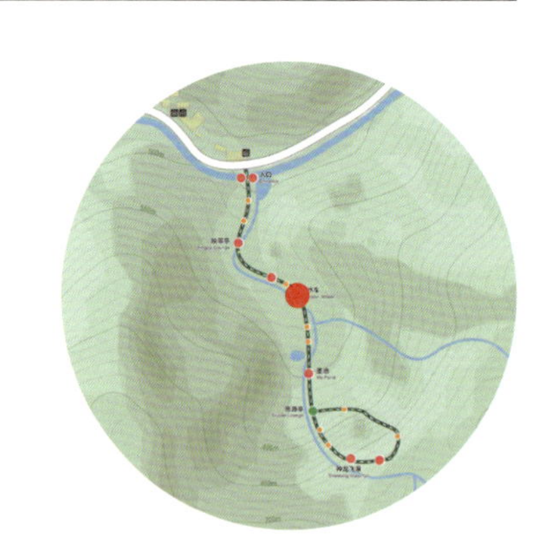

解说要点

1. 水车是当地山民因地制宜生活智慧的体现
2. 茶园是过去历史的见证者

场地概述

水车、木亭、茶园

解说方式 Interpretation Style

☑ 人员解说 Staff Interpretation　☐ 教育活动 Educational Activity
☑ 解说牌 Interpretation Signage　☐ 场馆解说 Hall Interpretation

解说季节 Interpretation Season

☑ 春季 Spring　☑ 夏季 Summer
☑ 秋季 Autumn　☑ 冬季 Winter

解说词

各位朋友，一路走来，我们欣赏到的大峡谷充满野性与自然之美，但是这里怎么会有一个大水车呢？其实在钱江源国家公园建成之前，这里曾经是人们从事农业活动的地方，这个水车即是公园建成后对传统生活方式的纪念。

钱江源国家公园山多地少的自然地理形成了“九山半水半分田”的独特农业形式，当地居民在认识自然的基础之上，因地制宜，借助水流落差产生的能量驱动水车，做成了这样一个装置，充分体现了人们古朴的生态智慧。

这边的茶园也是过去人类活动的见证，已建成的国家公园将这里划为核心保护区。为了保护这里的野生动植物以及它们的栖息地，不允许人类在此建设经济生产设施，它的存在代表着钱江源国家公园的过去，是历史的见证者。

除了水车和茶园之外，这里还有一棵高大的马尾松，是过去森林存在的证据。其实如果大家留心观察，可以挖掘很多这样的故事。

注意事项

- 这是沿途比较方便的一个休息点，可以让访客稍做停留，在亭子中休息的同时注意观察下水车的工作原理，解说水车背后的历史。

辅助道具推荐

- 标识牌：与山水共生的生活智慧
- 标识牌：江源自然的丰富馈赠

Qixi Town | Liyangtian-Grand Canyon | Canyon and Waterfall Experience Route

Interpretation No. QX-DXG-04

Stop 04: Water wheel

Interpretation theme

Wisdom of living with the nature

Key points

1. Water wheel reflecting the past living wisdom
2. The history of the tea garden

Site overview

Water wheel, water lounge, tea garden

Dear friends, we can see the Grand Canyon is full of wild and natural beauty along the way, but how can there be a big water wheel here?

In fact, before the Qianjiangyuan National Park was built, it used to be a place for people to engage in agricultural activities. This water wheel is the trace left by human activities after the completion of the park.

The natural geography of Qianjiangyuan National Park has formed the unique agricultural form of "nine mountains and half rivers and half fields". On the basis of understanding the nature, local residents drive the water wheel with the energy generated by the water to make such a device, which fully embodies the ancient ecological wisdom of people.

The tea garden here is also a witness to human activities in the past, which now is the core protected area of the National Park. In order to protect the wildlife and their habitats, human beings are not allowed to build economic production facilities here. Its existence represents the past of Qianjiangyuan National Park and is a witness to its history. In addition to water wheel and tea garden, a tall masson pine is standing here, which shows the existence of the forest. In fact, if you observe mindfully, you can dig up a lot of these stories.

Attentions

- This is a convenient rest point along the way, allowing visitors to stop for a while. Not only can you take a rest in the lounge, but also get to know the history of the water wheel.

Recommended auxiliary facilities and toolkits

- Interpretation signage: life wisdom inspired by the landscape
- Interpretation signage: generous gifts of nature

齐溪镇 | 里秧田—大峡谷 | 遇见峡谷飞瀑自然体验路线　　解说编号：QX-DXG-05

解说点 05 墨池

29° 23′ 45.57″ N，118° 13′ 5.65″ E
海拔：520m

解说主题

钱江源地区的森林氧吧

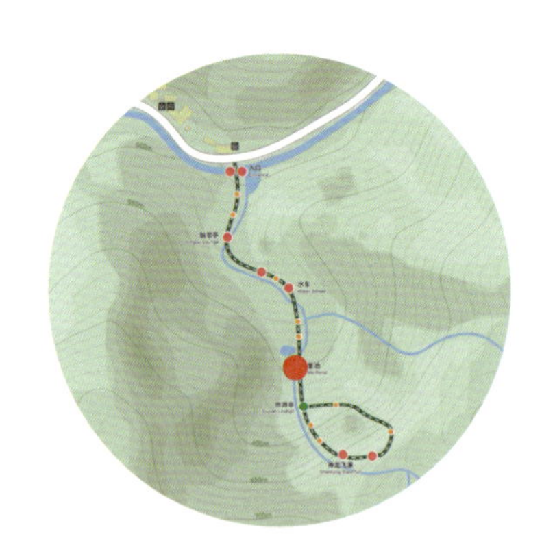

解说要点

1. 中国天然氧吧的评价标准
2. 钱江源负氧离子高的原因

场地概述

溪流边，可感受山间谷地清新的空气

解说方式 Interpretation Style

☑ 人员解说 Staff Interpretation　☐ 教育活动 Educational Activity
☑ 解说牌 Interpretation Signage　☐ 场馆解说 Hall Interpretation

解说季节 Interpretation Season

☑ 春季 Spring　☑ 夏季 Summer
☑ 秋季 Autumn　☑ 冬季 Winter

「中国天然氧吧的评价标准」

置身山林之中，并且有潺潺的溪水为伴，大家的各种感官是不是都被充分调动起来了？眼睛里看到的是巍巍青山，耳朵里听到的是潺潺水声，用手触摸岩石粗糙的肌理，感受清凉的水温，就连平时怠惰的嗅觉都变得灵敏，除了植物的芬芳之外，我们闻到的空气也是清新的。

行走在山林间，我们仿佛在经历一场森林的洗礼，身心舒畅，崇尚养生的中国人喜欢把这种地方叫作“森林氧吧”。“氧吧”这个词听起来非常主观，但是却有着实实在在的评判标准。被称作“氧吧”的地方不仅空气质量要高、水质要好、还要求有较高的森林覆盖率……其中最核心的指标是被称为“空气维生素”的负氧离子的含量。

举个例子，如果你居住在城市，你们家每立方厘米空气中的负氧离子含量约为 100 个。而世界卫生组织规定，当空气中负氧离子浓度不低于 1000～1500 个 /cm³ 时，才可视为清新空气。那我们这里的负氧离子含量有多少呢？大家可以猜一猜？平均值大约相当于你们家的 45 倍，是世界卫生组织给出的清新空气标准的 3 倍还多。开化县也因此在 2016 年就被评为首批“中国天然氧吧”第一名。

「钱江源负氧离子高的原因」

为什么开化县会获此殊荣呢？这还要从负氧离子的产生机理开始讲起。

负氧离子的形成需要一些特定的条件。自然界的闪电、光合作用、瀑布、喷泉以及一些植物分泌的芳香类物质都能促进负氧离子的产生。

我们来看看身边的大峡谷，这里不仅有奔腾不息的溪涧、声势浩大的瀑布，还有连绵茂密的原始森林，都为负氧离子的生成创造了优越的条件。同时，森林的滞尘作用、溅起的水花等还可以降低负氧离子的损耗。在产出较多，损失较少的情况下，这里的负氧离子含量自然也更高，空气也更清新，更有利于人类的健康。

注意事项

- 先请大家做一次深呼吸，引入森林氧吧的概念。
- 注意对负氧离子的介绍可以结合不同的受众群体选择解说内容。
- 比较一下自己家的空气指标与本地的差异（与身边建立关联）。
- 负氧离子浓度高与钱江源得天独厚的条件是密不可分的。

辅助道具推荐

- 标识牌：天然氧吧的最佳范本
- 不同环境下负氧离子浓度卡片

拓展知识点

「认识负氧离子」

- 负氧离子又被称为“空气维生素”，对人体健康十分有益。空气中的氧分子在瀑布水流冲击、紫外线照射、雷电等的诱导下，容易发生电离，生成负氧离子。一般水流速度越快，这种作用就越强烈。因此，在山间林地的瀑布附近空气中的负氧离子浓度相对也会比较高。

不同环境的负氧离子浓度及与人体健康的关系

环境	负离子浓度（个 /cm³）	与人体健康关系
森林瀑布	大于 3000	杀菌、减少疾病传染、提高人体免疫力
乡村田野	1000～3000	提高人体免疫力、抗菌力
城市公园	500～1000	提高人体健康
城市绿化带	200～400	提高人体健康功能较弱
城市住宅	100	诱发生理障碍，如头痛、失眠等
商业办公室	0～50	诱发生理障碍，如头痛、失眠等
工业区	0	诱发多种疾病，损害人体健康

Qixi Town | Liyang tian-Grand Canyon | Canyon and Waterfall Experience Route

Interpretation No. QX-DXG-05

Stop 05: Mo Pond

Interpretation theme

Forest oxygen bar in Qianjiangyuan National Park

Key points

1. Evaluation standard of natural oxygen bar in China
2. Reasons for the high negative oxygen ion in Qianjiangyuan

Site overview

By the stream, where you can feel the fresh air in the mountain valley

"Evaluation criteria of natural oxygen bar in China"

When in Qianjiangyuan National Park, do you feel that the air here in Qianjiangyuan National Park is extra fresh? We are now bathing in a natural "oxygen bar". But what is an oxygen bar? Is the oxygen content here higher? Congratulations, you are half correct. The oxygen bar actually means that the concentration of negative oxygen ions in the air is very high and reaches a certain standard.

What is negative oxygen ion? We will not explain it for the time being and leave it to the chemistry teacher. What we need to know is that negative oxygen ions are very beneficial to human health, also known as "longevity element" or "air vitamin". It is said that the concentration of negative oxygen ions in many longevity villages in China is very high.

Negative oxygen ions can eliminate smoke and dust, purifying the air. The higher the concentration of negative oxygen ions in the air, the better the air quality is.

The negative oxygen ion content of general urban residences is about $100/cm^3$. The World Health Organization (WHO) stipulates that it can be regarded as fresh air when the concentration of negative oxygen ions in the air is not less than 1000 – 1500 per cubic centimeter. Can you guess the amount of negative oxygen ions in Qianjiangyuan National Park? It is about three times the clean air standard given by WHO, or 39 times that of urban dwellings. As a result, Kaihua County was named the first place in the first batch of "China Natural Oxygen Bar" as early as 2016.

"Why is negative oxygen ion high in Qiangjiangyuan"

Why did Kaihua County get this honor? This Should start from the generation mechanism

of negative oxygen ions.

The formation of negative oxygen ions requires some specific conditions. Lightning, photosynthesis, waterfalls, fountains and aromatic substances secreted by some plants in nature can promote the production of negative oxygen ions.

Let's take a look at the Grand Canyon around us. Not only are there endless streams and magnificent waterfalls, but also the dense virgin forests have created excellent conditions for the generation of negative oxygen ions. At the same time, the dust-holding effect of the forest and splashing water can also reduce the loss of negative oxygen ions. In the case of more output and less loss, naturally the content of negative oxygen ions here is also higher, and the air is cleaner, which is more conducive to human health.

Attentions

- First please tell everyone to take a deep breath and then introduce the concept of forest oxygen bar.
- Conduct different interpretations of negative oxygen ions based on different audiences.
- Compare the difference between the air index of your home and Qianjiangyuan National Park (establish a connection with your surroundings).
- The high concentration of negative oxygen ions is inseparable from the unique conditions of Qian Jiangyuan.

Recommended auxiliary facilities and toolkits

- Interpretation signage: natural oxygen bar
- Concentrations of negative oxygen ion in various environments

Extra information

"Negative oxygen ion"

- Negative oxygen ion, also known as the "air vitamin", are beneficial to human health. The oxygen molecules in the air are easily ionized and generate negative oxygen ions under the induction of waterfall impact, ultraviolet radiation, lightning and so on. The faster the water flows, the stronger the effect, so the concentration of negative oxygen ion in the air is relatively high near the waterfall among the forest.

齐溪镇 | 里秧田—大峡谷 | 遇见峡谷飞瀑自然体验路线　　解说编号：QX-DXG-06

解说点 06 思源亭

29°23′43.94″ N，118°13′5.93″ E
海拔：528m

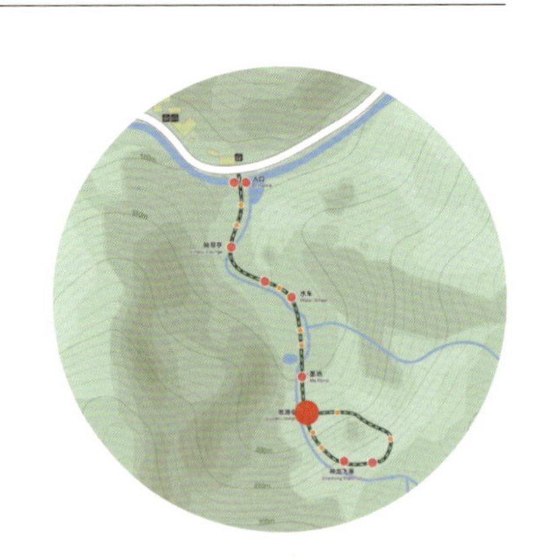

解说主题

石英脉的成因与历史见证

解说要点

1. 观察岩石，了解岩石的研究意义
2. 了解岩石见证的造山运动

场地概述

花岗岩石块，上有石英脉

解说方式 Interpretation Style

☑ 人员解说 Staff Interpretation　☐ 教育活动 Educational Activity
☑ 解说牌 Interpretation Signage　☐ 场馆解说 Hall Interpretation

解说季节 Interpretation Season

☑ 春季 Spring　☑ 夏季 Summer
☑ 秋季 Autumn　☑ 冬季 Winter

如果说峡谷的形成是山水共同作用的结果，那么一路上我们看到的大大小小各色各样的石头，则是亿万年来地质演变的见证者。

「观察岩石，了解岩石的研究意义」

可别小瞧你遇到的每块石头，它们动辄可能就已经是一位千万岁的老人了。我们眼前的这块大石头，就是其中一位。

请大家仔细观察一下它，然后告诉我你们的发现。它和我们常见的石头有什么不一样吗？表面有非常明显的纵横交错的灰白色纹理。那么，石头本来的颜色呢？由于布满苔藓和地衣，我们已经很难看清这块石头的本来面目了。不过，不要担心，我这里有一块石头可以给大家观察（可以提前准备一块未被风化的花岗岩，给访客展示）。从这块石头可以看出，它的质地坚硬且花纹细密，属于典型的花岗岩。上面那一条条白色的纵横交错的线条叫作石英脉，主要成分是石英，也叫作二氧化硅。石英是什么？你一定见过的。玻璃的成分就是石英。除此之外，我们经常看到的细沙中那些透明的颗粒也是由石英构成的。

「探求岩石见证的造山运动」

为什么一块石头上会出现两种截然不同的成分呢？背后其实是它们不同的形成过程。花岗岩是一种典型的岩浆岩，换句话说是由地底下的岩浆直接冷却形成的岩石。冷凝后的岩浆岩被抬升的地壳带出地表，这就是我们看到的花岗岩。在花岗岩形成之后，位于地壳更深处的岩浆熔化的石英渗入岩石的裂隙，冷却结晶之后，便形成了石英脉。

我们眼前这块填充着石英脉的花岗岩，可能就与 7000 万年前的地质运动有关，是久远地质历史的见证。

注意事项

- 注意观察石头时候提醒访客注意安全，且不要破坏石英脉。
- 思源亭是一个休息点，接着就要走上一段比较陡的铁梯，去观察瀑布和滴水崖壁等景观。因此，要提醒访客做好身体和心理上的准备。

辅助道具推荐

- 标识牌：岩石见证造山运动
- 捡拾的带石英脉的花岗岩石块

拓展知识点

- 花岗岩，英文名称是从拉丁文 *granum* 演绎而来，意指颗粒。而“花岗岩”一词是指“花”纹美丽又质地“岗”强的“岩”石。花岗岩经常用作建筑材料。

Qixi Town | Liyangtian Canyon | Canyon and Waterfall Experience Route

Interpretation No. QX-DXG-06

Stop 06: Siyuan Lounge

Interpretation theme

The formation and history of quartz veins

Key points

1. Observe the rock and understand the significance of rock research
2. Explore the orogeny witnessed by the rock

Site overview

The granite filled with quartz veins

Dear friends, along the way we have seen stones of different sizes and colors. They look ordinary, but these stones may easily have a history of hundreds of millions of years.

"Rock observation"

Please follow me to have a look on this big rock in front of us.

Look at it carefully and tell me what you have found. Is there any difference with common stones? Yes, there are obvious grey and white interlaced texture on the surface of this rock. What about the original color of the rock? Being covered by moss and lichen, we are difficult to see the true nature of the stone, but don't worry, I have prepared one for your observation (please bring a piece of un-weathered granite and show it to the visitors). We can see the stone is very hard with fine decorative pattern, belonging to the typical granite. The white crisscrossed lines above the surface are called quartz veins. Its main ingredient is quartz, also known as silica. What is quartz? You must have seen it. Glass is made of quartz. In addition, the transparent grains we often see in fine sand are also made of quartz.

"The orogeny witnessed by the rock"

Why do two very different ingredients appear on the same stone? That is because of their different formation processes. Can you guess which one was formed first, the granite or the quartz veins? Granite is a typical magmatic rock, in other words, this kind of rock was formed by direct cooling of magma from underground. The condensed magmatic rock was brought to the surface by the uplifted crust, where we see granite. After granite formed, magma deeper in the earth's crust melted quartz into cracks in the rock, cooled and crystallized, and finally turned to the quartz veins we see today.

The quartz veins of the granite in front of us may be related to the geological movement 70 million years ago, which is a witness of geological history.

Attentions

- Remind the tourists to pay attention to safety when observing stones.
- Remind tourists not to damage the quartz veins.
- After this stop, there is a steep and slippery trail to climb up to view the waterfall and the dripping cliffs. Remind the tourists to be prepared and careful.

Recommended auxiliary facilities and toolkits

- Interpretation signage: witness of orogeny
- A picked granite stone with quartz veins

Extra information

- Granite's English name is derived from the Latin word *granum*, meaning grain. And the word "granite" refers to the strong rocks with beautiful texture. Granite is often used as a building material.

齐溪镇 | 里秧田—大峡谷 | 遇见峡谷飞瀑自然体验路线　　解说编号：QX-DXG-07

解说点 07 峡谷瀑布

29° 23′ 42.25″ N，118° 13′ 6.05″ E
海拔：590m

● **解说主题**

峡谷瀑布的形成

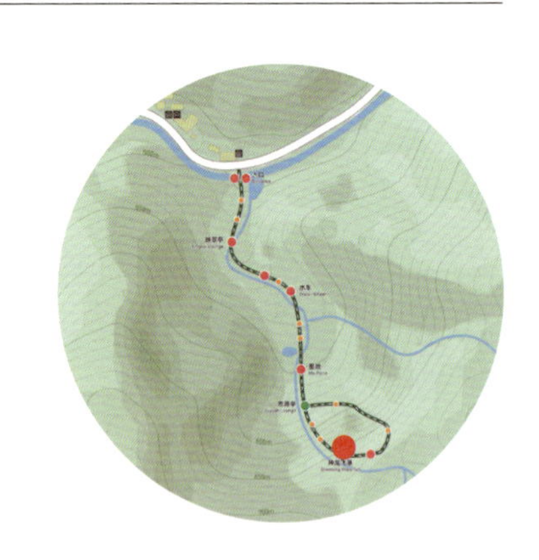

● **解说要点**

1. 认识峡谷地势与瀑布
2. 了解彩虹的形成原理

● **场地概述**

攀登的阶梯，较陡，有数个观景平台

解说方式 Interpretation Style

☑ 人员解说 Staff Interpretation　☐ 教育活动 Educational Activity

☑ 解说牌 Interpretation Signage　☐ 场馆解说 Hall Interpretation

解说季节 Interpretation Season

☑ 春季 Spring　☑ 夏季 Summer

☑ 秋季 Autumn　☑ 冬季 Winter

解说词

「认识峡谷飞瀑的形成与地形、流水的关系」

大家有没有发觉，顺着峡谷往上，水流越来越急、落差也越来越大、水流声音也越来越响亮。在水流落差较大的地方，容易形成瀑布。虽然沿途我们见过大大小小层层叠叠的瀑布，但是眼前的瀑布景观还是让大家忍不住赞叹！

这条瀑布发源于钱塘江源头主峰莲花尖，水流几乎呈90°直角垂直落下，落差超过120m，似一条飞龙从天上倾泻而下，因此本地人也称之为“神龙飞瀑”。有句俗话说得好“神龙见首不见尾”，如果说我们眼前这个瀑布是神龙的头，那么龙尾在哪儿呢？顺着瀑布往下，溪水流经茶园，流经长柄双花木群落，再到大峡谷的入口，最后会去往哪里？没错，正是钱塘江。

瀑布的形成是古老的山体运动与常年流水共同作用的结果，反过来，瀑布也在不断重塑峡谷地形。受到重力作用的影响，来自高处的瀑布水流会不断冲击、侵蚀河床，使得河床下切，这才形成了我们所在的这个“V”形峡谷。

瀑布之上是什么呢？大家一起来看下我手中这张航拍图，瀑布上方曾经是片农田，今天为了国家公园的保护已经渐渐恢复为森林，再次发挥其水源涵养地的功能。

「飞瀑彩虹」

（如果有彩虹，可以结合现象解说。）

大家平时看到过彩虹吗？（如果有的话，）是在什么情况下看到彩虹的呢？一般彩虹主要出现在雨后，这时候空气中有大量的小水滴且灰尘较少，太阳光会被反射或者折射，比较容易形成彩虹。但由于现在城市环境污染比较严重，我们已经很难在城市见到彩虹了。

不过，在大峡谷的瀑布附近随时都有充沛的水汽，因此只要天气晴好，有阳光直射，就很容易出现彩虹。只要我们做个有心人，一定有机会和彩虹来一次浪漫邂逅的。

注意事项

- 提醒游客攀爬时注意安全。
- 选择合适的地点开展人员解说，避免拥堵，注意安全。
- 关于彩虹的解说可以结合现场情况，选择性解说。

辅助道具推荐

- 标识牌：峡谷间的彩虹光影
- 无人机航拍的瀑布上方照片

Qixi Town | Liyangtian-Grand Canyon | Canyon and Waterfall Experience Route

Interpretation No. QX-DXG-07

Stop 07: Canyon Falls

● Interpretation theme

Formation of Canyon falls

● Key points

1. Get to know the waterfall and its formation
2. Introduce the formation of waterfall rainbow

● Site overview

Steep climbing stairs, where there are several landscape platforms

"Understanding the formation of Canyon Waterfall"

Have you noticed that as you go up the canyon, the water flow becomes more and more rapid, the drop becomes larger and larger, and the sound of the water flow becomes louder and louder. The sound that everyone now hears comes from a waterfall in front of us. Let's take a look together.

The waterfall is originates at Lianhua Tip, the main peak at the source of the Qiantang River. The water flows down almost at a 90° right angle with a drop of more than 120 meters. It looks like a flying dragon pouring down from the sky, so locally it is known as the "Magic Dragon Waterfall". As the saying goes, "One can see the head of a heavenly dragon, but not the tail." If we say that this waterfall is the head of the dragon, then where is the tail? Down the waterfall, past the tea plantation, past the home of rare and relic species, to the entrance of the Grand Canyon. Where does the stream end? Yes, it is the Qiantang River.

Waterfalls can be formed under two circumstances: first, the terrain needs to be steep with enough height to let the water drop, and second, there is enough water supply for the whole year. It is the complex mountainous terrain of Qianjiangyuan and sufficient precipitation that together created this waterfall. In turn, the waterfall is constantly reshaping the canyon terrain. Under the influence of gravity, the waterfall flow from high places will continue to impact and erode the river bed, undercutting the river bed, which forms the V-shaped canyon where we are.

What is above the waterfall? Let's take a look at this aerial picture in my hand. There was farmland above the waterfall. Today, for the protection of the national park, it has gradually

been restored to a forest, and once again plays the role of water conservation.

"Waterfall rainbow"
(If there is a rainbow, you can combine the phenomenon to interpret.)

Have you ever seen a rainbow? (If any) Under what circumstances did you see the rainbow? Generally, rainbows mainly appear after rain. At this time, there are a large number of small water droplets in the air and less dust. Sunlight will be reflected or refracted, and it is easier to form a rainbow. However, due to the serious environmental pollution in the city, it is difficult for us to see the rainbow in the city.

However, there is plenty of water vapor at any time near the waterfalls in the Grand Canyon, so as long as the weather is fine and the sun is shining towards the waterfall, it is easy for rainbows to appear. However, this requires us to have full patience and careful observation, so as to have the opportunity to encounter beautiful scenery.

Attentions

- Remind tourists to pay attention to safety when climbing.
- Choose a suitable place to conduct interpretation. Avoid the crowd and be safe.
- Interpretation of the rainbow can be selected according to the situation.

Recommended auxiliary facilities and toolkits

- Interpretation signage Rainbow beside the Waterfall.
- Aerial photo above the waterfall

齐溪镇 | 里秧田—大峡谷 | 遇见峡谷飞瀑自然体验路线　　解说编号：QX-DXG-08

解说点 08 滴水崖壁

29°23′42.25″ N，118°13′8.91″ E
海拔：602m

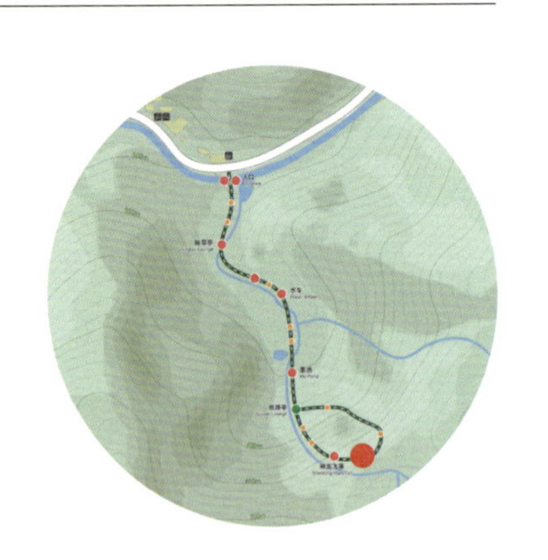

● 解说主题

滴水崖壁的生境与生物

● 解说要点

1. 认识滴水崖壁生境
2. 滴水崖壁上的独特植物

● 场地概述

攀登的阶梯，较陡，沿途有数个观景平台

解说方式 Interpretation Style

- ☑ 人员解说 Staff Interpretation
- ☐ 教育活动 Educational Activity
- ☑ 解说牌 Interpretation Signage
- ☐ 场馆解说 Hall Interpretation

解说季节 Interpretation Season

- ☑ 春季 Spring
- ☑ 夏季 Summer
- ☑ 秋季 Autumn
- ☐ 冬季 Winter

「观察滴水崖生境」

大峡谷地区充沛的降水和溪流、森林一起营造出一种湿润的环境。沿途大大小小的石头、树干上都长满了地衣、苔藓和蕨类等耐阴的植物。我们步道旁的这块大石头因为常年有流水的浸润，一直保持这种湿润的状态，被称为“滴水崖壁”。

形成滴水崖壁的石头成分为花岗岩，这种石头我们刚刚有提到过，因为是岩浆直接冷却形成的，所以质地坚硬，表面光滑平直，不易被风化。我们面前的这处滴水崖壁的侧面近乎垂直于地面，所以要想固定住土壤很难，再加上流水的冲刷，这里的土壤就更薄了。因此，对每一个选择在这里定居的生物来说，都是不小的挑战。

「滴水崖壁上的各种生物」

即使在这样贫瘠湿润的环境中，仍然生活着数十种生物。它们之所以能在这里定居，离不开“拓荒者”苔藓的功劳。作为植物界的“小矮人”，苔藓小小的身材里可是蕴藏着大世界。它们常常结伴生长，具有强大的吸水和除尘能力，同时在生长过程中会分泌酸性代谢物，促进岩石的风化。经过漫长的岁月，这些微小作用的积累会在原来裸露的崖壁上逐渐形成最初的土壤，成为减缓地表径流、防止风蚀和水土流失的保护层，从而让其他植物的生长成为可能。

另外，在潮湿的崖壁上常常会看到攀爬着一片片心形或戟形的绿色叶片。仔细观察，在叶片和叶柄连接处有个黄豆般大小的圆球，就像水珠滴上去了一样，这就是我国特有的植物——滴水珠。但这可不是真正的水珠，而是植物的育儿袋，脱落后就可以长成新的植株。这种独特的繁殖方式，是对潮湿、郁闭的生长环境的一种适应。

滴水崖壁上生长着一种身形迷你的食虫植物，它有一个非常可爱的名字 —— 圆叶挖耳草。这种植物看起来毫不起眼：叶片和花都很小，看起来非常娇弱纤细，惹人爱怜。但这可是一种地地道道的食虫植物，在它们普通的外表之下暗藏着杀机 —— 捕虫囊。如果不仔细观察，你甚至都看不到它们茎上的这些小球，这些实际上是为捕捉昆虫而布下的陷阱。

注意事项

- 滴水崖壁的植物种类有很多，除了苔藓作为必讲植物之外，其他植物可以根据实际情况进行选择性解说。
- 其他滴水崖壁植物除了解说词中提到的滴水珠和圆叶挖耳草之外，还有秋海棠、小沼兰等都是可以解说的对象，具体可以参考通用型生物多样性解说方案。

辅助道具推荐

- 标识牌：滴水崖壁上的自然天地

拓展知识点

「食虫植物」

- 能够捕虫的“天分”其实是食虫植物适应环境的一种生存智慧。在滴水崖壁这样潮湿、阳光微弱、土壤相对浅薄贫瘠的环境中生长的植物，往往可能缺少氮素等营养，于是，有些植物就演化出会捕获并消化动物的能力。达尔文早在 1875 年就发表了第一篇关于食虫植物的论文。现在人类已发现的能够捕捉动物并产生消化酶分解猎物再吸收其营养素的食虫植物超过了 600 种。

Qixi Town | Liyangtian-Grand Canyon | Canyon and Waterfall Experience Route

Interpretation No. QX-DXG-08

Stop 08: Dripping slope

Interpretation theme

Habitats and mysterious wildlifes in the dripping cliff

Key points

1. Understand the habitats in the dripping cliff
2. Unique plants in the dripping cliff

Site overview

Steep climbing stairs, where there are several Landscape platforms

"Observing the dripping slope"

Abundant rainfall combines with the streams and forests in the Canyon have created a humid environment. Thus, we can see along the trail almost all the stones and tree trunks are covered with shade plants such as lichens, mosses and ferns. Because of the constant water dripping, the rock beside the trail remains wet that it is called "wet and dripping slope".

The rock that forms the dripping slope is mainly made of granite, which is formed by direct cooling of magma. It's hard, smooth and straight in surface. It's also not susceptible to weathering. The slope in front of us are almost vertical, so you can imagine it's not easy to hold any soil on it. With the dripping water, the remained soil is even thinner. So it's a challenge for every creature that chooses to settle here.

"Biodiversity on the wet slope"

Even in such a barren and humid environment, there are still quite many creatures living on the rock slope, and the reason why they can settle here is inseparable from the contribution of the "pioneer" moss. As a "little dwarf" in the plant kingdom, the small body of moss contains the big world. They often grow together, having strong water absorption and dust removal capabilities, and secreting acidic metabolites during the growth process to promote the weathering of rocks. After a long period of time, the accumulation of these tiny effects will gradually form the initial soil on the original bare slope wall, which is also a protective layer to slow down the surface runoff and prevent wind erosion and soil erosion, thereby making the growth of other plants possible.

In addition, heart-shaped or halberd-shaped green leaves are often seen climbing on the damp cliffs. If you look closely, there is a bean-sized ball at the junction of the leaf and petiole, just like a water drop. It is the plant endemic to China — miniature green dragon. But this is not a real water drop, but a plant nursery bag, which can grow into a new plant after shedding. This unique gift of reproduction is an adaptation to this kind of humid and enclosed environment.

There is a miniature carnivorous plant growing on the wet slope, and it has a very cute name, i.e. striped bladderwort. This plant looks inconspicuous: their leaves and flowers are all very small with common color, which make it looks very delicate and fragile. But beneath their ordinary appearance, there is a hidden killer — the insect trap. If you don't observe carefully, you can't even see these small balls on their stems, which are traps laid for catching insects.

Attentions

- There are many types of plants on the drip cliffs. In addition to moss as a required plant, other plants can be selectively interpreted according to the actual situation.
- In addition to the miniature green dragon and striped bladderwort mentioned in the interpretation, other drip cliff plants, including hardy begonia, strawberry saxifrage, rattlesnake orchids and peacock orchid, can also be interpreted. Refer to the "General Interpretation of Biodiversity" for more details.

Recommended auxiliary facilities and toolkits

- Interpretation signage: micro-world on the wet rocks

Extra information

- The "genius" of carnivorous plants is actually a kind of survival wisdom for them to adapt to the environment. Plants growing in a humid, weak sunlight, relatively shallow and barren environment like drip cliffs often lack nutrients such as nitrogen, so some plants have evolved the ability to capture and digest animals. Darwin published his first paper on carnivorous plants as early as 1875. There are now more than 600 species of discovered carnivorous plants that can capture animals and produce digestive enzymes to break down prey and absorb its nutrients.

齐溪镇 | 里秧田—大峡谷 | 遇见峡谷飞瀑自然体验路线　　解说编号：QX-DXG-09

解说点 09 结束集散点

29°23′55.33″ N，118°13′2.44″ E
海拔：480m

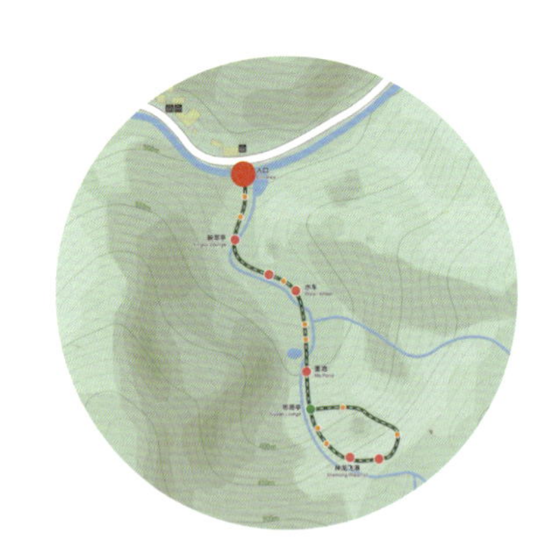

解说主题

本条游线总结与其他游线推荐

解说要点

1. 分享环节
2. 游线总结

场地概述

与步道的起点相同，延伸到周边的后山湾村与莲花溪

解说方式 Interpretation Style

- ☑ 人员解说 Staff Interpretation
- ☐ 教育活动 Educational Activity
- ☑ 解说牌 Interpretation Signage
- ☐ 场馆解说 Hall Interpretation

解说季节 Interpretation Season

- ☑ 春季 Spring
- ☑ 夏季 Summer
- ☑ 秋季 Autumn
- ☑ 冬季 Winter

解说词

各位朋友，我们的行程到这里就要结束了。虽然只有短短的一个多小时，但是相信大家此行的收获一定不小。

「游线总结」

亿万年前的造山运动和水流的不断冲刷共同铸就了我们今天看到的山高谷深的地形地势。溪水潺潺从高处流下，任由石头雕琢成不同的形态，有瀑布，有跌水，有湍流，也有深潭。溪流也是鸟类、昆虫、鱼类、植物等生物和谐共生的家园，在这里我们不仅可以看到雌雄异色的红尾水鸲、活泼好动的燕尾，还可以看到贴在水底捕食的虾虎鱼，在溪流中成群游动的石斑鱼。同时，钱江源国家公园还为长柄双花木等珍稀植物提供了栖身之所，使得它们得以捱过历史上的气候波动，存活至今。

一条短短的游线，就有那么多自然故事值得我们去发现。我们无法在短时间内看到国家公园的全貌，但可以将大峡谷线看作它的一个缩影，去探索看似平凡普通的石头、溪流背后的世界。

「其他游线推荐」

除了我们今天体验的这条游线之外，钱江源还有很多地方值得大家去探索。如果要继续玩水的话，建议你一定不能错过同样位于齐溪镇的莲花塘钱江源寻源之旅。当然，你也可以去苏庄镇的古田山走走，接受原始的森林氧吧的熏陶，体验一下当地的风土民情。

「游线分享」

接下来是我们的总结分享时间。我这里有两个简单的小问题想听听大家的答案。

1. 沿途令你印象最深的是什么？

2. 如果此行你打算带些东西回家与家人、朋友分享，那会是什么？

感谢大家的分享！这也给了我很多的启发。最后的小任务需要大家回去才能做。希望大家可以和家人朋友分享此行你的见闻，无论是照片还是亲自讲述，期待能把国家公园的精彩带给更多人。祝大家未来的旅途愉快，返程平安！

注意事项

- 召集大家集合，进行15分钟的小回顾，做些体力上的休整。
- 推荐其他游线。
- 邀请伙伴分享，并给予反馈，通过大家的反馈，延伸启发与思考，提出回去后的行为倡议。
- 感谢大家的到访，感谢大家聆听自己的解说（希望大家回去后向身边人宣传钱江源国家公园）。

Qixi Town | Liyangtian- Grand Canyon | Canyon and Waterfall Experience Route

Interpretation No. QX-DXG-09

Stop 09: Assembly ending Point

Interpretation theme

Summary and recommendation of other routes

Key points

1. Sharing
2. Route summary

Site overview

The same as the starting point of the trail, extending to the surrounding Houshanwan Village and Lotus Creek

Dear friends, our journey is coming to an end. Although it's only a short hour, we have achieved a lot.

"Travel summary"

The mountain-building movement hundreds of millions of years ago and the continuous scouring of the water flow together have created the deep terrain of the mountains and valleys we see today. The gurgling water flowed down from the mountain, and the stones were carved into different postures: there were waterfalls, there were falls, there were turbulences, and there were also deep pools. Streams are also home to the harmonious coexistence of birds, insects, fish, plants and other creatures. Here we can see not only dioecious red-tailed waterfowls, lively forktails, but also gobies attached to the bottom of water and hunter fishing spiders lying on the stone. At the same time, Qianjiangyuan National Park also provides shelter for rare plants such as *cercidifolius* var. *longipes* trees, allowing them to survive historical climate fluctuations until this day.

In a short tour, there are so many natural stories worth discovering, and the Canyon and Waterfall Nature Experience Route can be seen as a miniature. We cannot see the whole picture of the national park in a short time, but you can go through this tour to go behind the seemingly ordinary stones and habitats, there is an infinite world worth exploring.

"Recommended other tours"

In addition to the tour we took, Qian Jiangyuan has many places worth exploring. If you want to continue to play with water, you must not miss the source hunt for the Qianjiang source of the Lotus Pond, also located in Qixi Town. Of course, you can also go to Gutian Mountain

in Suzhuang Town to experience the original forest oxygen bar and the local customs.

"Travel sharing"

Next, is our summary sharing time. I have two simple questions here for everyone.

1. What did impress you the most along the way?

2. Share a new knowledge you learned on this trip.

Thank you all for sharing, that gave me a lot of inspiration. The last small task requires everyone to go back and do it. I hope you can share your experience with family and friends. Whether it is a photo or a personal story, I look forward to bringing the wonderful national park to more people. I wish you all a pleasant journey in the future and a safe return home!

Attentions

- Gather everybody, organize a 15-minute quick review, and take a break.
- Recommend other routes.
- Lighten inspiration and propose the environmental protection strategy.
- Thank them all for visiting and listening to your interpretation. (Wish everyone popularize the Qianjiangyuan National Park after returning home.)

03

通用型人员解说方案

General Interpretation

3.1 通用型生物多样性解说方案

General Interpretation of Biodiversity

认识钱江源国家公园的生物多样性

钱江源国家公园是万千生灵的家园。根据2018年有关钱江源国家公园的生物多样性调查报告，钱江源国家公园共有高等植物2062种，其中，种子植物1561种，蕨类植物176种，苔藓植物325种；已记录哺乳动物58种，鸟类237种，两栖爬行类77种和昆虫1156种，其中包括国家重点保护野生动物45种（其中，国家一级重点保护野生动物有黑麂、白颈长尾雉和穿山甲3种，国家二级重点保护野生动物有白鹇和黑熊等42种）。

据此，带领访客认识钱江源地区丰富的生物多样性，是钱江源国家公园解说工作的重要目标。从古老的孑遗植物到当地的本土物种，从微小的苔藓、地衣到高大的乔木，从备受关注的国家级保护物种，到随处可见的鸟类和昆虫……通过我们的解说为访客们呈现一个生机勃勃的自然世界。你可以带领访客穿越丛林，观察各种植物，并期待与林间出没的兽类和悠闲散步的雉鸡来一场偶遇，或顺着鸟鸣声寻找活跃在枝叶间隙的飞羽精灵，观察步道边的林下植被；也可以沿溪而行，在水边驻足，观赏水中的游鱼、两栖爬行动物和河岸边觅食的鸟类。

时间也是开展物种解说的重要线索。当夕阳落下，暗夜星空大幕拉开，夜晚的自然观察也是开展解说的好时机。在这里，你不仅可以带领访客听到蛙类和夏虫的合唱，或许还能与萤火虫来一次浪漫的邂逅，甚至一睹夜间狩猎者——猫头鹰的卓越捕食技巧。此外，笔者还挑选了一些比较有故事性和趣味性的生境和生态学概念使解说更加生动，同时也可以帮助解说员更加多角度地为访客讲述钱江源国家公园多样的物种之间以及物种与环境之间的关系。

请记住，国家公园的一草一木都受到保护，在解说中如果需要实物演示，请提前准备照片、标本、模型或捡拾自然物品。绝对不能随意采摘植物或捕捉动物，以免作出错误的示范，也要特别避免过多强调某些物种特别是保护物种的经济、药用等相关利用价值，以免对访客行为造成不必要的误导。

最后，请不要忘记我们对物种的解说不仅仅是一种知识的传递，更是希望访客在对物种更了解的基础之上，能够更加关注、珍视钱江源当地的生物多样性，进而愿意用实际行动保护更广义范围的野生动物和它们赖以生存的自然生态系统。

注：和自然体验路线不同，通用解说方案中的生物多样性部分的使用相对比较灵活，每一个解说点既与其他内容有关联又可以单独作为一个独立的解说内容，解说员在使用的时候可以根据情况灵活处理。

Understand the biodiversity of Qianjiangyuan National Park

In Qianjiangyuan National Park, the warm and humid climate, continuously distributed subtropical evergreen broad-leaved forests of large area, dense streams and abundant water resources and other conditions together have combined to shape the original nature here and make it home to thousands of creatures. According to current statistics, there are 2062 species of higher plants in the national park, including 1561 species of seed plants, 176 species of ferns and 325 species of bryophytes; 58 species of mammals, 237 species of birds, 77 species of amphibians and reptiles and 1156 species of insects have been recorded, including 45 species of state-level key protected wild animals (including 3 species of state first-level key protected wild animals including black muntjacs, Elliot's pheasants and pangolins, and 42 species of state second-level key protected animals including silver pheasants and black bears).

Guiding visitors to understand the rich biodiversity of Qianjiangyuan area is an important goal of the interpretation work of Qianjiangyuan National Park. From ancient relic plants to local native species, from tiny moss, lichens to tall trees, from state-level key protected species to birds and insects that can be seen everywhere… through our interpretation, we will present a vibrant natural world to the visitors. You can lead visitors crossing through the jungle, observe various plants, and look forward to encountering with land animals and pheasants walking leisurely in the forest, or follow the sound of birds to find the flying feathery elves active between branches and leaves, and observe the vegetation on the edge of the trail; you can also walk along the stream and stop by the water to watch the swimming fish in the water, the amphibians and reptiles, and the birds foraging on the river bank.

Time is also an important clue to species interpretation. When the sun sets, the dark night starry sky is on, it is a perfect time for night time natural observation interpretation. Here, you can not only lead visitors to listen to the chorus of frogs and summer insects, but also have a romantic encounter with fireflies, and even see the excellent hunting skills of night hunters, owls. In addition, we also selected some interesting habitats and ecological concepts to make the interpretation more vivid, and at the same time, it can also help the interpreter to explain the diverse species of Qianjiangyuan National Park from multiple angles, and the relationship between species and environment for visitors.

Please remember: every plant and tree in the national park is protected. If we need a physical demonstration during the interpretation, please prepare photos, specimens and models, or pick up natural objects in advance. It is absolutely not allowed to pick plants or capture animals at will for demonstration purposes, and avoid over emphasis on the economic and medicinal value of certain species especially protected species, so as not to cause unnecessary misleading to visitors.

Finally, please don't forget that our species interpretation is not only a transfer of knowledge, but also a hope that visitors can

pay more attention and cherish the local biodiversity of Qianjiangyuan and take practical actions to protect a wider range of wild animals and the natural ecosystems that wild animals depend on for survival.

Notes: Unlike the nature experience routes interpretation, the biodiversity section of general interpretation is relatively flexible. Each Interpretation material is not only related to other content, but also can be used as an independent interpretation content. The interpreter can flexibly handle it according to the situation when using it.

通用型解说 | 生物多样性解说　　　　解说编号：TY-WZ-01

01 四季常绿的阔叶树

● 解说要点

1. 从整体上认识钱江源国家公园的常绿阔叶林
2. 识别公园常见的常绿阔叶树：甜槠、木荷

● 观察地点推荐

1. S01 古田飞瀑：丛林飞瀑自然体验路线
2. S02 古田山—古田庙：探秘常绿阔叶林自然体验路线
3. S03 瞭望台：国家公园科研体验路线

木荷　木荷的咖啡色蒴果　甜槠林　甜槠果实

解说方式 Interpretation Style

☑ 人员解说 Staff Interpretation　☐ 教育活动 Educational Activity
☑ 解说牌 Interpretation Signage　☐ 场馆解说 Hall Interpretation

解说季节 Interpretation Season

☑ 春季 Spring　☑ 夏季 Summer
☑ 秋季 Autumn　☑ 冬季 Winter

解说词

「什么是常绿阔叶林」

大家在来的路上有远远地观察过钱江源国家公园的森林吗？如果有的话，你一定注意到了漫山遍野的树木似乎给大山盖了一层厚厚的绿色的“毛毯”。放大来看，每一个树冠似乎都被大自然的巧手精心修剪过：顶部浑圆，像一朵朵错落有致、色彩略异的“花椰菜”……这些常绿阔叶林中树木树冠的造型可以帮助植物最大限度地利用空间并接收阳光。

钱江源国家公园海拔 800m 以下的空间几乎全部被常绿阔叶林占据。常绿阔叶林是世界上生物种类最丰富的地区之一，从地被的苔藓到中层的灌木再到高大的林冠，森林的空间被充分利用，几乎毫无浪费。有些乔木不仅在数量上占据优势，同时树冠高度也位于森林的顶层，我们把它们称为森林的优势种。钱江源国家公园常绿阔叶林中常见的优势种有甜槠、木荷、青冈等。

「常见的常绿阔叶树：甜槠、木荷」

金秋时节，行走在钱江源国家公园的森林里，经常可见散落在地上的坚果。如果你看到一种很像电影《冰河世纪》中松鼠最喜欢囤积的橡子，有可能就是甜槠子。这种果实和板栗很像，也穿着一层刺猬状的苞衣，摸起来还有些扎手，但里面包裹着的甜槠果实却光滑无比，呈锥形，当地人将其称为甜槠子，味道十分甜美，是很多浙江人最难忘的童年味道之一。和大多数常绿阔叶树种一样，甜槠拥有椭圆形的革质叶片，表面有明显的光泽。每年的 4～6 月是甜槠的花季，当鹅黄色的花朵一起绽放，铺满整个树冠时，整棵树似乎都变成了一个花球。

另一种优势乔木——木荷拥有椭圆形的大叶片摸起来也如皮革般光滑，表面有一层薄薄的蜡质，可以有效地锁住水分。木荷是森林天然的“消防员”，整个植株的含水量超过 50%。不过大家不要轻易撕开木荷的叶子去观察哦，因为木荷汁液有微毒，如果不慎碰到，可能会引起皮肤不适。木荷的花为优雅的白色，与黄色的花蕊相映成趣。五枚花瓣中有一枚和其他不太一样，像风衣后面的帽子一样，所以也被称为“风帽花”。此外，木荷的果实也很特别，咖啡色，质地较坚硬，完全裂开前，顶端的裂缝颇似一颗五角星。

注意事项

- 从访客对森林的初步印象开始讲起，与他们的经验相连接。
- 可以邀请访客参与的部分：告诉你海拔高度，描述植被、植物特征。
- 可以根据情况，选择跳过前半部分，直接解说实际遇见的常绿阔叶树种，也可以根据情况，选择其他树种解说。

辅助道具推荐

- 捡拾的甜槠、木荷叶片、花朵或果实（根据季节定），或对应物种的照片

拓展知识点

- 常绿阔叶林在世界各地均有分布，但其中最为典型的，却在中国。为什么会这样呢？这是因为森林的生长离不开气候，而气候类型主要是由地理位置决定的。中国的北纬 30° 附近，由于受到亚热带季风气候的影响，夏季高温多雨、雨热同期，为常绿阔叶林的生长提供了极佳的条件，分布着世界上面积最大、生物多样性最高的常绿阔叶林。常绿阔叶林所在的区域也是我国人口密度最高的区域之一。几千年来，在人类与大自然的拉锯战中，森林节节败退，如今只在少数地区才能看到保存较为完好的原始森林。我们所在的钱江源国家公园，就是这样一片典型、珍贵的原真状态的常绿阔叶林。

General Interpretation | Biodiversity Section

Interpretation No. TY-WZ-01

01 Evergreen broad-leaved trees

Key points

1. Overall understand the evergreen broad-leaved forest of Qianjiangyuan National Park
2. Identify common evergreen broad-leaved trees in the park: the sweet oachestnut, the gugertree

Observation places recommended

1. S01 Jungle and waterfall experience route
2. S02 Evergreen broad-leaved forest exploration route
3. S03 Scientific research experience route

"What is an evergreen broad-leaved forest"
Did you observe the forests of Qianjiangyuan National Park from a distance on your way here? If so, you must have noticed that the trees cover all over the mountains like a thick green blanket. When you look closer, every tree canopy seems to have been carefully trimmed by nature: the top is round, like a bunch of "broccoli"··· The shape of the tree canopy in the evergreen broad-leaved forest can not only help plants maximize the use of space and sunlight, while at the same time it can effectively guide water to flow along the branches to the roots.

Almost all the space below 800 meters elevation in Qianjiangyuan National Park is occupied by evergreen broad-leaved forests. Evergreen broad-leaved forest is one of the regions with the richest biological species in the world. From moss on the ground to shrubs in the middle layer then to tall trees, the space of the forest is fully utilized and there is almost no waste. Some trees not only dominate in number, but also in height. We call them the dominant species of the forest. The common dominant species in the evergreen broad-leaved forests of Qianjiangyuan National Park are sweet oachestnuts, gugertree, ring-cupped oaks, *etc.*.

"Common evergreen broad-leaved trees: sweet oachestnuts, gugertrees"
In the autumn season, when walking in the forest of Qianjiangyuan National Park, you can often see nuts scattered on the ground. If you see a kind of acorn that squirrels like to hoard most in the movie *Ice Age*, it may be the sweet oachestnut. This kind of fruit looks very similar to chestnut, and it also wears a hedgehog-like bract, and it feels a bit pinchy to the touch, but the sweet oachestnut

fruit wrapped in it is very smooth with a cone-shaped. The locals call it the sweet oachestnut. Its taste is very sweet, which is one of the most memorable childhood tastes of many Zhejiang people. Like most evergreen broad-leaved trees, the sweet oachestnut has oval leathery leaves with obvious luster on the surface. The period from April to June each year is the flowering season of sweet oachestnuts. When the yellow flowers bloom together and cover the entire canopy, the whole tree seems to become a flower ball.

Another dominant species, the gugertree, has large oval leaves, which are as smooth as the leather and have a thin layer of wax on the surface. Its leaves can effectively lock in moisture. The gugertree is the natural "firefighter" of the forest, and the water content of the entire plant exceeds 50%. But don't tear leaves of the gugertree, because gugertree superba juice is slightly poisonous, if you accidentally touch it, it may cause your skin discomfort. Flowers of the gugertree are elegant white in contrast with the yellow stamens. One of the five petals is not the same as the rest, it looks like the hat behind the hoodie, so it is also called "the hood flower". In addition, fruits of the gugertree is also very special, with brown color and solid texture. Before the fruit is completely split, the crack at the top looks like a five-pointed star.

Attentions

- Start with the visitor's initial impression of the forest and connect with their experience.
- Invite visitors to participate: tell you the elevation, and describe characteristics of vegetation and plant.
- According to the situation, you can choose to skip the first half and directly interpret the evergreen broad-leaved tree species you actually encounter, or you can choose other tree species to interpret according to the situation.

Recommended auxiliary facilities and toolkits

- picked leaves, flowers or fruits (depending on the season) of sweet oachestnut and gugertree; or photos of the corresponding species

Extra information

- Evergreen broad-leaved forests are distributed all over the world, but the most typical one is in China. Why is that? This is because the growth of forests is inseparable from climate, and the type of climate is mainly determined by geographical location. The 30 degrees north latitude, the influence of the subtropical monsoon climate, hot and rainy summer time, provide excellent conditions for the growth of evergreen broad-leaved forests. It is the largest evergreen broad-leaved forest with the highest biodiversity in the world. The area where the forest is located is also one of the most densely populated areas in China. In the war of thousands of years between mankind and nature, the forest has been losing its survival ground. Nowadays, only a few areas can see relatively well-preserved virgin forest, the Qianjiangyuan National Park where we are located is such a typical and precious evergreen broad-leaved forest in its original state.

通用型解说 | 生物多样性解说 解说编号：TY-WZ-02

02 喜阳耐寒的针叶树

解说要点

1. 认识马尾松在群落演替早期作为“先锋树种”的作用
2. 识别公园常见的常绿针叶树：马尾松、黄山松、杉木

观察地点推荐

1. S02 古田山—古田庙：探秘常绿阔叶林自然体验路线
2. S03 瞭望台：国家公园科研体验路线
3. Q01 莲花塘：钱塘江寻源自然体验路线

解说方式 Interpretation Style

☑ 人员解说 Staff Interpretation ☐ 教育活动 Educational Activity
☑ 解说牌 Interpretation Signage ☐ 场馆解说 Hall Interpretation

解说季节 Interpretation Season

☑ 春季 Spring ☑ 夏季 Summer
☑ 秋季 Autumn ☑ 冬季 Winter

「认识“先锋树种”：马尾松」

刚刚我们在观察森林的时候有朋友注意到了，在一大片阔叶树中间，常有一些马尾松混生其间，它们树身高大，树干粗壮，看起来有些年头了。其实，这些马尾松的年纪有可能比其他胸径相似的阔叶树种更大。

马尾松是一种“先锋树种”，也就是说，在一片森林最开始形成的时候，马尾松常常是最早的拓荒者，它们根部有共生的根瘤菌，有固氮能力，对早期森林环境的改善起到重要的作用。但是，随着马尾松长成大树，林木下方由于缺少光照，其幼苗很难存活，这就使得一些阔叶树种有机会凭借其耐阴和生长速度快的特点，生根发芽，进而后来居上，占据森林上层空间。

「认识公园常见的针叶树：马尾松、黄山松和杉木」

在海拔较高的地方，由于温度较低，水分和养分条件相对较差，阔叶树生长受限，马尾松、黄山松和杉木等针叶树便长期占据统治地位。和阔叶树相比，钱江源国家公园针叶树的种类相对较少，也更容易识别。

钱江源国家公园的杉木常成片生长，组成杉木林。林下常堆满杉木的落叶，和凋落的松针不同，杉木一般是整枝凋落的。捡起一枝来看，会发现杉木叶片扁平，坚硬扎手，像一根根短剑一般依次生长在树枝两侧。枝干顶端是杉木锥形的球果。其鳞片像叶子一样坚硬，摸起来一定要小心。

正如大家所见，杉木的树干高大通直，且生长迅速，是“天生”的木材。早在明清时期，就有徽州商人将开化山中的杉木贩卖至全国各地，并因此富甲一方。中华人民共和国成立后，伐木业也曾是当地的支柱产业之一。后随着保护区和国家公园的建立，这里的伐木活动已被禁止，但这片杉木林所承载的记忆却永不会磨灭。

和杉木相比，松树可能更符合大家心中对于针叶树的认知。公园里常见的马尾松和黄山松都拥有针形的叶子和苍劲的树干，但它们却很少共同生活在一片区域。马尾松生活的海拔和阔叶树相近，一般在海拔 700m 以下，而黄山松适宜生长的区域海拔要略高于马尾松。

不同的生态位是马尾松和黄山松对环境长期适应的结果，其中最直接的体现就是二者外形的差异。我们不妨来推断一番。从海拔差异来看，谁更加耐寒呢？没错，是黄山松。谁对恶劣环境的适应能力更强呢？也是黄山松。谁的生长速度更快呢？是马尾松。大家知道吗？一棵 5m 高的黄山松可能动辄就有上百岁，但是同样高的马尾松可能才 5 岁不到。

除了树形，我们还可以从植物的局部特征来判断它们的差别。我这里刚好有黄山松和马尾松两种植物的叶子，请大家猜一猜，它们分别是属于哪种树木的？没错，这个短而坚硬的叶子是黄山松的，可以帮助它们适应高寒大风的环境；马尾松的叶子则相对长且软，也是所谓“马尾”的名字的由来。

General Interpretation | Biodiversity Section

Interpretation No. TY-WZ-01

02 Heliophilous and cold-resistant hardy conifer

Key points

1. Recognize the role of the masson pine as a "pioneer tree" in the early succession of the community

2. Identify common evergreen conifers in the park: masson pines, Huangshan pines, Chinese firs

Observation places recommended

1. S02 Evergreen broad-leaved forest exploration route

2. S03 Scientific research experience route

3. Q01 Qiantang River source exploration route

"Know the 'pioneer tree species': masson pine"

Just now when we were observing the forest, a friend noticed that in the middle of a large broad-leaved tree, some masson pines often intergrow. They are tall and their trunks are thick, and they look like some years old. In fact, these masson pines may be older than other broad-leaved species with the similar diameter at breast height. The masson pine is a kind of "pioneer tree species". When a forest is first formed, the masson pine is often the earliest pioneer. Their roots have symbiotic rhizobia and have the ability to fix nitrogen, which play an important role in the improvement of the early forest environment. However, as the masson pine grows into a big tree, the seedling is difficult to survive due to the lack of light under the forest. This allows some broad-leaved tree species to take advantage of their shading tolerance and fast growth characteristics to take root and sprout, and then come from behind and occupy the upper forest space.

"Know the common conifers in the park: masson pines, Huangshan pines and Chinese firs"

However, at higher elevation, the growth of broad-leaved trees is restricted due to low temperature, relatively poor water and nutrient conditions. Thus, conifers such as masson pines, Huangshan pines, and Chinese firs have long dominated. Compared with broad-leaved trees, the types of conifers in Qianjiangyuan National Park are relatively less and easier to be identified.

The Chinese firs at Qianjiangyuan often grow in groups, forming the fir forest. The forest is often piled with fallen leaves. Unlike the pine needles, the Chinese fir is generally fallen in branches. When you pick up a branch, you will find that the leaves of Chinese firs are flat and hard, growing on both sides of the branches like daggers. The cone-shaped cones of fir are on the tops of the branches, and the scales are as hard as the leaves. Be careful when you touch them.

As you can see, the trunk of Chinese fir is tall and straight, and it grows quickly, perfect for construction. As early as the Ming and Qing dynasties, Huizhou merchants sold Chinese firs in the Kaihua Mountain to all parts of the country, and they became rich. After the founding of the People's Republic of China, logging was once one of the local pillar industries. Later, with the establishment of protected areas and national parks, logging activities here have been banned, but the memory carried by this Chinese fir forest will never be erased.

Compared with Chinese firs, pine trees may be more in line with everyone's perception of coniferous trees. The common masson pines and Huangshan pines in the park have needle-shaped leaves and vigorous tree trunks, but they rarely live together in one area. Masson pine lives at an elevation similar to that of broad-leaved trees, generally below 700 meters elevation, while the area where, the Huangshan pine is suitable for growth is slightly higher than that of the masson pine.

The different ecological niches are the result of long-term adaptation of masson pines and Huangshan pines to the environment, and the most direct manifestation is the difference in their appearance. We might as well take a guess. From the perspective of elevation difference, who is more cold-resistant? Yes, it is the Huangshan pine. Who is more adaptable to harsh environments? It is also the Huangshan pine. Who does grow faster? the masson pine does. Do you know that a 5-meter-high Huangshan pine tree may be hundreds of years old, but a masson pine of the same height may be less than 5 years old?

In addition to tree shapes, we can also tell their differences from the local characteristics of plants. I happen to have the leaves of Huangshan pines and masson pines. Please guess what kind of trees they belong to? That's right, the short and hard leaves are from Huangshan pines, which can help them adapt to the environment of high cold and strong wind; the leaves of masson pines are relatively long and soft, which is also the origin of the so-called "horsetail".

Chinese firs

通用型解说 | 生物多样性解说　　　　解说编号：TY-WZ-03

03 古老的孑遗植物

解说要点

1. 孑遗植物的概念
2. 钱江源国家公园独特的地理环境守护的孑遗植物及其保护价值
3. 公园代表孑遗植物：青钱柳、杜仲、婺源安息香

观察地点推荐

1. S01 古田飞瀑：丛林飞瀑自然体验路线
2. S02 古田山—古田庙：探秘常绿阔叶林自然体验路线
3. Q01 莲花塘：钱塘江寻源自然体验路线

青钱柳（果序）　青钱柳（圆形翅果）　杜仲（叶片）　长柄双花木（花）　南方红豆杉　仲的“叶断丝连”现象

解说方式 Interpretation Style

☑ 人员解说 Staff Interpretation　☐ 教育活动 Educational Activity
☑ 解说牌 Interpretation Signage　☐ 场馆解说 Hall Interpretation

解说季节 Interpretation Season

☑ 春季 Spring　☑ 夏季 Summer
☑ 秋季 Autumn　☑ 冬季 Winter

「什么是孑遗植物」

各位朋友，大家知道地球有多少岁了吗？大约45亿年。那么人类呢？且不论人类是由哪种生物进化而来的，人类出现文明也不过一万年左右。在人类出现之前，地球曾经几度更换统治者，比如，其中大家最为熟悉的就是出现在距今约2亿~6500万年前的恐龙时代。

但是因为地球历史上的地质和气候剧变，很多生物都已灭绝，我们只能从埋藏在地球深处的化石中一窥它们的模样。但是，有一小群物种似乎受到了地球特别的眷顾，尽管它们的祖先早已变成了化石，它们却仍还坚强地活着。它们至今仍保留着化石中古老祖先的一些原始特征，千万年来似乎鲜有改变，它们是来自恐龙时代的孤独的旅行者，被形象地称为“孑（jié）遗植物”。

「国家公园守护孑遗植物」

如果把每种生物在地球上的存在过程比作一场旅程，孑遗植物就像是那种阅尽沧桑的孤独旅人。它们见识过地球历经的沧海桑田和巨大的历史变迁，曾几何时，地球上很多地区都有它们的分布，经过第四季冰期的生物大灭绝等灾难，与之同时期生存的物种，包括它们的很多近亲与同类都消失了，只有它们在一些“生物避难所”（Sanctuary）小环境中幸存下来，并存居至今，因此被称为“孑遗植物”。钱江源国家公园复杂的地理环境为杜仲、长柄双花木和青钱柳等众多孑遗植物提供了温暖的庇护所，它们和地层中的古老祖先化石一起从遥远的年代走来，见证了公园的过去与现在，也承载着公园的未来。

「孑遗植物的保护价值」

孑遗植物是穿越时空的古老物种，在它们身上，时间的魔力似乎暂时失效了。它们不仅容颜未改，身上也还流淌着远古的血液，保存着生命原始的基因密码和环境变化信息。这对研究植物系统发育、生物起源与进化以及保存生物基因多样性等具有重要意义。

注意事项

- 注意孑遗植物和稀有植物的区别（稀有植物的解说内容）。
- 孑遗植物非常珍贵，在解说的过程中可以通过对其古老历史的回顾，让访客了解到其保护价值，并在实际的旅行中践行这一点。

辅助道具推荐

- 地质年代表
- 孑遗植物银杏和其化石照片

拓展知识点

- 恐龙时代：恐龙最早出现在2亿3000万年前的三叠纪，曾统治地球陆地生态系统长达1亿4000万年之久，大部分恐龙于6500万年前的白垩纪大灭绝事件中灭绝，仅有鸟型恐龙中的鸟类存活下来，并繁荣至今。

注意事项

- 青钱柳的果实十分特别，像一串串的铜钱一样，可在7~9月观察到，可以在这个季节去搜集一些果实，以便在解说的时候当做道具。
- 最好能找到落在地上的杜仲叶片，亲自示范一下“叶断丝连”的现象。但如果没有，不要用采摘方式造成错误示范。
- 婺源安息香很难发现，因此可以提前在游线上确定此花的位置，以方便游客观察；花果期也非常短暂，注意安排观察时间，其他季节请以照片替代。
- 注意提醒访客不要乱折花草，可通过捡拾凋落物的形式弥补。

辅助道具推荐

- 捡拾的青钱柳的叶片、果实
- 青钱柳、枫杨的叶片和果实照片
- 杜仲“叶断丝连”现象的照片
- 杜仲的地史分布与现代分布示意图
- 婺源安息香的花、叶、果特写照片

「“摇钱树”青钱柳」

各位朋友，现在我们面前的这棵树名为青钱柳，也叫“摇钱树”。我们来摇一摇，看看能否落下钱来（假装轻摇树木，并悄悄拿出准备好的青钱柳果实）。瞧，钱就在这里！是不是像极了古代的铜钱，这其实是青钱柳的圆形翅果。

在历史上，人们真正利用过的是它的叶子。在湖南等地区曾有春季用青钱柳新鲜的芽叶制成青钱柳茶的习俗。现代医学研究发现，青钱柳叶片含有能降血糖以及抑制胆固醇的某些有效成分，但作为国家二级重点保护野生植物，野外的采集和利用是被绝对禁止的。也许未来这种植物有可能成为重要的药材原料，让我们把这份期待留给科学研究工作者们吧。

青钱柳和枫杨是近亲，都属于胡桃家族，比如，我们常吃的核桃也是胡桃家族的成员。不过，青钱柳是我国的特有种，枫杨、核桃却很普遍。作为一种古老的孑遗植物，青钱柳最早起源于北美大陆，曾经和恐龙一起生活在地球上的很多地方。但由于环境的变迁，大多数青钱柳家族的成员都已灭绝，如今只有中国的长江流域有青钱柳的分布，是一种典型的孑遗植物。

青钱柳是一种落叶乔木，常与青冈、楠木等常绿乔木组成常绿落叶阔叶混交林。我们面前这棵高大的树木看起来和普通树木没什么差别，但是它们的祖先却经历了青藏高原的隆起带来的地形和气候剧变，并且躲过了第四纪冰期的严寒。这除了得益于青钱柳本身对环境超强的适应能力外，更主要是因为钱江源国家公园为它们提供的避难所。

青钱柳的枝叶

「全身是宝的杜仲」

我们知道藕断会丝连，但是大家知道吗？有种植物的叶子断了，也会有“丝线黏连”现象。这种植物的名字说出来大家可能并不陌生：杜仲。不过大家对杜仲的认识，更多地可能是作为一种中药材。除此之外，杜仲还有哪些价值呢？我们不妨从面前的这棵杜仲树开始。

这是我刚在树底下捡到的一片杜仲叶，大家来看看有什么不一样吗？椭圆形，表面有磨砂的质感……似乎没什么突出的特征。那我们怎么知道这是不是杜仲叶呢？还记得我刚刚说的吗，杜仲体内因为含有杜仲胶，所以无论掰开新鲜的杜仲树叶、树枝还是树皮，都会有“丝连”现象。我们来试试看。果然有“丝”出现。杜仲全身都是宝，除了作为药材，杜仲胶还常被用来提取成硬质橡胶，是制造海底电缆等材料的重要原料。

科学家们通过对杜仲化石的研究发现，大约 500 万年前，杜仲还广泛分布在北半球各个大陆上，但因为第四纪冰期的影响，它们中的大多数从其他大陆消失。目前，只有生活在我国的杜仲由于青藏高原隆起带来的复杂地形与气候演变提供的特别庇佑，才幸运地逃过了这次劫难，存活至今。

「低调优雅的婺源安息香」

在钱江源国家公园的孑遗植物中，特别值得一提的是婺源安息香。这是一种非常低调的灌木，平时喜欢生活在比较阴湿的环境中，4 月开花，花朵为白色，花梗纤细，花开时一串串像玉坠一样并排垂在一根花茎上，结成的果实也是这么排列着的。卵圆形的果实顶端有短短的尖头，煞是可爱。

婺源安息香是中国特有的一种植物，数量稀少，分布区域有限，仅限于中国的江西、安徽和浙江的部分地区。在浙江，婺源安息香仅分布于古田山周边地区。大家在徒步的过程中可要留心看看会不会发现它们的身影。

General Interpretation | Biodiversity Section

Interpretation No. TY-WZ-03

03 Ancient Relic plants

● Key points

1. Concept of relic plants

2. The relic plants protected by the unique geographical environment of Qianjiangyuan National Park and their protection value

3. Representative relic plants in the park: wheel wingnuts, hardy rubber trees, Wuyuan Snowbell (*Styrax wuyuanensis*)

● Observation places recommended

1. S01 Jungle and waterfall experience route
2. S02 Evergreen broad-leaved forest exploration route
3. Q01 Qiantang River source exploration route

"What is a relic plant"

Dear friends, do you know how old the earth is? Approximately 4.5 billion years. What about humans? Regardless of what kind of organisms human beings evolved from, human civilization has only been about 10,000 years old. Before the emergence of human beings, the earth had changed its ruler several times.

What everyone is most familiar with is that it appeared in the age of dinosaurs about 200 to 65 million years ago.

But because of the drastic changes in geology and climate in the history of the earth, many organisms have become extinct. We can only get a glimpse of their appearance from the fossils buried deep in the earth. However, there is a small group of species that seem to be particularly favored by the earth. Although their ancestors have long been fossilized, they are still alive. They still retain some of the original features of the ancient ancestors in the fossils, and they seem to have little changed over the years. They are lonely travelers from the age of dinosaurs and have been vividly called "relic plants".

"Qianjiangyuan National Park: a guardian of relic plants"

If you compare the existence of each creature on earth to a journey, the relic plants are like the kind of lonely traveller who has read through the book of life. They have seen the huge

wheel wingnut

Eucommia ulmoides

wheel wingnut

geological and historical changes of the earth's vicissitudes of life. They were once prosperous and widely distributed in history, but their territories have been shrinking since then, and many of them are now only a small corner. Therefore they are called "relic plants". The complex geographical environment of Qianjiangyuan National Park provides a warm shelter for many relic plants such as *Eucommia ulmoides*, *Disanthus cercidifolius* var. *longipes* and wheel wingnuts together with the strata, they come from ancient geological movements and witness the National Park's past. And now, it also carries the future of the Park.

"The conservation value of relic plants"

Relic plants are travelers who have travelled through time and space. The magic of time seems to have temporarily expired on them. Not only have their appearances unchanged, they also have ancestors' blood flowing in their bodies, preserving the original genetic code of life and information about environmental changes. This is of great significance for studying plant phylogeny, biological origin and evolution, and preserving biological genetic diversity.

Attentions

- Pay attention to the difference between relics and rare plants (interpretation content of rare plants).
- Relic plants are very precious. During the interpretation, visitors can learn about its conservation value by reviewing its ancient history, and practice this in actual travel.

Recommended auxiliary facilities and toolkits

- Geological Time Scale
- Photographs of the relic plant—ginkgo trees and its fossils

Extra information

- The Age of Dinosaurs: Dinosaurs were originated during the Triassic Period about 230 million years ago. They have dominated the terrestrial ecosystems on the earth for over 140 million years. Most dinosaurs died out during the Cretaceous Extinction 65 million years ago. Only a few bird-like dinosaurs survived and evolved into later birds, which thrived and then flourished today.

Attentions

- The fruit of the wheel wingnut is very special, like a string of copper coins, which can be observed from July to September. Some fruits can be collected during this season, which can be used as a tool in the interpretation.
- It is best to find the leaves of *Eucommia* that fall on the ground and personally demonstrate the phenomenon of "fiber connection". But don't pick the leaves on the tree to cause wrong demonstrations.
- It is difficult to find *Styrax wuyuanensis*, you can locate this flower in advance to help the visitors; the flowering and fruit period is also very short, pay attention to arrange the observation time, please replace them with photos in other seasons.
- Remind visitors not to fold flowers and plants, and you can make up for it by picking up litter.

Recommended auxiliary facilities and toolkits

- Picked leaves and fruits of the wheel wingnut
- Photos of the leaves and fruits of Wheel wingnut and Chinese wingnut
- Photos of *Eucommia* "fiber connection" phenomenon
- Schematic diagram of the geographical history and modern distribution of *Eucommia*
- Close-up photos of flowers, leaves and fruits of *Styrax wuyuanensis*

"Money tree" wheel wingnuts

Dear friends, the tree in front of us is called wheel wingnut, also called "money tree". Let's shake it to see if it can drop the money (pretend to shake the tree lightly and quietly take out the prepared wheel wingnut fruit). Look, the money is here! Does it look like an ancient cash coin? This is actually the round samara of the wheel wingnut.

In history, what people have really used is its leaves. In Hunan and other regions, there was a custom of making tea with the fresh buds of wheel wingnut in spring—wheel wingnut tea. Modern medical research has found that Wheel wingnut leaves contain certain active ingredients that can lower blood sugar and inhibit cholesterol. However, as a state second-level key protected plant, collection and use in the wild is absolutely prohibited. Perhaps this plant may become an important raw material for medicinal materials in the future. Let us leave this expectation to scientific researchers.

Wheel wingnut and Chinese wingnut are close relatives and both belong to the walnut family. For example, the walnuts we often eat are also members of the walnut family. However, the wheel wingnut is endemic to our country, while maple and walnut are common. As an ancient relic plant, wheel wingnut originated in North America and once lived with dinosaurs in many places on earth. However, due to changes in the environment, most members of the wheel wingnut family have become extinct. They are now only distributed in the Yangtze River valley in China.

Wheel wingnut is a deciduous tree, which often forms an evergreen and deciduous broad-leaved mixed forest with evergreen trees such as ring-cupped oak and *Eucommia*. The tall tree in front of us looks no different from ordinary trees, but their ancestors experienced drastic changes in terrain and climate brought about by the uplift of the Tibetan Plateau, and escaped the cold of the Quaternary glaciation, which is not only because of its strong adaptability to the environment, but also the refuge provided by Qianjiangyuan National Park.

"*Eucommia*, a treasure itself"

We know that when the lotus root is broke, the fiber is still connected. But do you know same thing also applies to

Disanthus cercidifolius var. *longipes*

the leaves of some plants? Everyone may be familiar with the name of this plant: *Eucommia*. However, what everyone knows about *Eucommia* is more likely to be a kind of Chinese herbal medicine. In addition, what value does *Eucommia* have? We might as well start with the *Eucommia* tree in front of us.

This is a piece of *Eucommia* leaf I just picked up under the tree. Come and see what is different? Oval, with a frosted texture on the surface··· it seems that there are no outstanding features. So how do we know if this is *Eucommia* leaf? Remember what I said just now, because *Eucommia* contains eucommia gum, no matter if you break fresh *Eucommia* leaves, branches or bark, there will be "silky" fibers. Let's try it. *Eucommia* is a treasure itself. In addition to being used as a medicinal material, *Eucommia* is often used to extract hard rubber, which is an important raw material for making submarine cables and other materials.

Scientists have discovered through research on *Eucommia* fossils that *Eucommia* were widely distributed on all continents in the northern hemisphere 5 million years ago, but most of them were slowly retreated from other continents due to the influence of Quaternary glaciers until they disappeared. Only the *Eucommia ulmoides* living in our country escaped this catastrophe due to the complex terrain and climate evolution brought about by the uplift of the Tibetan Plateau.

"Low-profile and elegant *Styrax wuyuanensis*"

Among the relic plants in Qianjiangyuan National Park, *Styrax wuyuanensis* is particularly worth mentioning. This is a very low-profile shrub. It usually likes to live in a relatively humid environment. It blooms in April. The flowers are white with slender pedicels. When the flowers bloom, clusters of flowers hang side by side on a flower stem like jade pendants. The fruits are arranged in this way with an ovoid shape and have short pointed tips, which is really cute.

Styrax wuyuanensis is scarce with a limited distribution area. Only in parts of Jiangxi, Anhui, and Zhejiang Province in China, can you see the natural *Styrax wuyuanensis*. In Zhejiang, *Styrax wuyuanensis* is only distributed in the surrounding areas of Gutian Mountain. You can pay attention to see if you will find them during the hike.

通用型解说 | 生物多样性解说　　　　解说编号：TY-WZ-04

04 特立独行的长柄双花木

解说要点

1. 长柄双花木的主要特征与繁殖方式
2. 长柄双花木的分布状况与珍贵价值

观察地点推荐

Q02 里秧田—大峡谷：遇见峡谷飞瀑自然体验路线

长柄双花木（花）　长柄双花木（叶）

解说方式 Interpretation Style

☑ 人员解说 Staff Interpretation
☐ 教育活动 Educational Activity
☑ 解说牌 Interpretation Signage
☐ 场馆解说 Hall Interpretation

解说季节 Interpretation Season

☑ 春季 Spring
☑ 夏季 Summer
☑ 秋季 Autumn
☑ 冬季 Winter

「长柄双花木的主要特征与繁殖方式」

钱江源国家公园为各种生命的演绎提供了一个大舞台，一年四季精彩不断，你方唱罢我登场。相对于春的繁华，夏的热闹，秋、冬季节的钱江源可能相对冷清，但如果你选择在这个时候来拜访钱江源国家公园，也许会有不期而遇的惊喜，其中，长柄双花木就是适宜秋冬季节观赏的代表性植物。

长柄双花木，单听名字就觉得这个植物不一般。大家来看下这张图，首先从外形上看，它们的花柄和叶柄都很细长，而且一根花柄之上生有两朵背对背紧贴着的并蒂花，这也是它得名的原因。

长柄双花木是一种落叶灌木。每年的秋季，叶子会完成它们的使命，纷纷由绿转红，直至凋落。此时，你若留心，会发现长柄双花木凋零的枝干上，紫红色的花朵不知何时已悄然爬上枝头。大多数花朵都会选择在温暖的春夏季节开花，长柄双花木却选择了不走寻常路。它们体内似乎有一个自己的生物节拍，不仅花开得晚，就连结果也慢慢悠悠的，从冬到春，再从春到夏，直到来年的秋季，新一轮的花苞立上枝头的时候，黑色的蒴果才算成熟。

有很多植物知道自己能力有限，会借助风、动物甚至流水等帮助自己传播种子，以便最大限度地延续自己家族的基因。但是长柄双花木却选择"自力更生"，经过漫长时节好不容易长成的果实在成熟后会自动炸裂，然后种子就依靠果实爆裂的力量被弹射出果壳，落在了地上。这还没完，落地后的种子不会立刻长出新芽，而是要经过漫长的休眠期，再遇到合适的生长条件才会发芽、长叶，变成小双花木。自然界的时间稍纵即逝，在这个过程中还可能伴随着病虫害、自然灾害和人为干扰等因素的影响。所以你看，一棵长柄双花木的长成是多么不容易啊！

「长柄双花木的珍贵价值与分布状况」

作为一种古老的孑遗植物，最早的植物化石表明，长柄双花木在白垩纪的恐龙时代曾繁盛一时。地质时期的环境变化使得恐龙等大型动物从地球上消失，长柄双花木却幸运地逃过了一劫。不过，现今长柄双花木的分布区域狭窄，只在浙江、湖南和江西三省的部分区域以孤岛状存在着。钱江源国家公园温凉湿润的气候为长柄双花木的生长提供了优越的条件，是长柄双花木仅存的孤岛之一。这里的长柄双花木面积高达约3.33km²，是全球长柄双花木天然连片分布面积最大的种群地之一。我们对面的这片林子，就集中生长有大量的长柄双花木。这里也是长柄双花木自然海拔分布的下限（480m）和最北的居群。

长柄双花木虽然美丽，也很脆弱，钱江源国家公园已经这样静静守候了它们千万年。让我们就这样隔着溪流，保持适当的距离，不去打扰它们，远远地欣赏它们的美好。

注：解说员需要了解长柄双花木的特征和位置，并根据合适的季节选择是否进行本点位的人员解说。

General Interpretation | Biodiversity Section

Interpretation No. TY-WZ-04

04 Maverick *Disanthus cercidifolius* var. *longipes*

● Key points

1. The main characteristics and reproduction methods of *Dianthus cercidifolius* var. *longipes*
2. The precious value and distribution of *Disanthus cercidifolius* var. *longipes*

● Observation places recommended

Q02 Canyon and waterfall experience route

"The main characteristics and reproduction methods of *Disanthus*"

Qianjiangyuan National Park provides a big stage for the interpretation of various lives. It is wonderful throughout the year. Compared with the bustling spring and summer, autumn and winter in Qianjiangyuan may be relatively deserted, but if you choose to visit Qianjiangyuan National Park at this time, you may have unexpected surprises. *Disanthus cercidifolius* var. *longipes* is representative plants suitable for viewing in autumn and winter.

Disanthus cercidifolius var. *longipes*, just hearing the name, you can tell it is unusual. Let's take a look at this picture. First of all, from the appearance, their flower stalks and petioles are very slender, and on a flower stalk, there are two flowers back to back closely.

Disanthus cercidifolius var. *longipes* is a deciduous shrub. In autumn, the leaves will complete their mission, turning from green to red until they fall. At this time, if you pay attention, you will find that purple-red flowers have quietly climbed onto its withered branches. Most of the flowers will choose to bloom in the warm spring and summer, but this kind of chooses to take an unusual path. It seems to have its own biological rhythm in its body. Not only its flowers bloom late, but also bears fruit late, from winter to spring, then to summer, until the autumn of the following year, when a new round of flower buds stand on the branches, the black capsules are considered mature.

Many plants know that their abilities are limited, and will use wind, animals, and even running water to help them spread their seeds in order to maximize their territory. However, the *Disanthus cercidifolius* var. *longipes* chooses to depend on itself. Its fruit that grows after a long period of time will automatically burst when it matures, and then the seeds will be ejected out of the shell

flower of *Disanthus cercidifolius* var. *longipes*

by the force of the bursting of the fruit and fall to the ground. This is not over yet. The seeds that fall to the ground will not grow new sprouts immediately, but will have to go through a long dormant period before they encounter suitable growth conditions before they germinate, grow leaves, and become seedlings. The time in nature is fleeting, and in this process it may be affected by factors such as plant diseases and insect pests, natural disasters and human interference. So you see, how difficult it is to grow a tree!

"The precious value and distribution of *Disanthus cercidifolius* var. *longipes*"

As an ancient relic plant, the earliest plant fossils indicate that the *Disanthus cercidifolius* var. *longipes* flourished during the Cretaceous dinosaur era. Environmental changes during the geological period caused dinosaurs and other large animals to disappear from the earth, but the *Disanthus cercidifolius* var. *longipes* was lucky to escape. However, its living area is narrow nowadays, and it only exists as isolated islands in parts of Zhejiang, Hunan and Jiangxi provinces. The warm and humid climate in Qianjiangyuan National Park provides excellent conditions for its growth. It is one of the only islands left. Its area here is as high as 3.33km^2, and it is one of the largest populations of *Disanthus cercidifolius* var. *longipes* in the world. In the forest opposite to us, there are a large number of its trees. This is also the lowest (480m) and the northernmost population of the natural elevation distribution of *Disanthus cercidifolius* var. *longipes*.

Although the *Disanthus cercidifolius* var. *longipes* are beautiful, it is also very fragile. Qianjiangyuan National Park has been guarding them quietly for thousands of years. Let us just keep a proper distance across the stream, don't disturb them, and appreciate their beauty.

Notes: The interpreter needs to understand the characteristics and location of the *Disanthus cercidifolius* var. *longipes*, and choose whether to conduct interpretation according to the appropriate season.

通用型解说 | 生物多样性解说　　　　解说编号：TY-WZ-05

05 稀有植物知多少

解说要点

1. 稀有植物的概念
2. 钱江源国家公园的稀有植物和稀有植物群落

观察地点推荐

1. S02 古田山—古田庙：探秘常绿阔叶林自然体验路线
2. Q01 莲花塘：钱塘江寻源自然体验路线

南方红豆杉
野含笑
香果树

解说方式 Interpretation Style

- ☑ 人员解说 Staff Interpretation
- ☐ 教育活动 Educational Activity
- ☑ 解说牌 Interpretation Signage
- ☐ 场馆解说 Hall Interpretation

解说季节 Interpretation Season

- ☑ 春季 Spring
- ☑ 夏季 Summer
- ☑ 秋季 Autumn
- ☑ 冬季 Winter

「什么是稀有植物」

我们常说“物以稀为贵”，但是对于一个物种来说，稀有则意味着生存受到威胁甚至具有灭绝的危险。大家觉得什么是稀有植物呢？是数量稀少吗？

稀有植物包含有两种情况。一种是仅存在于非常有限的区域内，另一种则是虽然分布范围较广，但是只是零星存在于一些区域。这两种情况下的植物都有可能陷入濒危或者灭绝。从我国的珍稀濒危植物分类体系来看，稀有植物的受危等级最低，排在渐危种之后，因此受到较少的关注。稀有植物目前虽不至于灭绝或者濒危，但是如果没有充分有效的保护策略，一些稀有种如果遇到突发的气候变化或者人为干扰，极有可能造成其野生种群的破坏甚至消失。

「钱江源国家公园的稀有植物和稀有植物群落」

钱江源国家公园特殊的地形和气候条件，成为南方红豆杉、香果树、野含笑等众多稀有植物的庇护所。中国科学院等研究机构通过对古田山大样地中的生物多样性调查，发现了大量稀有种的存在，达到了总物种数的 1/3 以上。

这些珍稀或者濒危的植物本身分布区域狭窄，个体数量稀少，大多仅零星分布于各类植物群落中，其中只有少数种类能作为优势种或者次优势种形成特定的群落，我们将其称为稀有植物群落。

钱江源国家公园内就有多处稀有植物群落，如野含笑群落、紫茎群落、香果树群落、南方红豆杉群落等，这在全国范围内都十分罕见。

注意事项

- 注意稀有植物和孑遗植物的区别。
- 无论是公园的稀有植物还是稀有植物群落，都与当地的环境条件密不可分。

辅助道具推荐

- 部分稀有植物和及群落图片

拓展知识点

- 大部分孑遗植物受气候变化、人为干扰等因素影响，都已经成为稀有植物，但孑遗植物不一定都是稀有植物，比如银杏。同时，稀有植物也不一定都是孑遗植物，比如兰花。另外，相对于孑遗植物的不变性，稀有植物是可以随着种群数量的变化而变化的。同一物种在不同的地区可能分别属于稀有或非稀有物种。

General Interpretation | Biodiversity Section

Interpretation No. TY-WZ-05

05 Rare plants

Key points

1. The concept of rare plants
2. Rare plants and rare plant communities in Qianjiangyuan National Park

Observation places recommended

1. S02 Evergreen broad-leaved forest exploration route
2. Q01 Qiantang River source exploration route

"What is a rare plant"

We often say: "Rare things are precious", but for a species, being rare means that its survival is facing threats or even extinction. What does determine a rare plant? By its quantity?

There are two situations in rare plants. One is that it only exists in a very limited area, and the other is that although the distribution is wide, it only exists in some areas sporadically. In both cases, plants may be endangered or extinct. From the perspective of my country's rare and endangered plant classification system, rare plants have the lowest endangered level, ranking behind the increasingly endangered species, so they have received less attention. Although rare plants are not currently extinct or endangered, if there are no adequate and effective protection strategies, some rare species may face destruction or even disappearance of their wild populations if they encounter sudden climate change or human disturbance.

"Rare plants and rare plant communities in Qianjiangyuan National Park"

The special topography and climatic conditions of Qianjiangyuan National Park have become a refuge for many rare plants such as Chinese yews, *Emmenopterys henryi*, and *Michelia skinneriana*. Research institutions such as the Chinese Academy of Sciences have investigated the biological diversity in the Gutian Mountain plot and discovered the existence of a large number of rare species, reaching more than one third of the total number of species.

These rare or endangered plants have a narrow distribution area and few individuals. Most of them are scattered in various plant communities. Only a few species can be considered as dominant or sub-dominant species to form specific communities. We call them rare plant communities.

Taxus chinensis

There are many rare plant communities in Qianjiangyuan National Park, such as wild *Michelia* community, Chinese stewartia community, *Emmenopterys henryi* community, Chinese yew community, *etc*., which are very rare across the country.

Attentions

- Pay attention to the difference between rare and relict plants.
- The existence of both rare and relic plants are inseparable from the effective protection of local environment as a sanctuary during the natural disasters.

Recommended auxiliary facilities and toolkits

- Pictures of some rare plants and their communities (such as the communities of Chinese yews, Chinese stewartias, and Chinese anises)

Extra information

- Most relic plants have become rare plants due to factors such as climate change and human disturbance, but the relic plants are not necessarily rare plants, such as ginkgo. At the same time, rare plants are not necessarily relic plants, such as orchids.
- In addition, different from the invariance of relic plants, rare plants can change with the change of population, that is, the same species may belong to rare or non-rare species in different regions.

通用型解说 | 生物多样性解说 解说编号：TY-WZ-06

06 小苔藓，大世界

● 解说要点

1. 了解苔藓的分类，认识钱江源国家公园常见的藓类
2. 了解苔藓对于环境的重要价值

● 观察地点推荐

1. S02 古田山—古田庙：探秘常绿阔叶林自然体验路线
2. S03 瞭望台：国家公园科研体验路线
3. Q01 莲花塘：钱塘江寻源自然体验路线

解说方式 Interpretation Style

☑ 人员解说 Staff Interpretation ☐ 教育活动 Educational Activity
☑ 解说牌 Interpretation Signage ☐ 场馆解说 Hall Interpretation

解说季节 Interpretation Season

☑ 春季 Spring ☑ 夏季 Summer
☑ 秋季 Autumn ☑ 冬季 Winter

解说词

钱江源国家公园温暖湿润的气候为众多生物提供了生存的场所，其中有一种生物虽无处不在，却也最容易被大家忽视，这就是苔藓。清代诗人袁枚将苔藓的这种处境用诗句形象地表达了出来："白日不到处，青春恰自来。苔花如米小，也学牡丹开。"但是如果遇到一个较真的植物学家，他可能要说这首诗虽美，却犯了一个科学性的错误，因为苔藓没有花和种子，它们主要是靠孢子繁殖的。但这却不妨碍苔藓被归类为高等植物，而同样容易被忽视的地衣和藻类却属于低等植物。

作为一种高等植物，苔藓身上有哪些"高明"之处呢？我们不如从这块石头(或这个树干上、脚下)的苔藓开始讲起。

「了解苔藓的分类，认识钱江源国家公园常见的藓类」

首先，大家要知道苔藓中的苔和藓实际上不是一个物种。苔藓实际上是一类植物的统称，包括苔、藓和角苔三个大的类群，其中，藓类植物最多。藓类植物中最为大家熟悉的又有葫芦藓、金发藓和泥炭藓等。

我们现在观察的这块石头上的苔藓就是葫芦藓。大家可以告诉我它们为什么会被叫作葫芦藓吗？只要凑近看就会发现，这片苔藓丛中长有许许多多突出的"小葫芦"。不知道这些"葫芦"里卖的是什么药。还记得我们说苔藓靠什么繁殖吗？没错，是孢子。那么孢子藏在哪儿呢？实际上就在这样一颗颗的小葫芦里。

不过，不是所有的苔藓都是那么矮小。这是金发藓，最高可达 40cm，堪称苔藓中的"巨人"。不过金发藓名字中虽带个"发"字，却一点也不似头发般柔软飘逸。它们的叶片摸起来微硬，仿佛是由杉木树枝扦插而成的微缩景观。那么金发藓是如何得名的呢？大家看它们的孢蒴，表面由一层金黄色的绒毛包裹着，孢柄修长，像不像一位亭亭玉立的金发女郎？金发藓也因此得名。

泥炭藓是公园中常见的一种沼泽藓类，多生活在滴水崖壁或者湿地沼泽等水分充足的地方。泥炭藓植株柔软，色泽偏灰，呈灰绿色或者灰白色。具有强大的吸水能力，可以蓄积自身重量 20～25 倍的水分。此外，泥炭藓还具有强大的固碳能力，在全球碳循环中发挥着重要的作用，可以说是名副其实了。

注意事项

- 在介绍不同苔藓的时候，最好可以结合实际的场景。
- 苔藓的优点可以通过提问互动的形式，让访客说出来。
- 提醒访客在游览钱江源国家公园的过程中，可以试着观察下遇到的苔藓，看看能否分辨出来遇见的具体是哪种苔藓？

辅助道具推荐

- 角苔、苔和葫芦藓、金发藓、泥炭藓的照片

「了解苔藓对环境的重要价值」

苔藓和其他高等植物一样，会通过光合作用吸收二氧化碳，释放氧气。但是在森林的下层，作为光合作用最重要的一味原料——光却十分稀缺。这时候，小小的苔藓就发挥作用了。它们虽然身形不大，但凭借着对极端环境惊人的适应能力和数量优势，占据了其他高等植物（如蕨类植物和种子植物）无法立足之地，最大限度地“成就”了叶绿体的光合作用“事业”。

但苔藓的功能还不止于此。作为一种由水生向陆生过渡的植物类型，苔藓就像海绵一样，具有极强的吸水、储水能力，在水体净化和水源涵养方面发挥着重要的作用。苔藓的健康与否，还被用作森林生态系统是否优质的重要指标。一般环境质量越高，苔藓的长势也越好。

苔藓常成丛生长，每丛苔藓都是一个小小的王国。除了苔藓之外，这里还时常是藻类、地衣、细菌和真菌等不易察觉的微小生物和蜗牛、蛞蝓、蜘蛛等昆虫的栖居之所。在这个小小的王国里，它们不仅可以自给自足，制造氧气，留住水分，还会通过吸收大气中的水分、尘土等不断改善自己的生存环境，同时还会分泌出一种酸性物质促进岩石的风化。经过漫长的岁月，这些微小作用积累后在原来裸露的石块表面逐渐形成的最初的土壤，能够减缓地表径流、防止风蚀和水土流失，从而为其他植物的生长创造可能。因此，苔藓也被称为“拓荒者”。

拓展知识点

「苔藓为什么是高等植物」

- 高等植物和低等植物的区别在于高等植物有胚，且具有根、茎、叶的分化，能适应陆地的生活环境。苔藓植物被划分为高等植物正是因其具有胚的分化。

「苔藓的繁殖方式」

- 苔藓植物是依靠孢子繁殖的，既不开花也不结果。苔藓植物成熟的孢子囊会释放出里面的孢子，当孢子在适宜的环境下萌发时可形成原叶体（也有叫原丝体，单倍配子体）。原叶体的腹面上长有雌、雄生殖器官，当原叶体被水浸湿时，精子游到雌性生殖器官里与卵细胞结合完成受精作用，形成受精卵（二倍孢子体），由受精卵再发育成新的植物体。

「苔藓的种类」

- 苔藓的种类非常丰富，全球已知超过 20000 余种，中国分布 3000 余种。据统计，钱江源国家公园的古田山自然保护区已发现苔藓植物 55 科 142 属 325 种，分别占全国苔藓植物科的 45.83%、属的 24.61%、种的 9.68%，其中包含 17 个中国特有种，这从侧面反映了古田山苔藓植物区系的古老性。

General Interpretation | Biodiversity Section

Interpretation No. TY-WZ-06

06 Microcosm in the mosses

Key points

1. Understand the classification of mosses and the common mosses in Qianjiangyuan National Park
2. Understand the important value of mosses to the environment

Observation places recommended

1. S02 Evergreen broad-leaved forest exploration route
2. S03 Scientific research experience route
3. Q01 Qiantang River source exploration route

The warm and humid climate of Qianjiangyuan National Park provides a place for many creatures to survive. One of them is ubiquitous but also the easiest to be overlooked, which is the moss. The poet Yuan Mei of the Qing Dynasty vividly expressed the situation of mosses in verses: "The day is not everywhere, but youth comes from it. The moss flowers are as small as rice, and you can learn how to bloom as peonies do." But if you encounter a more serious botanist, he might say that although this poem is beautiful, it made a scientific mistake because the moss has no flowers or seeds. They reproduce mainly by spores. But this does not prevent moss from being classified as higher plants, while lichens and algae, which are also easily overlooked, belong to lower plants.

As a higher plant, what are the "smart" aspects of the moss? We might as well start with the moss on this stone (or on the trunk or under the feet).

"Understand the classification of mosses and recognize the common mosses in parks"

First of all, everyone should know that the moss is actually a collective name for a group of plants, including three large groups. The most familiar mosses include bonfire mosses, haircap mosses and blunt-leaved bogmosses.

The moss on this stone we are observing now is bonfire moss. Can you tell me why they are called bonfire moss? If you look closer, you will find that there are many "small spheres" growing in this moss bush. So what's in it? Do you remember how the moss reproduced we said? Yes, it is a spore. So where are the spores hidden? It's actually in these little spheres.

Not all mosses are so short. This kind of moss is called haircap moss, which can reach up to 40cm in height and can be called the "giant" of mosses. Although there is the word hair in its name, the haircap moss is not soft at all, the leaves are slightly hard, like a miniature landscape made from cuttings of fir branches. So how did the haircap moss get its name? Look at their spore capsules. Their surface is covered by a layer of golden fluff and their spore trunks are slender. Does it look like a cap made of hair? This is why the haircap moss gets its name.

Blunt-leaved bog moss is a kind of swamp moss commonly found in the national park. It lives in places with sufficient water such as dripping cliffs or wetland swamps. Blunt-leaved bog mosses are soft, with grayish green or grayish white body. It has strong water absorption capacity, which can store 20 – 25 times its own weight in water. In addition, the blunt-leavcd bog moss has strong carbon sequestration ability and plays an important role in the global carbon cycle.

Attentions

- When introducing different mosses, it is best to combine actual scenes.
- You can use interactive questions to let visitors talk about the advantages of mosses.

Recommended auxiliary facilities and toolkits

- Photos of hornworts, mosses and bonfire mosses, haircap mosses, blunt-leaved bog mosses

Extra information

"Why is the moss higher plant"

- The difference between higher plants and lower plants is that higher plants have embryos, and have root, stem and leaf differentiation, which can adapt to the terrestrial living environment. The moss is classified as the higher plant because of its embryonic differentiation.

"The way of reproduction of moss"

- The moss relies on spores to reproduce and neither bloom nor bear fruits. The mature sporangia of mosses will release the spores. When the spores germinate in a suitable environment, they can form a prothallium (also called a prothion, a haploid gametophyte). There are female and male reproductive organs on the ventral surface of the prothallium. When the prothallium is soaked in water, the sperm will swim into the female reproductive organs and combine with the egg cells to complete fertilization, forming a fertilized egg (diploid sporophyte). The eggs then develop into new plants.

"Types of moss"

- The species of moss is very rich, more than 20,000 species are known worldwide, and more than 3,000 species are distributed in China. According to statistics, in the Gutian Mountain Nature Reserve of Qianjiangyuan National Park, 55 families, 142 genera, and 325 species of mosses have been discovered, accounting for 45.83% of the families, 24.61% of the genera, and 9.68% of the species in the country. 17 species of them are endemic to China, which reflects the ancient nature of the moss flora of Gutian Mountain.

"Understand the important value of mosses to the environment"

Like other higher plants, the moss absorbs carbon dioxide and releases oxygen through photosynthesis. But in the lower part of the forest, light that is the most important raw material for photosynthesis is very scarce. At this time, the little mosses come into play. Although they are not big in size, with their amazing ability to adapt to extreme environments and their quantitative advantages, they occupy a place where other higher plants (such as ferns and seed plants) cannot live, and contribute to the photosynthesis.

But the function of moss does not stop there. As a plant type that transitions from aquatic to land, the moss is like a sponge, with strong water absorption and storage capacity, and plays an important role in water purification and water conservation. The health of mosses is also used as an important indicator of the quality of forest ecosystems. Generally, the higher the environmental quality, the better the growth of mosses.

The moss often grows in clumps, and each clump of moss is a small kingdom where not only mosses grow, but also tiny creatures such as algae, lichens, bacteria and fungi often live. It is also home to snails, slugs, spiders and other creatures. In this small kingdom, mosses can not only be self-sufficient, produce oxygen, retain water, but also continuously improve their living environment by absorbing moisture and dust in the atmosphere. At the same time, they can secrete an acidic substance to promote weathering of rocks. After a long period of time, the initial soil gradually forms on the surface of the exposed rocks after these tiny effects have accumulated, which can slow down surface runoff, prevent wind erosion and soil erosion, and create the possibility for the growth of other plants. Therefore, the moss is also called "pioneer".

通用型解说 | 生物多样性解说　　解说编号：TY-WZ-07

07 地衣：菌藻共生体

● 解说要点

1. 地衣的概念
2. 钱江源国家公园的三种地衣种类

● 观察地点推荐

1. S02 古田山—古田庙：探秘常绿阔叶林自然体验路线
2. Q01 莲花塘：钱塘江寻源自然体验路线
3. Q02 里秧田—大峡谷：遇见峡谷飞瀑自然体验路线

解说方式 Interpretation Style

- ☑ 人员解说 Staff Interpretation
- ☐ 教育活动 Educational Activity
- ☑ 解说牌 Interpretation Signage
- ☐ 场馆解说 Hall Interpretation

解说季节 Interpretation Season

- ☑ 春季 Spring
- ☑ 夏季 Summer
- ☑ 秋季 Autumn
- ☑ 冬季 Winter

解说词

各位朋友，在徒步的过程中，我们发现有些岩石或树干上常有一些色彩灰白色或艳丽的斑块，就像石头或者树皮上的装饰一般，它们的名字叫作地衣。

地衣和苔藓生长环境相似，它们常常相伴而生，以至于很多人误认为它们是一种生物，或干脆分不清哪个是苔藓，哪个是地衣。大家看，这块石头（树桩）上生长的，是苔藓还是地衣？这里我教给大家一个分辨苔藓和地衣的小窍门。一般来说，苔藓都是绿色的，且有比较明显的茎、叶的分化，而地衣的颜色相对来说会比较丰富一些。另外，相对苔藓来说，地衣植株较为矮小，也更加耐寒、耐旱、耐贫瘠，以及适应更极端的环境。

「什么是地衣」

地衣和苔藓在本质上也是有区别的。从生物学分类的角度来看，苔藓是一种植物，地衣却不是。难道地衣是动物？可能性也不大。实际上地衣既不是植物也不是动物，而是由藻类和真菌共生的生物复合体。

在这个“组合家庭”里，真菌扮演着“守护者”的角色，一般我们看到的部分就是地衣中的真菌，藻类则被它们用菌丝严严实实地包裹在身体里，保护它们免受强光、干旱等恶劣环境的影响，此外，真菌还会吸收环境中的水分、无机盐和二氧化碳等物质，为藻类输送养分。反过来，藻类也会通过光合作用提供真菌所需的营养物质。如果把这两种生物分开，藻类可艰难存活，而真菌则会被活活饿死。

地衣体内藻类和真菌的这种优势互补、互利共生的关系已经维系了亿万年之久。远在苔藓和蕨类植物登陆之前，地衣可能就已经开始改造地球表面的岩石圈了。它们是比苔藓更早的拓荒者和先锋物种，为地球早期环境的塑造作出了了不起的贡献。

「常见的地衣种类」

地衣是一种非常坚强的生物，从干燥缺水的热带荒漠，到南北两极这样的酷寒之地，它们的身影遍布全球。但是，它们又十分敏感脆弱，环境中的微小污染对地衣来说都有可能是致命伤害。因此，如今我们在城市中已经很难觅得地衣的踪迹。

地衣的形态有很大不同，大致可分为壳状地衣、叶状地衣和枝状地衣三大类型。其中，最低矮也是最常见的为壳状地衣，它们像是龟裂的油漆般牢固地附着在岩石或者树皮上，很难剥离；叶状地衣外形似叶片，有明显的上下表面，比较容易剥落；枝状地衣形态最为复杂，看起来似乎有枝叶一般，长得也更加高大立体。

拓展知识点

「可食用的地衣」

- 有些地区的人们将地衣当作一种食物，比如石耳、肺衣、松萝、树花等。同时，地衣也是某些野生动物重要的食物来源。比如，滇金丝猴就偏爱一种叫作松萝的地衣，这是它们在冬季食物匮乏的时节最主要的食物。还有北方的驯鹿，冬季也是靠着刨开冰雪之下的石蕊等地衣度过漫长的严寒的。

General Interpretation | Biodiversity Section

Interpretation No. TY-WZ-07

07 Lichen: Symbiosis of bacteria and algae

Key points

1. Concept of the lichen
2. Three lichen species in Qianjiangyuan National Park

Observation places recommended

1. S02 Evergreen broad-leaved forest exploration route
2. Q01 Qiantang River source exploration route
3. Q02 Canyon and waterfall experience route

Dear friends, during the hike, we found that some rocks or tree trunks have some brightly colored or gray patches, which are like wearing floral clothes on the earth. Their names are called lichens.

The growth environment of lichens and mosses is similar, and they often grow together, so that many people mistakenly believe that they are a kind of organism, or simply cannot tell which is the moss and which is the lichen. Everyone, is it moss or lichen growing on this stone (tree stump)? Here I teach you a little trick to distinguish between mosses and lichens. Generally speaking, the moss is green, and there are obvious differentiation of stems and leaves, and the color of lichen is relatively richer. In addition, compared with the moss, the lichen is more resistant to cold, drought, and barrenness, and can adapt to more extreme environments.

"What is the lichen"

Next, I want to ask you a question: Is the lichen a plant? In fact, the lichen is neither a plant nor an animal, but a biological complex composed of algae and fungi.

In this "combined family", fungi play the role of "guardian". Generally, the part we see is the fungus in the lichen, and the algae are tightly wrapped in the body with hyphae to protect them from the effects of harsh environments such as strong light and drought. In addition, fungi can also absorb water, inorganic salts, carbon dioxide and other substances in the environment to transport nutrients to the algae. In turn, algae will also provide nutrients needed by fungi through photosynthesis. If these two organisms are separated, the algae can survive hard, but the fungus will starve to death.

This complementary and mutually beneficial symbiosis relationship between algae and fungi in lichens has been maintained for hundreds of millions of years. Long before mosses and ferns landed, lichens may have begun to transform the lithosphere on the earth's surface. They are pioneers and pioneer species earlier than mosses, and they have made great contributions to the formation of early earth soil.

"Common lichen types"

The lichen is a very strong creature, you can see it from dry tropical deserts to extreme cold places like the north and south poles. But it is very sensitive and fragile. Small pollution in the environment can be fatal to lichens. Therefore, it is difficult to find lichens in cities today.

But when you come to Qianjiangyuan National Park, you will find lichens everywhere, whether on the rocks or on the tree trunks. If you observe carefully, you will find that their morphologies are quite different, which can be roughly divided into three types: shell-like lichens, leaf-like lichens and branch-like lichens. Among them, the shortest and most common ones are shell-like lichens, which are firmly attached to rocks or barks like cracked paint and are difficult to peel off; leaf-like lichens are like leaves, with obvious upper and lower surfaces, which are easier to peel off; branch-like lichens have the most complex morphology, they seem to have branches and leaves, and they grow taller and three-dimensional.

Extra information

"Edible lichen"

- People in some areas regard lichen as a kind of food, such as stone fungus, lung clothing, tree flowers and so on. At the same time, the lichen is also an important food source for certain wild animals. For example, Yunnan golden monkey prefers a lichen called usnea, which is their main food in winter when food is scarce. There are also the reindeer in the north. In winter, they also survive the long and severe cold by cutting away the litmus and other lichens under the ice and snow.

通用型解说 | 生物多样性解说　　　　解说编号：TY-WZ-08

08 认识模式标本

● 解说要点

1. 模式标本的概念
2. 钱江源国家公园的植物与昆虫模式标本

● 观察地点推荐

1. S02 古田山—古田庙：探秘常绿阔叶林自然体验路线
2. S03 瞭望台：国家公园科研体验路线

美丽马醉木　蚁墙蜂　粉花出蕊四轮香　浙江红花油茶

解说方式 Interpretation Style

☑ 人员解说 Staff Interpretation　☐ 教育活动 Educational Activity

☑ 解说牌 Interpretation Signage　☐ 场馆解说 Hall Interpretation

解说季节 Interpretation Season

☑ 春季 Spring　☑ 夏季 Summer

☑ 秋季 Autumn　☐ 冬季 Winter

解说词

大家知道地球上有多少种生物吗？这个问题其实没有答案。但目前已经被科学家正式定名的物种有130多万个，这还只是地球物种数量的冰山一角，据估计，地球上可能还有数以百万的物种未被发现。每年科学家都会发现新的物种，仅2014年一年，全球科学家就命名了18000个新物种。

「模式标本的概念」

不过，你怎么知道或者证明你发现的是不是新物种呢？很简单，和现有的物种比较差异。但是，现有物种这么多，我们需要选取一个能够代表该物种的形象，这时候就该“模式标本”登场了。

大家或许以前从未听说过模式标本这个词，但对科学家而言，这个词具有特别重要的意义。它是一种植物在世界上首次被发现时，植物学家用以作为命名证据的那份标本。模式标本是对发现者和发现地的一种认定。而钱江源所在的开化县，就是很多动植物物种模式标本的采集地。

「钱江源国家公园的植物与昆虫模式标本」

这里发现的植物模式标本有11种，比如，古田山鳞毛蕨、古田山黄精、开化鳞毛蕨、浙江红花油茶等。仔细观察，有些物种的拉丁名里就包含有拼音“gutianshan”（古田山）或“kaihua”（开化）这样的信息。

大家浏览一下表格中模式标本的采集时间一栏会发现，这些模式标本的采集和命名时间自1955年开始，一直到2017年都有新的模式物种被发现，这也从侧面说明了钱江源国家公园所保护的原真生态系统的科研价值历久弥新，仍然具有无限可挖掘的空间。

除了植物，钱江源国家公园也是众多动物模式种的采集地。其中，以古田山为模式产地的昆虫就有164种。2014年，中外科学家还在钱江源国家公园发现一种会用蚂蚁尸体封住自身巢穴来保护其后代的新物种：蚁墙蜂。它们这种独特的行为在动物界是首次发现。

拓展知识点

- 所谓模式标本，即是某种植物在世界上首次被发现时，植物学家用以作为命名证据的那份标本。模式标本一般包括一个物种典型的根、茎、叶、花、果实、种子的形态特征，能更加全面地从分类学上体现该物种。它的引入使得一个物种不再只是抽象的名称或符号，而是看得见、摸得着、可参考的实物，对物种的识别具有重要意义。
- 世界上很多自然博物馆都收藏有大量物种的模式标本，其中不乏珍稀或者已灭绝的物种，这对于我们研究物种的分类和起源等具有重要意义。

附表 模式标本采自开化县的主要解说植物

中文名	拉丁名	科	属	采集地点	采集时间	采集人
古田山鳞毛蕨	*Dryopteris gutishanensis*	Dryopteridaceae	*Dryopteris*	开化（古田山）	1981.10	郑朝宗、丁炳扬
开化鳞毛蕨	*Dryopteris kaihuaensis*	Dryopteridaceae	*Dryopteris*	开化（古田山）	1979.10	洪利兴、丁炳扬
浙江红花油茶	*Camellia chekiangoleosa*	Theaceae	*Camellia*	开化（古田山）	1955.4	王景祥
粉花广东蔷薇	*Rosa kwangtungensis* f. *roseoliflora*	Rosaceae	*Rosa*	开化（古田山）	1990.4	徐耀良，等
齿叶石灰花楸	*Sorbus folgneri* var. *duplicato-dentata*	Rosaceae	*Sorbus*	开化	1964	王景祥
粉花出蕊四轮香	*Hanceola exserta* f. *subrosa*	Lamiaceae	*Hanceola*	开化（南华山）	2016.9	丁炳扬，等
洪林龙头草	*Meehania hongliniana*	Lamiaceae	*Meehania*	开化（南华山）	2017.4	丁炳扬
洪林薹草	*Carex honglinii*	Cyperaceae	*Carex*	开化	1987.5	洪林
黄鞍竹	*Phyllostachys chlorina*	Poaceae	*Phyllostachys*	开化	1982	占荣富
古田山黄精	*Polygonatum cyrtonema* var. *gutianshanicum*	Liliaceae	*Polygonatum*	开化（古田山）	2002	金孝锋
短茎萼脊兰	*Hygrochilus subparishii*	Orchidaceae	*Hygrochilus*	开化	1959.5	浙江植物资源普查队

美丽马醉木

General Interpretation | Biodiversity Section

Interpretation No. TY-WZ-08

08 Type specimen

Key points

1. The concept of type specimen
2. Type specimen of plants and insects in Qianjiangyuan National Park

Observation places recommended

1. S02 Evergreen broad-leaved forest exploration route
2. S03 Scientific research experience route

Do you know how many kinds of living things there are on earth? There is no answer at all. We can only say that there are more than 1.3 million species officially named by scientists, but this is only the tip of the iceberg of the number of species on the earth. According to estimates, there may be millions of species on the earth that have not been discovered. Scientists discover new species every year. In 2014 alone, scientists around the world named 18,000 new species.

"Concept of type specimen"

But how do you know or prove whether you have discovered a new species? It's very simple, compared with existing species. But there are so many existing species, we need to choose an image that can represent that species, and it's time for the "type specimen" to appear.

You may have never heard of the term "type specimen" before, but for scientists, this term has a particularly important meaning. It is the specimen used by botanists as the evidence of naming a plant when it was first discovered in the world. The type specimen is an identification of the discoverer and the place of discovery. Kaihua County, where Qianjiangyuan is located, is the collection site for type specimens of many animal and plant species.

"Type specimen of plants and insects in Qianjiangyuan National Park"

There are 11 types of plant type specimens found here, such as Gutian Mountain lepidoptera (Dryopteris gutishanensis), Ancient pteridophyte (*Polygonatum cyrtonema* var. *gutianshanicum*), Kaihua lepidoptera (*Dryopteris kaihuaensis*), Zhejiang safflower camellia (*Camellia chekiangoleosa*). Observe carefully, the Latin names of some species include the spelling "gutianshan" or "kaihua".

If you look at the collection time column of the

Extra information

- The so-called type specimen is the specimen used by botanists as the evidence of naming a certain plant when it was first discovered in the world. Type specimens generally include the morphological characteristics of typical roots, stems, leaves, flowers, fruits, and seeds of a species, which can reflect the species more comprehensively. Its introduction makes a species no longer just an abstract name or symbol, but a visible, tangible and referable object, which is of great significance for species identification.
- Many natural museums in the world have collections of type specimens of a large number of species, many of which are rare or extinct. It is of great significance for us to study the classification and origin of species.

Camellia chekiangoleosa

type specimens in the table, you will find that the collection and naming time of these type specimens started in 1955 and until 2017, new type species have been discovered, which also explains Qianjiangyuan from the side that the scientific research value of the original real system protected by the national park has lasted forever, and there is still unlimited space for excavation.

In addition to plants, Qianjiangyuan National Park is also a collection site for many animal type species. Among them, there are 164 species of insects with Gutian Mountain as the type producing area. In 2014, Chinese and foreign scientists also discovered a new species in Qianjiangyuan National Park that uses ant carcasses to seal its nests to protect their offsprings. This unique behavior is the first discovery in the animal kingdom.

Attached list typical plants for interpretation whose type specimens collected from Kaihua County

Latin name	Family	Genus	Gathering place	Gathering time	Gathering person
Dryopteris gutishanensis	Dryopteridaceae	*Dryopteris*	Kaihua (Gutian Mountain)	1981.10	Chaozong Zheng, Bingyang Ding
Dryopteris kaihuaensis	Dryopteridaceae	*Dryopteris*	Kaihua (Gutian Mountain)	1979.10	Lixing Hong, Bingyang Ding
Camellia chekiangoleosa	Theaceae	*Camellia*	Kaihua (Gutian Mountain)	1955.4	Jingxiang Wang
Rosa kwangtungensis f. *roseoliflora*	Rosaceae	*Rosa*	Kaihua (Gutian Mountain)	1990.4	Yaoliang Xu, *etc.*
Sorbus folgneri var. *duplicato-dentata*	Rosaceae	*Sorbus*	Kaihua	1964	Jingxiang Wang
Hanceola exserta f. *subrosa*	Lamiaceae	*Hanceola*	Kaihua (Nanhua Mountain)	2016.9	Bingyang Ding, *etc.*
Meehania hongliniana	Lamiaceae	*Meehania*	Kaihua (Nanhua Mountain)	2017.4	Bingyang Ding
Carex honglinii	Cyperaceae	*Carex*	Kaihua	1987.5	Lin Hong
Phyllostachys chlorina	Poaceae	*Phyllostachys*	Kaihua	1982	Rongfu Zhan
Polygonatum cyrtonema var. *Gutianshanicum*	Liliaceae	*Polygonatum*	Kaihua (Gutian Mountain)	2002	Xiaofeng Jin
Hygrochilus subparishii	Orchidaceae	*Hygrochilus*	Kaihua	1959.5	Zhejiang Plant Resources Survey Team

Camellia chekiangoleosa

通用型解说 | 生物多样性解说　　　　解说编号：TY-WZ-09

09 丛林中出没的兽类

● 解说要点

钱江源国家公园代表性的野生哺乳动物：黑麂、小麂、毛冠鹿、中华鬣羚、野猪、亚洲黑熊、云豹

● 观察地点推荐

1. S02 古田山—古田庙：探秘常绿阔叶林自然体验路线
2. Q01 莲花塘：钱塘江寻源自然体验路线

黑麂　小麂　毛冠鹿　中华鬣羚　黑熊　野猪　云豹

解说方式 Interpretation Style

☑ 人员解说 Staff Interpretation　☐ 教育活动 Educational Activity
☑ 解说牌 Interpretation Signage　☐ 场馆解说 Hall Interpretation

解说季节 Interpretation Season

☑ 春季 Spring　☑ 夏季 Summer
☑ 秋季 Autumn　☑ 冬季 Winter

解说词

各位朋友，不知道大家在公园中行走的时候有没有这样的疑惑：公园里的植物种类丰富多样，森林里、溪流边偶尔也能见到一些鱼类、鸟类和两栖类、爬行类动物的身影，但是唯独缺少了大型的哺乳动物。作为钱江源国家公园的明星物种，哺乳动物们都去哪儿了呢？要知道答案的话，我们得先从认识公园的哺乳动物开始。

钱江源国家公园已经记录到的哺乳动物有 58 种，其中，最为珍贵的是黑麂（jǐ），属于国家一级重点保护野生动物。说到麂大家可能有些陌生，但是说起它所属于的鹿类大家族，每个人脑海里或许都会有一些关于这种动物的印象。麂是一种小型的鹿类，也是最古老的鹿类。中国一共有三种麂类，除了我们刚刚提到的黑麂之外，还有赤麂和小麂。小麂也是钱江源国家公园分布非常广泛的一种野生动物，数量远在黑麂之上。

我们知道，植食性的鹿类都比较温和胆小，黑麂和小麂也不例外。平时它们都躲在深山老林之中，很少出没在森林边缘或者开阔的区域，更别说是人类活动的区域了。此外，为了躲避天敌的追踪，它们进化出了非常灵敏的听觉，稍微有点风吹草动，都逃不过它们的耳朵，哪怕是轻踩落叶的声音，都可能使得它们受到惊吓，立刻逃窜开。因此，一般我们只能通过密布在丛林深处的能够 24 小时进行实时监测的红外相机窥见它们的身影。

大家看，这两张黑麂和小麂的照片就是红外相机拍摄到的。大家可以通过照片来判断两种麂类的差别吗？首先，我们从体形上可以看出，黑麂的个头比小麂略大，体长可达 1m 以上，而一般小麂体长都不到 1m。其次，从颜色上来看，虽然它们身体都为棕色，但是黑麂相比小麂体色更深一些，这也是黑麂名字的来源。

大家再仔细观察一下，这两种动物有没有哪些细节不太一样呢？是的，黑麂的头顶有一丛十分个性的棕黄色长毛。有朋友还注意到了黑麂下部的毛呈醒目的白色尾巴，这也是我们野外识别黑麂的重要标志。此外，雄性成年黑麂的嘴角还露出一对獠牙，但在它们那双无辜的大眼睛的衬托下，倒显得可爱。和黑麂相比，小麂虽没有那么突出的外貌，但我们却可以通过小麂犬吠般的叫声很容易地认出它们。

注意事项

- 通过红外相机照片，来讲述代表性野生哺乳动物的故事。
- 鹿科动物重点讲解其对于环境的适应性，注意几种麂的区别。
- 提醒访客观察动物的时候一定要动作轻缓、保持安静，不要让小动物受到惊吓，小声讲话，尽量使用手势交流。
- 不要试图干涉动物的生活，别给它们喂食物，也别去帮助它们，更不要触碰它们的洞穴或者窝巢。

辅助道具推荐

- 红外相机拍摄到的哺乳动物的物种照片
- 纪录片《秘境之眼》相关动物视频
- 音频：吴金黛《森林狂想曲——山羌（小麂）声音》

拓展活动推荐

- 解说员可以考虑打扮成黑熊或者云豹的样子，为访客解说，增加趣味性。

黑麂是我国的特有物种，现存数量不足6000只，又因为其雌雄个体的DNA数量不同，具有特别的研究价值，被称为“世界上最神秘的鹿科动物”。野生的黑麂一共有两个主要的分布区，一个在安徽南部，另一个就在我们这里，这里也是黑麂最大的一个集中分布区。尽管如此，我们却依然很难在野外见到它们的身影。

除了麂类，公园里还生活着一种正宗血统的鹿类：毛冠鹿。从名字就可以看出来，毛冠鹿的头顶也有一撮长毛，只不过和黑麂的“杀马特”黄毛不同，毛冠鹿头顶长毛为黑色，形状类似马蹄，远看就像两耳间戴了一顶小尖帽一样。它们的短角则被遮在“尖顶”之下若隐若现。另外，毛冠鹿还有个别名叫“青鹿”，这是因为它们的身体颜色偏青灰色，可以和麂类的棕色明显区分开来。

「“四不像”中华鬣（音 liè）羚」

钱江源国家公园的森林里，生活着这么一种动物：它们的角像鹿而不是鹿，蹄像牛而不是牛，头像羊不是羊，尾像驴不是驴，总体来说是“四不像”，这种动物就是中华鬣羚。大家看，这是鬣羚的照片，它们的上身黑色，四肢棕黄色，头上有一对短角。但是第一眼就被大家注意到的可能还是它们脖颈上方那又长又蓬松的白色鬃毛，潇洒飘逸。森林是中华鬣羚的家，它们常单独活动，喜欢在早晨或者傍晚出来觅食，也是素食主义者。鬣羚遇到人类惊扰时反应特别大，常乱蹦乱跳地逃跑。

「最熟悉的陌生“猪”」

钱江源国家公园的森林里，还生活有大量的野猪。和家猪相比，野猪的体形更加健壮，成年野猪毛色更深，幼猪身上则有明显的条纹。野猪背上长有长而硬的鬃毛，头部较大，四肢短粗，雄猪嘴边还有两对不断生长的獠牙露在外面，仿佛全身都在透露一个信息“不要惹我”。不过，野猪也的确不好惹，它们性情粗暴、攻击性强，受到威胁时，公猪会用獠牙来保护自己，没有獠牙的母猪则用咬的。大家如果见到野猪，一定要“敬而远之”。

如今关于野猪和人类的矛盾是人兽冲突中最典型的案例。由于野猪的食性广泛，且和人类一样，偏爱淀粉、糖和脂肪含量高的食物，因此在栖息地被破坏以后，它们常走出森林，来到人类的果园、田园寻找食物。历史上，人们通过“保苗节”等温和而有效的方式驱散它们，今天，钱江源国

拓展知识点

《IUCN红色名录》

- 《国际自然保护联盟濒危物种红色名录》（或称《IUCN红色名录》，简称红皮书）于1963年开始编制，是全球动植物物种保护现状最全面的名录，被认定为对生物多样性状况最具权威的指标。
- IUCN根据物种数目下降速度、物种总数、地理分布、群族分散程度等准则将物种分为九类。
- 绝灭（EX, extinct）
- 野外绝灭（EW, extinct in the wild）
- 极危（CR, critically endangered）
- 濒危（EN, endangered）
- 易危（VU, vulnerable）
- 近危（NT, near threatened）
- 无危（LC, least concern）
- 数据缺乏（DD, data deficient）
- 未评估（NE, not evaluated）

「国家重点保护野生动物」

- 《中华人民共和国野生动物保护法》第九条：国家对珍贵、濒危的野生动物实行重点保护。
- 国家重点保护的野生动物分为一级重点保护野生动物和二级重点保护野生动物。

家公园设立了生态补偿制度，确保人与动物能在此和谐共生。

「黑熊出没」

各位朋友，如果我们在公园中游览时突然遇到一只熊，该怎么办呢？大家不要紧张，上一次有人直接遇到熊已经是几十年前了，但 2009 年和 2014 年，钱江源国家公园的红外相机确实记录到了黑熊的出没踪迹。亚洲黑熊是钱江源国家公园目前监测到的体形最大的食肉兽类。不过，大家却不用过于担忧，因为黑熊的食谱中有 98% 是植物，只有 2% 是动物。黑熊可以直立行走，站立时高度可达近 2m，体重接近于 4 个成人加起来的重量（240kg）。黑熊最显著特征是胸部有“V”形的白斑，形似月牙，故又称“月牙熊”。同时，由于视力不佳，黑熊又得了个“黑瞎子”的外号，但其嗅觉和听觉却很灵敏。黑熊会随着食物和季节的变化而迁移，夏季它们会在高山避暑，入冬前则由海拔较高的地方转移到较低处避寒，并准备冬眠。

黑熊过去分布在中国西北的大部分地区，然而由于人类的过度捕猎（为了获取皮毛和熊胆）和栖息地的丧失，种群数量急剧减少，在野外已经很少能见到黑熊的身影。这里已是目前为数不多的黑熊可以自在生长的地方了。

「树栖生活的云豹」

猫科动物因为它们独立的个性、矫健的身形等特征，深受人类的喜爱。人们把一些体形较大的猫科动物称为大猫，比如，雪豹、老虎、美洲狮、猎豹、金钱豹……接下来我们一起认识一种不太常见且不为大家熟知的“大猫”：云豹。

“它们是非凡的运动员。在其他猫科动物当中，我没见过能像它们那样攀爬的。它们可以用单爪悬吊，可以倒挂。我见过它们做一些着实令人惊叹的事情！”来自美国动物园的一位工作人员如此赞叹道。这是一种了不起的猫科动物。

云豹的“云”字可以有两个含义：一是指它们身上有着美丽的深色云状斑纹，这种斑纹为丛林生活的云豹提供了很好的伪装，无论是吃别人还是被吃，都具有天然的优势。另一个“云”是指什么呢？不得不提到云豹的“云上生活”，当然此云并非天上飘浮的白云，而是指云豹的一生，无论是捕食还是休息，大都是在树上进行的，也难怪它们能练就倒挂树梢的本领。多亏了云豹那又长又粗的大尾巴保持着身体的平衡，它们才能在树上行走自如。

云豹名字中虽带有一个“豹”字，但和豹的关系并不亲近，是现存猫科动物中比较古老的类型。除了树栖生活之外，云豹的行踪也极为隐蔽，人们对云豹在野外的行为所知甚少。直到新技术，包括红外相机监测和无线电项圈出现以后，科学家们才有办法对云豹有了更多的了解。但当人们知道的时候，已经有些晚了。云豹数量非常稀少，分布范围也十分狭窄（只分布在南亚和东南亚部分地区）。在 20 世纪 90 年代古田山曾有关于云豹捕食家畜的记录，但此后一直未再监测到云豹活动。

General Interpretation | Biodiversity Section

Interpretation No. TY-WZ-09

09 Beasts that haunt the jungle

● Key points

Representative wild mammals in Qianjiangyuan National Park: black muntjacs, Chinese muntjacs, tufted deer, Chinese hyenas, wild boars, Asian black bears, snow leopards

● Observation places recommended

1. S02 Evergreen broad-leaved forest exploration route
2. Q01 Qiantang River source exploration route

Dear friends, I don't know if you have such doubts when you walk in the park: there are many kinds of plants in the park, and you can occasionally see some fish, birds, amphibians and reptiles in the forest and by the stream, but only large mammals are missing. As the star species of Qianjiangyuan National Park, where have all the mammals gone? To know the answer, we have to start by getting to know the mammals in the national park.

There are 58 species of mammals recorded in Qianjiangyuan National Park, among which the most precious is the black muntjac, which belongs to the state first-level key protected animals. The muntjac may sound a little strange to everyone, but when it comes to the deer family, everyone may have some impressions of this animal in their minds. The muntjac is a kind of small deer and the oldest deer kind. There are three types of muntjacs in China. In addition to the black muntjac we just mentioned, there are also the southern red muntjac and the Chinese muntjac. The Chinese muntjac is also a very widely distributed wild animal in Qianjiangyuan National Park, far more than the black muntjac.

We know that the herbivorous deer are docile and shine, and the black muntjac and the Chinese muntjac are no exception. They usually hide in deep mountains and forests, and rarely appear in forest edges or open areas, not to mention areas with human activity. In addition, in order to avoid being tracked by natural enemies, they have evolved a very sensitive hearing, nothing can escape their ears, even the sound of lightly stepping on fallen leaves may startle them so that they run away immediately. Therefore, in general, we can only see them through the infrared cameras that can monitor them in real time in the deep jungle for 24 hours.

You see, these two photos of black muntjac and Chinese muntjac were taken by an infrared camera. Can you tell the difference between the two muntjac species through photos? First of all, we can see from the body shape that the black muntjac is slightly larger than the Chinese muntjac, and the body length can reach more than 1 meter, while the general Chinese muntjac body is less than 1 meter long. Secondly, from their color, although their bodies are brown, the black muntjac is darker than the Chinese muntjac, which is also how it got its name.

Let's take a closer look and see if there are any differences between these two animals. That's right, the black muntjac has a very characteristic yellow long hair on its head. Some friends also noticed that the black muntjac has a white tail, which is very eye-catching. This is also an important sign for us to identify the black muntjac in the wild. In addition, a pair of fangs are exposed at the corners of the male adult black muntjac's mouth, but they look cute against their big innocent eyes. Compared with the black muntjac, although the Chinese muntjac does not have the prominent appearance, we can easily recognize it by its barking.

Chinese muntjac

Attentions

- Tell the story of representative wild mammals through infrared camera photos.
- The interpretation of deer is focused on their adaptability to the environment and the difference between several deer.
- Remind visitors to move gently and quietly when observing animals, and do not scare small animals. Speak quietly and try to communicate using gestures.
- Don't try to interfere with the lives of animals, don't feed them, don't help them, do not touch their burrows or nests.

Recommended auxiliary facilities and toolkits

- Photos of mammals from the infrared camera
- Animal videos related to the documentary *Eye of Secrets*
- Photo comparison of wild boars and domestic pigs
- Audio: Wu Jindai *Forest Rhapsody—Sound of Shan Qiang* (*Chinese muntjac*)

Recommended activities

- The interpreter may consider dressing up as a black bear or a snow leopard to explain to visitors and add interest.

Extra information

IUCN Red List

- Extinct (EX) – beyond reasonable doubt that the species is no longer extant.
- Extinct in the wild (EW) – survives only in captivity, cultivation and/or outside native range, as presumed after exhaustive surveys.
- Critically endangered (CR) – in a particularly and extremely critical state.
- Endangered (EN) – very high risk of extinction in the wild, meets any of criteria A to E for Endangered.
- Vulnerable (VU) – meets one of the 5 red list criteria and thus considered to be at high risk of unnatural (human-caused) extinction without further human intervention.
- Near threatened (NT) – close to being endangered in the near future.
- Least concern (LC) – unlikely to become endangered or extinct in the near future.
- Data deficient (DD). Not evaluated (NE).

"Wild animals under priority conservation"

- According to the *Wild Animal Conservation Law of the People's Republic of China*, species of wild animals under state priority conservation are divided into wild animals under Grade I conservation and wild animals under Grade II conservation.

The black muntjac is a unique species in our country, and the number is less than 6000. It has special research value because of the different amounts of DNA of male and female individuals. It is called "the most mysterious deer animal in the world". There are two wild black muntjac main distribution areas, one is in southern Anhui and the other is here. This is also the largest concentrated distribution area of black muntjac. Nevertheless, it is still difficult for us to see them in the wild.

In addition to the muntjac, there is also a pure blood deer in the park: the tufted deer. As you can tell from the name, the tufted deer also has a bunch of long hair on the top of its head, but it is different from the black muntjac's yellow hair. The hair is black, like a small pointed hat. Their short horns are hidden under it. In addition, the tufted deer are also called "green deer" because their body color is greenish-gray, which can be clearly distinguished from the brown of the muntjac.

"Multiple faces 'Chinese hyena'"

In the forests of Qianjiangyuan National Park, there lives such an animal: their horns are like deer but not deer, their hooves are like cows but not cows, their heads are like sheep but not sheep, and their tails are like donkeys but not donkeys. In general, they are "multiple faces". This kind of animal is the Chinese hyena (sounding liè). Everyone, this is a photo of hyenas. Their upper body is black, their feet are brown, and they have a pair of short horns on their heads. But what everyone noticed at first glance may be the long and fluffy white mane on the top of their neck, cool and elegant. The forest is the home of Chinese hyenas, they often move alone, like to come out for food in the morning or evening, and they are also vegetarian. When hyenas are disturbed by humans, they react very strongly, and often run away.

"The most familiar stranger 'pig'"

There are a large number of wild boars living in the forest of Qianjiangyuan National Park. Compared with domestic pigs, wild boars are more robust, growing pigs have darker coats, and young pigs have obvious stripes. The wild boar has a long stiff mane on its back, a large head, stubby feet, and two pairs of growing fangs around the male boar's mouth, as if the whole body is revealing a message "don't mess with me". However, wild boars are tough to deal with. They are rough and aggressive. When threatened, male boars use fangs to protect themselves, while females like to bite. If you see a wild boar, you must stay away from it.

The wild boar is a state second-level key protected wild animal, but the contradiction between wild boar and humans is now the most typical case of human-beast conflict. Because wild boars have a wide range of food habits, and like humans, they prefer foods with high starch, sugar and fat content. Therefore, after their habitat is destroyed, they often go out of the forest and come to human orchards and gardens to find food. Historically, people used mild and effective ways to drive them away, such as the "Bud Conservation Festival". Today, the national park has established an ecological compensation system to ensure that humans and animals can live in harmony here.

"Black bear haunted"

Dear friends, what should we do if we suddenly encounter a bear while visiting the national park? Don't be nervous! The last time someone encountered a bear directly was decades ago, but in 2009 and 2014, infrared cameras in the national park did recorded the presence of black bears. The Asian black bear is the largest carnivorous animal currently monitored in Qianjiangyuan National Park. But don't worry too much, because 98% of the black bear's diet is plants and only 2% are animals. Black bears can walk upright, stand up to nearly two meters high, and weigh close to the combined weight of four adults (240kg). The most distinctive feature of the black bear is the "V"-shaped white spot on the chest, which resembles a crescent moon, so it is also called "crescent bear". At the same time, due to poor eyesight, the black bear got the nickname "black blind", but his sense of smell and hearing is very sensitive. Black bears will migrate with changes in food and seasons. In summer, they will escape the heat in the mountains. Before winter, they will move from higher elevation to lower places to avoid the cold and prepare for hibernation.

Black bears used to be distributed in most areas of northwestern China. However, due to human overhunting (for fur and bear bile) and loss of habitat, the population has declined sharply, and black bears are rarely seen in the wild. This is currently one of the few places where black bears can live freely.

"Clouded leopards living on the tree"

Cats are deeply loved by humans because of their independent personality, vigorous body shape and other characteristics. People call some large cats in the wild as big cats, such as snow leopards, tigers, cougars, cheetahs, leopards… Next we will get to know a less common big cat: clouded leopard .

"They are extraordinary athletes. Among other cats, I have never seen one that can climb like them. They can hang on the tree with a single claw or upside down. I have seen them do some truly amazing things! " A staff member from a zoo in the US was so amazed. This is an amazing cat.

The "cloud" of the name of clouded leopards has two meanings: one means that they have beautiful dark cloud-like markings on their bodies, which provides a good camouflage for clouded leopards living in the jungle, whether they are eating others or being eaten. What does the other "cloud" mean? I have to mention the clouded leopard's "life on the cloud". Of course, this cloud is not a white cloud floating in the sky, but refers to the clouded leopard's life. Whether they are predating or resting, most of them do it on the tree. No wonder they can practice hanging on the tree upside down. Thanks to the clouded leopard's long and thick tail for keeping the balance of the body, they can walk freely on the tree.

Although the clouded leopard has the word "leopard" in its name, it is not closely related to the leopard and is an older type of cat in existence. In addition to their life on the trees, the clouded leopard's traces are also extremely hidden, and people know very little about the clouded leopard's behavior in the wild. It was not until the advent of new technologies, including infrared camera monitoring and radio collars, that scientists had a way to learn more about clouded leopards. But when people know it, it's already a bit late. The number of clouded leopards is very little, and their distribution range is very limited (only distributed in parts of South Asia and Southeast Asia). There were records of clouded leopards preying on domestic animals in Gutian Mountain in the 1990s, but clouded leopards have not been monitored since then.

通用型解说 | 生物多样性解说 解说编号：TY-WZ-10

10 穿山甲：生存还是毁灭

解说要点

1. 穿山甲的外形特征与生活习性
2. 穿山甲的保护现状与保护价值

观察地点推荐

1. S02 古田山—古田庙：探秘常绿阔叶林自然体验路线
2. Q01 莲花塘：钱塘江寻源自然体验路线

解说方式 Interpretation Style

- ☑ 人员解说 Staff Interpretation
- ☐ 教育活动 Educational Activity
- ☑ 解说牌 Interpretation Signage
- ☐ 场馆解说 Hall Interpretation

解说季节 Interpretation Season

- ☑ 春季 Spring
- ☑ 夏季 Summer
- ☑ 秋季 Autumn
- ☑ 冬季 Winter

解说词

说起穿山甲，大家应该并不陌生。这是一种已在世界上存在超过 4000 万年的古老生物，也是现存唯一身披鳞甲的哺乳动物，同时也是世界上非法走私最严重的野生哺乳动物。

「穿山甲的外形特征与生活习性」

虽然身披坚硬的铠甲，但是穿山甲却远没有人们想象的那么强大。它们性情胆小，一旦受到惊吓，就会把自己卷成球形，自我保护。但“穿山”这种“硬挖”的技能倒不是浪得虚名。穿山甲很擅长挖土打洞，它们尖锐的大长爪就是为此而生的。在没有亲眼看到之前，你几乎无法想象穿山甲打洞的速度有多快，它们 10 分钟左右就可以在土中打出一个 1m 多深的洞穴。

接下来，它们会将细长的头部和长长的舌头伸进洞里，迅速将一个蚁穴的蚂蚁或者白蚁扫荡一空。穿山甲白天常在地下的洞穴里休息，夜晚才出来活动。它们的眼睛很小，视力不佳，主要依靠敏锐的嗅觉寻找食物。

「穿山甲的保护现状与保护价值」

全球有 8 种穿山甲均受到威胁，其中，亚洲的物种要么濒危，要么极其濒危。20 多年前，随处可见穿山甲在中国南方诸多有野生穿山甲分布的省份活动。但是现在，在我国，穿山甲几乎已经在野外绝迹。

2020 年 6 月，国家林业和草原局发布公告，将穿山甲由国家二级重点保护野生动物调整为国家一级重点保护野生动物。保护等级提升意味着相应的合法猎捕审批权限也将上调，非法猎杀交易等惩处将更严格。

注意事项

- 穿山甲均为古老的野生动物，具有非常独特的外貌特征和生活习性。
- 也可以结合 2020 年的新冠疫情介绍穿山甲，加深访客对于珍稀物种的理解，进而采取实际行动保护它们，比如，不吃野味。

辅助道具推荐

- 穿山甲照片

拓展知识点

- 像其他食蚂蚁和白蚁的动物一样，穿山甲没有牙齿。穿山甲的鳞甲如同人的手指甲和犀牛角，其成分是角蛋白。过去十年中，大约有 1,000,000 只穿山甲在野外被捕。
- 据不完全统计，20 世纪 60 年代至今，我国野生穿山甲的数量已经下降了 88.88%~94.12%。我国华南地区原有的穿山甲分布区，至少有 50% 以上已经成为罕见或濒危绝迹的地区。

General Interpretation | Biodiversity Section

Interpretation No. TY-WZ-10

10 Pangolins: survival or destruction

Key points

1. The appearance characteristics and life habits of pangolins
2. The protection status and value of pangolins

Observation places recommended

1. S02 Evergreen broad-leaved forest exploration route
2. Q01 Qiantang River source exploration route

Speaking of the pangolin, everyone should be familiar with it. This is an ancient creature that has existed in the world for more than 40 million years, and it is also the only mammal in existence in armor. It is also the most illegally smuggled wild mammal in the world.

"The appearance and life habits of pangolins"
Although wearing hard armor, pangolins are far from being as strong as people think. They are timid, and when they are frightened, they will roll themselves into a ball. But pangolins are very good at digging holes. Their sharp, long claws are made for digging holes. Before you see it with your own eyes, you can hardly imagine how fast pangolins can make holes. They can make a hole more than one meter deep in the soil in about 10 minutes.

Next, they will stick their slender heads and long tongues into the holes, and quickly sweep away the ants or termites in an ant nest. Pangolins often rest in underground caves during the day and only come out at night. They have small eyes and poor eyesight, and they rely on a keen sense of smell to find food.

"The protection status and value of pangolin"
Eight species of pangolins are threatened globally, and Asian species are either endangered or extremely endangered. More than 20 years ago, pangolins were seen everywhere in many provinces in southern China where wild pangolins were distributed. But now, in our country, wild pangolins have almost disappeared in the wild.

In June 2020, the National Forestry and Grassland Administration issued an announcement to adjust pangolins from the state second-level key protected wild animals to the state first-level key protected wild animals. The increase in the protection level means that the authority

of corresponding legal hunting permission will also be increased, and the punishment for illegal hunting transactions will be stricter.

Attentions

- Pangolins are ancient wild animals with very unique appearance characteristics and living habits.
- You can also introduce pangolins with the Coronavirus epidemic in 2020, deepen visitors' understanding of rare species so as to make them take practical actions to protect pangolins, such as not eating wild animals.

Recommended auxiliary facilities and toolkits

- photos of pangolins

Extra information

- Like other ant-eaters, pangolins have no teeth. The scales of pangolins are like human fingernails and rhino horns, and their ingredients are keratin. In the past ten years, approximately 1,000,000 pangolins have been captured in the wild.
- According to incomplete statistics, since the 1960s, the number of wild pangolins in my country has dropped by 88.88% – 94.12%. At least 50% of the original pangolin distribution areas in South China have become rare or endangered areas.

通用型解说 | 生物多样性解说　　　　解说编号：TY-WZ-11

11 与溪流相伴而生的鸟类

● 解说要点

1. 常见溪流鸟类：白冠燕尾、灰背燕尾、小燕尾、红尾水鸲、褐河乌、紫啸鸫、白鹡鸰、灰鹡鸰

2. 溪流与溪流之间相互依存的关系

● 观察地点推荐

1. Q01 莲花塘：钱塘江寻源自然体验路线

2. Q02 里秧田—大峡谷：遇见峡谷飞瀑自然体验路线

白冠燕尾　小燕尾　灰背燕尾　红尾水鸲（雄）　红尾水鸲（雌）　紫啸鸫　褐河乌　白鹡鸰　灰鹡鸰

解说方式 Interpretation Style

☑ 人员解说 Staff Interpretation　☐ 教育活动 Educational Activity

☑ 解说牌 Interpretation Signage　☐ 场馆解说 Hall Interpretation

解说季节 Interpretation Season

☑ 春季 Spring　☑ 夏季 Summer

☑ 秋季 Autumn　☑ 冬季 Winter

解说词

各位朋友，当我们在森林中行走的时候，偶尔会听到几声鸟叫传到我们耳边，但是顺着鸟叫的声音望去，却很难在繁茂的枝叶中寻觅到它们的身影。但是，在我们面前的溪流边，要观察鸟儿就容易一些。在这里，我们经常可以看到一些与湍急的溪流相伴而生的鸟类：比如，灰白两色的燕尾、长着红尾巴的红尾水鸲、全身乌褐色的褐河乌以及身披华丽紫色羽毛的紫啸鸫等。

「三种燕尾」

想必大家已经注意到了，和平原地区的大江大河不同，山间溪流大都水流湍急、水质清澈，且河底多碎石，这样的环境是燕尾属鸟类的乐园。从名字就可以看出来，燕尾长着燕子一样剪刀似的尾巴，除此之外，燕尾的羽毛也和家燕一样，偏黑、白、灰三色。

钱江源的溪流边常出现的燕尾有白冠燕尾、灰背燕尾和小燕尾三种。那么我们怎么区分这三种燕尾呢？咱们不妨从名字开始。白冠燕尾的前额和顶冠为白色，有时会耸起成小凤头状；灰背燕尾和其他燕尾的区别在于头顶及背部为灰色；小燕尾的得名是因为其体形较小，体长约 13cm，比常见的麻雀小一些。

尾巴也是我们区分三种燕尾的一种重要特征。白冠燕尾体形略大，尾羽修长，有明显的分叉，尾尖为白色；灰背燕尾体形比白冠燕尾略小，尾羽相较于白冠燕尾更为整齐，近似梯形，尖端白色且有些微分叉；小燕尾的体色和白冠燕尾相似，但尾巴较短且分叉很浅。

为了方便在溪流浅水区行走，燕尾家族的鸟儿们大都长着长腿长爪，颜色近似鲜艳的肉粉色。燕尾常与溪流相伴而生，尤其是小燕尾，它们一般生活在海拔 1000～3500m 的山区溪流边，会随着季节上下迁徙。小燕尾对溪流环境非常挑剔，容不得一丝污染，因此小燕尾也常被用作优良水质的指示物种。

注意事项

- 这里提到的溪流鸟类众多，在实际解说时，需结合具体场景，有选择性地进行解说，并根据情境，对解说词进行适当修改。

辅助道具推荐

- 各种溪流鸟类图片

「红尾水鸲」

大家如果在公园里看到一种红尾巴、蓝身子，体形和麻雀差不多大小的鸟类，那八成就是红尾水鸲了。红尾水鸲是一种雌雄异色的鸟类，栗红色的尾巴是雄鸟最为抢眼的标志，和它们灰蓝色的身体形成鲜明的对比。雌鸟全身布满灰褐色，腹部有鳞状白色斑纹，和麻雀还有几分相似。红尾水鸲很少集群，常独自或成对活动于山泉溪涧或者村庄附近的池塘堤岸边。以昆虫为食，也吃少量的植物果实和种子。红尾水鸲的雄鸟不仅羽色更艳丽，性格相对于雌鸟也更为活泼大胆一些，比较容易被观察到。在岩石上停歇或者快速移动时，红尾水鸲尾巴经常摆动，就像手中挥舞着一把扇子一样不停地打开合拢。

「褐河乌」

褐河乌是钱江源国家公园里的常住居民，如果你在水边看到一种全身深褐色、羽色暗沉，喜欢贴着水面飞行的小型鸟类，可能就是它了。褐河乌具有高超的游泳与潜水技能，能在水底行走，常捕食水底的昆虫或者小鱼、小虾。当然，褐河乌对生活环境也十分挑剔，它们生活的溪流必须水质清澈无污染、水流不能太急或太深。褐河乌终年活动于溪流中的大石、倒木或者河岸树枝上，寻觅水中的昆虫或者蚯蚓等小型无脊椎动物。

「紫啸鸫」

紫啸鸫全身的羽毛都是蓝紫色，泛着金属光泽，在阳光的照耀下非常绚丽夺目，因此又名“琉璃鸟”。紫啸鸫是一种体形较大的鸟类，身长可超过30cm。它们常出双入对，在灌丛中互相追逐，边飞边鸣，声音洪亮短促，音调高且尖锐。紫啸鸫是一种比较亲人的鸟类，一般栖息于中低山区的森林溪流沿岸，以昆虫、虾蟹为食，也会吃浆果和其他植物。

「两种鹡鸰」

和那些因外形特征而命名的鸟类不同，鹡（jí）鸰（líng）是因为叫声与“jí líng”近似而得名的。从外形上看，鹡鸰无论是体形还是尾羽都比较纤细修长，在地上行走时，尾羽会一上一下地摆动，飞行时则呈波浪状起伏且边飞边叫。公园里有白鹡鸰与灰鹡鸰两种，颜色是区分它们最好的办法。白鹡鸰全身由黑、白、灰三色构成，上身羽毛为灰色，下体白色，翅膀和尾巴则是黑白相间的颜色；灰鹡鸰的上背为灰色，腰黄绿色，下体黄色，飞行时会显现白色的翼斑和黄色的腰部。

山间的溪流水流湍急、水质清澈、环境多变，能为鸟儿提供的食物并不多，因此鸟种稀少。这些与溪流相伴而生的鸟类，早已与溪流融合为一个整体，既不能离开，又必须选择清澈干净的溪流。水质、流速乃至水的深浅都有可能会影响到它们的生存。反过来，如果没有鸟类，溪流也必将成为一片死寂的世界。

「普通翠鸟」

翠鸟又叫“鱼狗”，两个名字分别代表了它们典型外貌和突出习性。普通翠鸟的翠来自于它们蓝绿色的羽毛，在阳光的照耀下，泛着金属光泽，十分华美；普通翠鸟的蓝绿色羽毛上还点缀有白色点斑，好像画家不经意泼上去的一样；此外，橙棕色的下体也是我们辨别翠鸟的一个重要特征之一。

翠鸟是一种广泛分布于全球各地的鸟类，生活在各种各样天然的或人工的水域环境，比如，淡水湖泊、溪流、鱼塘等。翠鸟常把家安在水边的土崖壁上，既可以躲避一些地面捕食者，也可以方便捕食。翠鸟喜欢单独行动，常独立地停栖在近水边的树枝或者岩石上，等待时机，迅速地扎进水中捕鱼，即使在水中，也能保持极佳的视力，具有高超的捕鱼本领，“鱼狗”一名即由此得来。实际上翠鸟食性广泛，除了小鱼之外，还会捕食虾蟹、昆虫、蛙类和少量水生植物。

普通翠鸟

General Interpretation | Biodiversity Section

Interpretation No. TY-WZ-11

11 Birds that accompany streams

● Key points

1. Common stream birds: the white-crowned forktail, slaty-backed forktail, little forktail, plumbeous water redstart, brown dipper, blue whistling thrush, white wagtail and grey wagtail

2. The interdependent relationship between streams and streams

● Observation places recommended

1. Q01 Qiantang River source exploration route

2. Q02 Canyon and waterfall experience route

Dear friends, when we are walking in the forest, we occasionally will hear birds chirping, but it is difficult to find them among the lush foliage. But beside the stream in front of us, the birds have nowhere to hide. Here, we can often see some birds that live with the turbulent stream: for example, gray-white forktails, plumbeous water redstarts with red tails, brown dippers, and gorgeous blue whistling thrushes, *etc.*.

"Three types of forktails"

Everyone must have noticed that unlike the big rivers in the plains, the mountain streams are mostly turbulent, clear, and there are gravels at the bottom. This environment is a paradise for forktail birds. It can be seen from the name that the forktail has a fork-like tail like a swallow. In addition, the feathers of the forktail are the same as the swallow, which are black, white and gray.

There are three types of forktails, white-crowned forktail, gray-backed forktail and little forktail. So how do we distinguish these three forktails? Let's start with the name. The forehead and crown of the white-crowned forktail are white, and sometimes rise into a small crested head; the difference between the gray-backed forktail and other forktails is that the top and back of the head are gray; the little forktail is named because of its small size and body. It is about 13cm long and smaller than the common sparrow.

The tail is also an important feature that distinguishes the three types of forktails. The white-crowned forktail is slightly larger, its tail feathers are slender, with obvious bifurcation, and the tail tip is white; the gray-backed forktail is slightly smaller than the white-crowned forktail, and its tail feathers are more neat than that of the white-crowned

White-crowned forktail
White wagtail
Plumbeous water redstart (female)
Plumbeous water redstart (male)
brown dipper
blue-whistling thrush

forktail, similar to a trapezoid, with a slightly white tip forked; the body color of the little forktail is similar to the white-crowned forktail, but its tail is shorter and its forks are very shallow.

In order to facilitate walking in the shallow waters of streams, most of the forktail family birds have long legs and claws with vibrant pink color. Forktails often live along streams, especially little forktails. They generally live near mountain streams at an elevation of 1,000 to 3,500 meters and migrate up and down with the seasons. The little forktail is very picky about the stream environment and cannot tolerate any pollution, so the little forktail is often considered as an indicator species for good water quality.

Attentions

- There are many stream birds mentioned here. During the actual interpretation, please make appropriate modifications according to the situation.

Recommended auxiliary facilities and toolkits

- Photos of various stream birds

"Plumbeous water redstart"

If you see a red-tailed, blue-bodied bird in the national park that is about the size of a sparrow, it is probably plumbeous water redstart. The plumbeous water redstart is a dioecious bird. The red tail is the most eye-catching sign of a male bird, which forms a sharp contrast with their gray-blue body. The female bird is covered with gray-brown body and has scaly white markings on the abdomen, which is somewhat similar to sparrows. Plumbeous water redstart rarely colonizes, and often lives alone or in pairs on the banks of mountain springs, streams or ponds near villages. They feed on insects, but also eat a small amount of fruits and seeds. The male plumbeous water redstart not only has a brighter plumage, but also more active than the female, and is easier to be observed. When resting or moving fast on a rock, the plumbeous water redstart's tail often swings, opening and closing like a fan.

"Brown dipper"

The brown dipper is a resident of Qianjiangyuan National Park. If you see a small bird with dark brown plumage by the water, it may be it. The brown dipper has superb swimming and diving skills, which makes it can walk underwater, and often prey on underwater insects or small fishes and shrimps. Of course, brown dipper is also very picky about the living environment. The streams they live in must be clean and pollution-free, and the water must not be too rapid or too deep. The brown dipper lives on the rocks, fallen trees, or riverbank branches by the streams all year round, looking for small invertebrates such as insects or earthworms in the water.

The plumage of the blue whistling thrush is blueish-purple with a metallic luster. It is very dazzling under the sunlight, so it is also known as the "glass bird". The blue whistling thrush is a large bird, up to 30cm in length. They often come and go in pairs, chasing each other in the bushes, whistling while flying, their voices are loud and short, and the pitch is high and sharp. Blue whistling thrush is a kind of relatively human friendly bird, generally inhabiting along the forest stream in the middle and low mountains. It feeds on insects, shrimps and crabs, and also eats berries and other plants.

"Two wagtails"

Unlike the birds that are named because of their appearance, the wagtail is named because it sounds similar to pronunciation of the word "wagtail" in Chinese. From their appearances, the wagtail is relatively slender, both in size and tail feathers. When walking on the ground, the tail feathers will swing up and down, and when flying, they will undulate in a wavy shape and whistle while flying. There are two kinds of wagtails, white wagtail and grey wagtail in the national park. Color is the best way to distinguish them. The whole body of the white wagtail is black, white and gray. The upper body feathers are gray, the lower body is white, and the wings and tail are black and white. The gray wagtail has gray upper back, yellowish green waist, and yellow lower body, which will appear white when flying with white spots on the wings.

The mountain streams are turbulent, the water quality is clear, and the environment is changing all the time. There is not much food for the birds, so bird species are scarce. These birds that accompany the stream have long been integrated with the stream. They cannot leave, and must choose a clear and clean stream. Water quality, flow rate and even water depth may affect their survival. *Vice versa*, if there were no birds, the stream would surely become a dead world.

Common kingfisher

"Common kingfisher"

Kingfisher is also called "fish dog", and the two names represent their typical appearance and outstanding habits. The green of the common kingfisher comes from their blue-green feathers, which are glowing with a metallic luster under the sun, looking gorgeous; the blueish-green feathers of the common kingfisher are also dotted with white spots, as if the painter accidentally splashed it on; in addition, the orange-brown lower body is also one of the important characteristics for us to distinguish kingfishers.

Kingfisher is a bird that is widely distributed all over the world, living in various natural or artificial water environments, such as freshwater lakes, streams, and fish ponds. Kingfishers often place their homes on the cliffs near the water, which can help them avoid some ground predators and also find food easily. Kingfishers like to act alone, often staying by themselves on branches or rocks near the water, waiting for the opportunity, and diving into the water quickly, even in the water, they can maintain excellent eyesight and have superb fishing skills. The word, "Fish Dog", is derived from this. In fact, the kingfisher has a wide range of food. In addition to small fish, it also preys on shrimps, crabs, insects, frogs and a small amount of aquatic plants.

通用型解说 | 生物多样性解说　　解说编号：TY-WZ-12

12 在林中生活的鸟类

解说要点

1. 树林中华丽的地栖鸟类：白颈长尾雉、白鹇、仙八色鸫
2. 其他林鸟：红嘴相思鸟

观察地点推荐

1. S01 古田飞瀑：丛林飞瀑自然体验路线
2. S02 古田山—古田庙：探秘常绿阔叶林自然体验路线
3. S03 瞭望台：国家公园科研体验路线
4. Q01 莲花塘：钱塘江寻源自然体验路线

白鹇　白颈长尾雉　仙八色鸫

解说方式 Interpretation Style

- ☑ 人员解说 Staff Interpretation
- ☐ 教育活动 Educational Activity
- ☑ 解说牌 Interpretation Signage
- ☐ 场馆解说 Hall Interpretation

解说季节 Interpretation Season

- ☑ 春季 Spring
- ☑ 夏季 Summer
- ☑ 秋季 Autumn
- ☑ 冬季 Winter

解说词

行走在钱江源国家公园的茂密山林间，我们会听到各种各样的鸟叫声，但大家的第一反应往往是抬头看，试图在层层叠叠的枝叶间寻找到它们的身影。但是，你知道吗？森林里从地表到树干再到树顶，都是鸟儿们的安身之所。我们把生活在森林最底层，也就是地上的鸟类叫作地栖型鸟类。

据说鸟类的祖先是恐龙，现在的很多地栖型鸟类还保留着恐龙不善飞行善行走的一些特征，比如，家鸡。生活在钱江源国家公园里久负盛名的白鹇、白颈长尾雉等与家鸡就是近亲，但个头却比家鸡大得多，体态更优雅，羽色也更华美，令许多观鸟爱好者趋之若鹜。

「珍贵稀有的华丽林鸟：白颈长尾雉」

白颈长尾雉是钱江源国家公园最为珍稀的鸟类，属于国家一级重点保护野生动物，也是我国特有种，野外种群数量仅 1 万多只。钱江源国家公园里生活的白颈长尾雉约有五六百只，是全国野生白颈长尾雉种群密度较高的区域之一。

白颈长尾雉常三五成群，集群生活在 800m 以下的常绿阔叶林或者混交林中，为了躲避天敌的追捕，它们常在晨昏或者阴雨天出来活动、觅食，生性机警，行踪隐蔽，再加上它们不喜欢鸣叫的个性与稀少的种群数量，很难在野外被直接观察到。为了了解白颈长尾雉的种群数量和生存现状，公园内现已覆盖了全部区域的红外相机网格化监测网络，人们才得以逐渐揭开白颈长尾雉生活的神秘面纱。

（展示一张红外相机拍摄到的白颈长尾雉的照片）请大家看看这张红外相机拍摄的照片，告诉我图上有几只鸟？没错，是两只。但是乍一看，还以为只有一只呢。这只羽色华丽、颈部白色、尾羽修长的是雄鸟，旁边那只布满棕白色斑点的是雌鸟。如果大家在野外有幸遇到白颈长尾雉，相信大家一眼就可以将它们认出来。

注意事项

- 林鸟不易观察，且不易靠近；
- 条件允许的话，可佩带几个望远镜。
- 从色彩、典型特征、观察场所等角度重点阐述。

辅助道具推荐

- 各种鸟类照片
- 成年人的手掌张开约有 20cm 左右宽，可以作为参考来比对鸟类的大小

拓展活动推荐

- 鸟类涂色游戏（可选白鹇、白颈长尾雉、红嘴蓝鹊等）。

「优雅的林中隐士：白鹇」

在古田山中行走，特别在清晨和傍晚，不经意就可能遇到一种身披白衣的大鸟：白鹇。它们虽不如白颈长尾雉羽色鲜艳，但是对于偏好素雅的中国人来说，这绝对是一种怎么赞美都不为过的鸟类。大诗人李白曾用“朝步落花闲”形容这种鸟类超凡脱俗的品性，清代更是将白鹇图案作为五品文官补服的特定品级徽识，取其清廉、正直之意。

我们在野外虽可以远观白鹇，却很难近距离观赏它们，这张照片为大家清晰地展示了白鹇的样貌。这里的两只鸟都是白鹇，一雄一雌。雄鸟上身披着件一雪白色的素衣，一直延伸到长长的尾羽，头部和下体则为黑色，眼周和双脚为鲜艳的红色。雄性白鹇的羽色并非只有单纯的白，仔细观察会发现上面有极细的黑色纹理，宛如云纹一样华美而秀丽，十分令人惊艳。白鹇的雌鸟体形只有雄鸟的一半大小，羽色为朴素的橄榄色，和周围的环境巧妙地融为一体，很难被发现。不过，白鹇夫妇常成队活动。白天，它们常在林间悠闲地散步、觅食，并时不时发出“咕咕咕”的叫声，到了夜晚它们则会成群结对地在树上睡觉。

「羽色华丽的林中仙子：仙八色鸫」

除了白鹇、白颈长尾雉这样的大鸟，钱江源国家公园的森林中还生活着许多体色艳丽的小鸟。其中，色彩最丰富的非仙八色鸫莫属。从名字就可以看出来，它们羽毛的颜色多达八种，如果鸟类也有选美大赛的话，那么仙八色鸫一定是名列前茅的。高颜值的仙八色鸫深受观鸟爱好者的喜欢，被他们亲切地称为“小八”。仙八色鸫是一种非常机敏、胆小的地栖型鸟类，很难靠近，另外，由于它们行踪隐秘，常独自在林灌丛中活动，不仔细观察，是很难发现它们的。目前，仙八色鸫全球数量已不足一万只，被评为国家二级重点保护野生动物、IUCN 易危物种，由于森林砍伐和人为干扰造成的栖息地破碎化正在不断威胁着这种美丽的小鸟，国内稳定的繁殖地屈指可数。

白鹇

「红嘴巴的小鸟：红嘴相思鸟」

相较于仙八色鸫的鲜少露面，同样身材娇小、羽色艳丽的红嘴相思鸟就亲民多了。它们的嘴巴红红的，上体羽毛为橄榄绿色，下体橙黄色，翅膀尖端的红色和黄色在歇息时十分显眼。红嘴相思鸟不仅外形美丽，而且叫声响亮婉转，生性活泼喧闹，常集群活动于次生林的林下植被中，主要以毛毛虫、甲虫、蚂蚁等昆虫为食，也吃果实、种子等植物性食物。

红嘴相思鸟

红嘴相思鸟

General Interpretation | Biodiversity Section

Interpretation No. TY-WZ-12

12 Birds living in the forest

Key points

1. Gorgeous ground-dwelling birds in the woods: the Elliot' s pheasant, silver pheasant and fairy pitta
2. Other forest birds: red-billed leiothrix

Observation places recommended

1. S01 Jungle and waterfall experience route
2. S02 Evergreen broad-leaved forest exploration route
3. S03 Scientific research experience route
4. Q01 Qiantang River source exploration route

Walking in the dense mountains and forests of Qianjiangyuan National Park, we will hear all kinds of birds calling, and our first reaction is often to look up and try to find them among the layers of branches and leaves. But do you know from the surface to the trunk then to the top of the tree, the forest is a shelter for the birds? We call the birds that live at the bottom of the forest the ground-dwelling birds.

It is said that the ancestors of birds are dinosaurs, and many land-dwelling birds still retain some of the characteristics that dinosaurs have, such as not good at flying and walking, like farm chickens. The well-known silver pheasants and Elliot' s pheasants living in Qianjiangyuan National Park are close relatives of domestic chickens, but they are much larger than domestic chickens, their bodies are more elegant and their plumages are more beautiful, which make many bird lovers scramble for watching them.

"Precious and rare gorgeous forest bird: Elliot's pheasant"

The Elliot' s pheasant is the rarest bird in Qianjiangyuan National Park. It is a state first-level key protected animal and is also endemic to our country. The wild population is only more than 10,000. There are about 500 or 600 Elliot' s pheasants living in the national park. Qianjiangyuan National Park is one of the areas with a higher population density of wild Elliot' s pheasants in the country.

Elliot' s pheasants often live in groups of three to five. They live in evergreen broad-leaved forests or mixed forests below 800 meters. In order to avoid hunting by natural enemies, they often come out for food in the morning and dusk or on cloudy and rainy

silver pheasant

days. They are vigilant and good at hiding, coupled with their silent personality and small population.

They are difficult to be directly observed in the wild. In order to understand the population and survival status of the Elliot's pheasant, the infrared camera grid monitoring network has covered all areas in the national park, and people can gradually uncover the mystery of Elliot's pheasant's life.

(Show a photo of Elliot's pheasant taken by an infrared camera) Please take a look at this photo taken by an infrared camera. Tell me how many birds are in the picture? Yes, there are two. But at the first glance, I thought there was only one. This gorgeous plumage, with a white neck and slender tail feathers, is a male bird, and the one with brownish white spots next to it is a female bird. If you are lucky enough to encounter Elliot's pheasants in the wild, you can recognize them at the first sight.

Attentions

- Forest birds are not easy to be observed and not easy to be approached.
- If conditions permit, you can wear several telescopes.
- Emphasize on the perspectives of color, typical characteristics, observation places, *etc.* to conduct interpretation.
- The palm of an adult is about 20cm wide, which can be used as a reference to compare the size of birds.

Recommended auxiliary facilities and toolkits

- Photos of various birds

Recommended activities

- Bird coloring game (silver pheasant, Elliot's pheasant, red-billed blue magpie, *etc.*).

"The graceful hermit in the wood: silver pheasants"

Walking in the Gutian Mountains, especially in the early morning and evening, you may inadvertently encounter a big bird in white: the silver pheasant. Although they are not as bright as the Elliot's pheasants, they are definitely famous birds for the Chinese who prefer simple elegance. The great poet Li Bai used to describe the extraordinary qualities of this bird as "walking among the flowers in the morning". In the Qing Dynasty, the silver pheasant pattern was used as a special grade emblem for the five-rank government officers, which means being clean and honest.

Although we can see the silver pheasants from a distance in the wild, it is difficult to see them up close. This photo clearly shows the appearance of the silver pheasants. The two birds here are silver pheasants, one male and one female. The male bird wears a snow-white plain clothes on the upper body, extending to the long tail feathers, the head and lower body are black, and the eyes and feet are bright red. The plumage color of the silver pheasant is not just pure white.

If you look carefully, you will find that there are very fine black textures on the male bird, which is as gorgeous and beautiful as a cloud pattern, truly amazing. The female of the pheasant is only half the size of a male bird, and the plumage is a simple olive color, which is ingeniously integrated with the surrounding environment and is difficult to be discovered. However, the couples often move in pairs. During the day, they often wander in the woods, foraging for food, and make "cuckling" sounds from time to time. At night, they would sleep in groups on trees.

"The fairy in the forest with gorgeous feathers: the fairy pitta"

In addition to the big birds such as the Silver pheasant and the Elliot's pheasant, there are many beautiful small birds living in the forest of Qianjiangyuan National Park. Among them, the most colorful is the fairy pitta. As you can see from the name, they have eight colors all over their bodies. If there are also beauty pageants for birds, the fairy pitta must be among the best. The good looking fairy pitta is very popular among birdwatchers, and is affectionately called the "little eight". The fairy pitta is a very alert and timid ground-dwelling bird, which is difficult to be approached. In addition, because of their secret traces, they often move alone in the bushes. It is difficult to find them without careful observation. At present, the number of fairy pittas in the world is less than 10,000, and it has been rated as a state second-level key protected animal and an IUCN vulnerable species. The fragmentation of the habitat caused by deforestation and human disturbance is constantly threatening this beautiful bird. There are only a handful of stable breeding places in China.

"Red-billed bird: red-billed leiothrix"

Compared with the seldom appearance of the fairy pitta, the red-billed leiothrix with its petite figure and bright plumage is much more friendly to human beings. Their mouths are red, the upper body feathers are olive green, the lower body orange-yellow, and the red and yellow on the tips of the wings are very prominent when resting.

Red-billed leiothrixes are not only beautiful in appearance, but also have a loud voice, lively and vivid, and they often cluster in the understory vegetation of the secondary forest. They mainly feed on insects such as caterpillars, beetles, and ants. They also eat plant foods such as fruits and seeds.

fairy pitta

通用型解说 | 生物多样性解说　　　　解说编号：TY-WZ-13

13 “森林医生”：啄木鸟

● 解说要点

1. 国家公园代表性啄木鸟——星头啄木鸟及其对森林的贡献
2. 啄木鸟特殊的生理构造及其功能
3. 其他鸟种介绍：大拟啄木鸟、红嘴蓝鹊

● 观察地点推荐

1. S02 古田山—古田庙：探秘常绿阔叶林自然体验路线
2. S03 瞭望台：国家公园科研体验路线
3. Q01 莲花塘：钱塘江寻源自然体验路线

星头啄木鸟的树洞　　星头啄木鸟

解说方式 Interpretation Style

☑ 人员解说 Staff Interpretation　☐ 教育活动 Educational Activity
☑ 解说牌 Interpretation Signage　☐ 场馆解说 Hall Interpretation

解说季节 Interpretation Season

☑ 春季 Spring　☑ 夏季 Summer
☑ 秋季 Autumn　☑ 冬季 Winter

解说词

各位朋友，我们知道森林中的很多鸟儿都以昆虫为食，但是说到“森林医生”，大家却不约而同地会想到啄木鸟。这是为什么呢？有朋友说啄木鸟啄木时会制造声响，比较容易引起大家的注意。但是，啄木鸟怎么知道树洞里有没有害虫呢？其实，啄木鸟在疯狂啄木之前，还有“听诊”的本领，它们会通过用嘴巴敲打树干为树木进行“身体检查”，听声音不对后就一阵猛啄，直至捉到害虫为止。

「国家公园的星头啄木鸟及其对森林的贡献」

星头啄木鸟为钱江源国家公园最常见的啄木鸟类之一，这种啄木鸟体形和麻雀大小差不多，全身具黑白条纹。它们个头虽小，但是啄起树干来可是一点也不含糊。它们常单独或成对活动，多在树木中上部活动和取食，偶尔也到地面倒木取食。星头啄木鸟主要以天牛、刺蛾、毒蛾等昆虫为食，且食量惊人，一只啄木鸟每天能吃掉 1000~1400 条害虫的幼虫，在一块 $100hm^2$ 的树林里，如果有 3 对啄木鸟栖居，就可以控制住害虫的蔓延了。

「啄木鸟特殊的生理构造及其功能」

既然叫啄木鸟，哪能没有点看家本领呢。首先，啄木鸟能够在树上打洞，这离不开它们又长又硬的嘴巴。但这只是冰山一角。有数据表明，啄木鸟平均每天要啄木 12000 次，每秒钟可以啄 20 次，在这样的高频率冲击下，啄木鸟小小的头是如何承受如此的重击而没有发生脑震荡呢？这还要归功于啄木鸟大脑周围特殊的保护层：含有液体的海绵状的骨骼，再加上头骨外侧具有减震作用的肌肉，一起共同塑造了一个非常强大的头部防震系统，可以有效地缓冲和消解来自外部的冲击力。不过，要说啄木鸟身上最奇妙的结构，非舌头莫属。我们可以想象，为了抓到树洞里的食物，啄木鸟的舌头必须很长，而且能够分泌黏液，以便能牢牢地抓住虫子。不过这么长的舌头，平时放在哪里呢？你知道吗？啄木鸟的舌根居然是长在右鼻孔中的！这条有弹性的舌头会从脑袋的前半部分的顶上伸展向后，绕过后脑壳，再从下颌伸出进入口腔，整整绕了头骨一圈，待到需要时伸出捕食，不用时卷起放好，还能起到减震的效果。

注意事项

- 从对啄木鸟的既定印象出发逐步引导，解释敲打树干的独特生理结构的奇妙。
- 选取两种体形、习性差别明显的啄木鸟进行比较。

辅助道具推荐

- 啄木鸟生理结构图片
- 啄木鸟的啄木声
- 成年人的手掌张开约有 20cm 左右宽，可以作为参考来比对鸟类的大小除此之外，啄木鸟的“站姿”也很特别，它们不像其他鸟类，是站立在树枝上的，而是攀援在直立的树干上的。这样难道不会滑下来吗？我们来继续看看啄木鸟的身体构造：为了“抓”住树干，啄木鸟的鸟爪不仅锐利，而且四趾中有两趾在前，两个趾在后，和一般鸟类三趾在前而一趾在后的结构稍有差异。此外，它们的尾巴呈现楔形，羽轴坚硬且富有弹性，攀爬时可以很好地支撑住身子。因为这样特殊的身体结构，啄木鸟可以非常灵活地在树干上迅速移动，无论是向上跳跃还是向下反跳，抑或是向两侧转着圈爬行，似乎都不是什么难事儿。

「大拟啄木鸟」

大拟啄木鸟名字中虽有“啄木鸟”三字，但却主要以马桑、五加科植物以及其他植物的花、果实和种子为食，偶尔才会取食昆虫。它们之所以被叫作“啄木鸟”，其实是因为它们那粗厚的大嘴巴。大拟啄木鸟体形相比一般啄木鸟较大，颜色也更为绚丽。全身色彩大致可以分为 3 段：靠近头、颈的为蓝色，中间偏暗褐色，尾巴为草绿色。大拟啄木鸟常单独或成对活动，在食物丰富的地方有时也成小群，常栖于高树顶部，会啄出树洞用于筑巢。这些树洞巢通常不会连续 2 年被使用，于是废弃的巢穴可能被领鸺鹠、领角鸮或红背鼯鼠利用。大拟啄木鸟通常不易被看见，却容易被听见，因为它的叫声非常嘹亮，像婴儿的哭声一样，容易被察觉。

大拟啄木鸟

「美丽却强悍：红嘴蓝鹊」

红嘴蓝鹊是国家公园里一种较容易被观察到的体形较大，身形优美的鸟类。它们鲜红的喙与黑色的头部形成鲜明的对比，长长的蓝灰色尾羽和红色的脚让它们无论是飞翔还是停落都十分显眼。自古以来，在中国的山水画中就常见它们的优雅身影。红嘴蓝鹊广泛分布于中国东部和南部中低海拔的各种生境，甚至包括城市。它们不喜欢独居，常三五成群，结伴生活。别看红嘴蓝鹊外形华美，个性却十分强悍。它们团结、领域性强，有时会主动结群攻击其他鸟类，包括猛禽，甚至侵入它们的领地去威胁雏鸟的人类。

红嘴蓝鹊

General Interpretation | Biodiversity Section

Interpretation No. TY-WZ-13

13 "Forest doctor" —— Woodpecker

● Key points

1. The representative woodpecker of Qianjiangyuan National Park—The grey-capped pygmy woodpecker and its contribution to the forest
2. The special physiological structure and function of woodpecker
3. Introduction to other birds: the great barbet and red-billed blue magpie

● Observation places recommended

1. S02 Evergreen broad-leaved forest exploration route
2. S03 Scientific research experience route
3. Q01 Qiantang River source exploration route

Dear friends, we know that many birds in the forest feed on insects, but when it comes to "forest doctor", everyone will think of woodpeckers. Why is this? Some friends say that woodpeckers make noises when pecking wood, which is easier to attract everyone's attention. But how does a woodpecker know if there are pests in the tree hole? In fact, woodpeckers have the ability to "authenticate" trees before pecking wildly. They will "check the body" of trees by tapping the trunk with their mouths. After hearing the wrong sound, they will peck fiercely until they catch the pests.

"The national park's representative woodpecker, the grey-capped pygmy woodpecker and its contribution to the forest"

The most common woodpecker in Qianjiangyuan National Park is the grey-capped pygmy woodpecker. This woodpecker is about the size of a sparrow and has black and white stripes. Although they are small, they are strong when they peck at the trunk. They often move singly or in pairs. They mostly move and feed on the middle and upper parts of trees, and occasionally on fallen wood on the ground to feed. Grey-capped pygmy woodpeckers mainly feed on longhorn beetles, spiny moths, tussock moths and other insects, and their food intake is amazing. A woodpecker can eat 1,000 to 1,400 pest larvae every day. In a 100-hectare forest, you can control the spread of pests with three pairs of woodpeckers.

"The special physiological structure and function of woodpeckers"

Since it is called a woodpecker, you can imagine how talented it is. First of all, woodpeckers can make holes in trees without their long and hard mouths. But this is only the tip of the iceberg.

Data shows that woodpeckers peck wood 12,000 times a day on average, and they can peck 20 times per second. Under such high-frequency impacts, how can the small woodpeckers withstand such heavy blows without having a concussion? This is also due to the special protective layer around the woodpecker's brain: the liquid-containing spongy bones, together with the shock-absorbing muscles on the outside of the skull, together create a very powerful head shock-proof system that can effectively absorb and dispel the impact from the outside.

But the most wonderful structure of woodpeckers is the tongue. We can imagine that in order to catch the food in the tree hole, the tongue of the woodpecker must be very long and able to secrete mucus so that it can hold the bug firmly. But where do you usually put such a long tongue? Do you know the root of the woodpecker's tongue actually grows in the right nostril? This elastic tongue will stretch back from the top of the front half of the head, go around the back of the skull, and then extend from the lower jaw into the mouth, making a full circle around the skull. When needed, stretch out to prey, roll up and put away when not in use, it can also have the effect of shock absorption. What a wonderful design of nature!

In addition, the "standing posture" of woodpeckers is also very special. Unlike other birds, which stand on branches, it climbs on upright tree trunks. Wouldn't it slide down like this? Let's continue to look at the body structure of woodpeckers: in order to "catch" the trunk, the claws of woodpeckers are not only sharp, but also have two toes in the front and two toes at the back, different from most of the birds with three toes in the front and one toe: at the back. In addition, their tails are wedge-shaped, and the feather shafts are hard and flexible, which can support their bodies well when climbing. Because of this special body structure, woodpeckers can move on the trunk very quickly and flexibly, whether it is jumping up or down, or crawling in circles to the sides, it seems that it is not a difficult task.

"Great barbet"

Although the name "woodpecker" is used in the name of the great barbet, it mainly feeds on the flowers, fruits and

Attentions

- Starting from the stereotypes of woodpeckers, we will gradually guide and explain their unique physiological structure for beating tree trunks.
- Select two woodpeckers with obvious differences in body shape and habits for comparison.

Recommended auxiliary facilities and toolkits

- The picture of woodpecker physiological structure
- Voice of woodpecker pecking the wood
- The palm of an adult is about 20cm wide, which can be used as a reference to compare the size of birds

great barbet

great barbet

seeds, and occasionally eats insects. They are called "woodpeckers" because of their thick and big mouths. The great barbet is larger in size and more colorful than the average woodpecker. The color of the whole body can be roughly divided into three sections: the one near the head and neck is blue, the tail is grass green, and the middle is dark brown. Large woodpeckers often move alone or in pairs, and sometimes in small groups where food is abundant. Often living on top of tall trees, it will peck out tree holes for nesting. These tree hole nests are usually not used for two consecutive years, so abandoned nests may be used by collared owlets, collared scops owls or red-backed squirrels. It is a lot easier to hear the great barbets instead of seeing it. Because it has a loud voice like the cry of a baby.

"Beautiful but powerful: red-billed blue magpie"

The red-billed blue magpie is a relatively large and graceful bird that is easier to be observed in the national park. Their bright red beaks are in sharp contrast with their black heads, and their long blue-gray tail feathers and red feet make them very conspicuous whether they fly or stop. Their graceful figures have been common in Chinese landscape paintings since ancient times. The red-billed blue magpie is widely distributed in various habitats at low and middle elevation in eastern and southern China, even in cities. They don't like to live alone, and often live in groups. Regardless of its gorgeous appearance, the red-billed blue magpie has a very strong personality. They are united and territorial, and sometimes actively attack other birds in groups, including raptors, and even humans who invade their territories to threaten nestlings.

red-billed blue magpie

通用型解说 | 生物多样性解说　　解说编号：TY-WZ-14

14 天空王者：猛禽

● 解说要点

钱江源国家公园常见的猛禽：蛇雕、林雕、赤腹鹰

● 观察地点推荐

1. S02 古田山—古田庙：探秘常绿阔叶林自然体验路线
2. S03 瞭望台：国家公园科研体验路线
3. Q02 里秧田—大峡谷：遇见峡谷飞瀑自然体验路线

蛇雕

赤腹鹰

林雕

解说方式 Interpretation Style

☑ 人员解说 Staff Interpretation　☐ 教育活动 Educational Activity
☑ 解说牌 Interpretation Signage　☐ 场馆解说 Hall Interpretation

解说季节 Interpretation Season

☑ 春季 Spring　☑ 夏季 Summer
☑ 秋季 Autumn　☑ 冬季 Winter

解说词

在公园的开阔地带，大家经常可以看到翱翔在天空中的猛禽。它们是天空中的绝对王者，不仅拥有非常高超的飞行技巧，同时也是位于食物链顶端的凶猛的猎食者。

「捕蛇能手：蛇雕」

蛇雕上体暗褐色，下体土黄色，头上顶着一个别致的圆形羽冠，黑中带点白色，尤其在站立时十分显眼。蛇雕在天空中飞行的时候，双翼和尾巴下方均有一条宽阔的白色横带，是其识别的重要标志。

天气晴朗的春日，钱江源的山林里经常传来蛇雕嘹亮的“呼—呼—悠悠阿”的鸣叫声，再加上它们硕大的身影，让人很难不注意到它们的存在。蛇雕属于大中型猛禽（体长约61～73cm），是天生的捕蛇好手，也吃蛙类、鸟类和甲壳类和其他小型哺乳类动物。蛇雕常停留在开阔地区的枯树顶端枝杈上，以守株待兔的方式等待捕食机会。

「滑翔高手：林雕」

林雕是体形中形偏大的猛禽（体长约70cm），主要活跃于中低山地区的阔叶林和混交林地区，从不远离森林，是一种完全以森林为栖息地的猛禽，因此得名林雕。

林雕全身暗褐色，翅膀伸展后可达170cm左右，飞行时7枚张开的翼指清晰可见，尾羽则呈方形。林雕堪称开化县天空的王者，晴天正午前后，留意天空和山脊，常可见到它们单独或成对在空中翱翔，飞行时两翅煽动缓慢，显得相当从容不迫和轻而易举。而其招牌的大波浪展示飞行更是一绝，仿佛是在高空中乘坐着云霄飞车般，高超的飞行技巧让人赞叹不已。

林雕的食物以栖息在树上的松鼠等小型兽类为主，一般不吃蛇类和两栖类，因此它们才能和同样体形高大且生态位近似的蛇雕共用一片林子和天空。

注意事项

- 注意理清猛禽的概念和分类，帮助访客认识猛禽的不同类别和习性。
- 猛禽虽大，但因距离较远，不易被观察到，观察时注意识别主要特征，必要时可借助望远镜等工具。
- 秋季是观赏赤腹鹰的最佳时机，它们常常在天空中集群飞翔，形成“鹰球”“鹰河”等景观。

辅助道具推荐

- 各种鸟类照片（最好有站立和高空飞翔两种）
- 成年人的手掌张开约有20cm左右宽，可以作为参考来比对鸟类的大小

拓展活动推荐

- 猛禽的生物链：自主选择一种猛禽作为线索，将捕食者与被捕食者的关系描述出来，比一比谁的食物链画的最长。

赤腹鹰/John Gerrard Keulemans于1874年绘制

「会迁徙的猛禽：赤腹鹰」

赤腹鹰是浙江省较为常见的一种小型猛禽（体长30cm左右），在钱江源国家公园常见其在空中盘旋觅食。从名字就可以看出来，赤腹鹰有着偏红色的腹部，身体的其他部分则呈淡蓝灰色。因外形像鸽子，也叫“鸽子鹰”。

赤腹鹰飞行时翅膀羽毛会张开，翅尖为黑色，可见到4枚翼指，成年赤腹鹰的腹部没有什么纹路，比较干净，幼鸟的腹部才有横纹。

赤腹鹰是一种会迁徙的候鸟，我国一般在夏季可以观察到它们，等到冬季，它们就会举家搬迁到更暖和的东南亚菲律宾和印度尼西亚等国家过冬。

拓展知识点

「猛禽的主要外形特征与习性」

- 猛禽不仅在居住环境上占据高位，同时还位于食物链的顶端，扮演着高级消费者的角色。作为生态系统中的顶级捕食者，猛禽在生活环境与习性上均十分接近，尽管它们大小不一，却表现出一些相似的外形特征。大家能想到它们有哪些共同特征吗？嘴巴坚韧，视力极佳，爪子很有力量，除此之外，它们的听觉也非常敏锐且非常善于飞行。
- 猛禽主要以蛙、蜥蜴为食，也食小型鸟类、鼠类和昆虫，它们数量虽然不多，却能有效控制自然界中小动物的种群数量，在生态系统中占有重要的地位。
- 值得一提的是，所有猛禽的保护级别均在国家二级或之上。

「观鸟守则」

- 观鸟是国家公园自然体验活动的重要内容。通过观察鸟类的形态、动作、飞行姿态、取食行为和独特叫声，我们可以更直观真切地理解自然万物的独特和精彩。观鸟要以不干扰鸟类的正常生活为基本原则，包括不投喂诱拍、不靠近打扰、不破坏鸟儿的栖息地等。观鸟时，我们的衣帽不宜过于艳丽以免惊扰鸟类，应持望远镜观鸟，并对所观所学有所记录。在繁殖季节，观鸟时尤其要小心谨慎，避免发生惊扰，避免造成亲鸟的弃养行为。

General Interpretation | Biodiversity Section

Interpretation No. TY-WZ-14

14 King of the sky: raptor

Key points

Common Raptors in Qianjiangyuan National Park: crested serpent eagle, black eagle, Chinese sparrowhawk

Observation places recommended

1. S02 Evergreen broad-leaved forest exploration route
2. S03 Scientific research experience route
3. Q02 Canyon and waterfall exploration route

In the open area of the park, you can often see raptors flying in the sky. They are the absolute kings in the sky. They not only have very superb flying skills, but also are fierce predators at the top of the food chain.

"Snake catcher: crested serpent eagle"

The upper body of the crested serpent eagle is dark brown, and the lower body is ocher. There is a unique round crown with a little white in black on its head, which is very conspicuous when standing. When a crested serpent eagle is flying in the sky, there is a broad white horizontal band under both wings and tail, which is an important sign to identify it.

On a sunny spring day, in the mountains and forests of Qianjiangyuan, there are often the loud calls of crested serpent eagles, "Huh－Huh－Yoyo", and their huge figures make it hard not to notice their existence. Crested serpent eagles are large and medium-sized raptor (about 61-73cm in length). They are natural snake catchers. They also eat frogs, birds, crustaceans and other small mammals. Crested serpent eagles often stay on top branches of dead trees in open areas, waiting for predation opportunities.

"Gliding master: black eagle"

The black eagle is a medium to large raptor (about 70cm in length). It is mainly active in broad-leaved forests and mixed forest areas in the middle and low mountain areas. It is never far away from the forest. It is a raptor that completely takes forest as its habitat.

The whole body of the black eagle is dark brown, and the wings can reach about 170cm after being extended. The 7 open wing fingers are clearly visible when flying, and the tail feathers are square. Black eagles can be called the king of the sky in Kaihua

crested serpent eagle

County. At noon on a sunny day, pay attention to the sky and ridges, you can often see them soaring in the air alone or in pairs. When flying, their two wings flapping slowly, making them look quite calm and easy. And its signature big wave flying technique is even more amazing, as if riding a roller coaster high in the sky.

The food of black eagles is mainly small beasts such as squirrels living on trees. They generally do not eat snakes and amphibians. Therefore, they can share the forest and sky with crested serpent eagles of the same size and similar ecological niche.

Attentions

- Pay attention to clarify the concept and classification of raptors to help visitors understand the different categories and habits of raptors.
- Although the raptor is large, it is not easy to be observed due to the long distance. Pay attention to identifying the main features when observing, and use telescopes and other tools if necessary.
- Autumn is the best time to watch Chinese sparrowhawks. They often fly in groups in the sky, forming landscapes such as "eagle ball" and "eagle river".

Recommended auxiliary facilities and toolkits

- Photos of various birds [preferably close view (standing) and far view (flying)].
- The palm of an adult is about 20cm wide, which can be used as a reference to compare the size of birds.

Recommended Activities

- Raptor's biological chain: choose a raptor as a clue, describe the relationship between the predator and the predator, and compare whose food chain is the longest.

"Migrating Raptor: Chinese sparrowhawk"

The Chinese sparrowhawk is a small raptor (about 30cm in length) commonly seen in Zhejiang Province. It is common in Qianjiangyuan National Park hovering in the air for food. The Chinese sparrowhawk has a red abdomen, and the rest of its body is light blue-gray. It is also called "pigeon eagle" because it looks like a pigeon.

When the Chinese sparrowhawk is flying, its wings and feathers will spread. The tips of the wings are black, and four wing fingers can be seen. The abdomen of the adult Chinese sparrowhawk has no patterns and is relatively clean. Only the young bird has horizontal stripes on the abdomen.

The Chinese sparrowhawk is a migratory bird that can be observed in summer in China. When it comes to winter, they will move their families to warmer countries such as the Philippines and Indonesia to spend the winter.

Chinese sparrowhawk

Extra information

"The main features and habits of the raptor"

- Raptors not only occupy a high position in the living environment, but also at the top of the food chain, playing the role of high-end consumers. As the top predator in the ecosystem, the raptors are very similar in living environment and habits. Although they are different in size, they show some similar appearance characteristics. Can you think of what they have in common? They have tough mouths, excellent eyesight, and powerful claws. In addition, they have a very keen sense of hearing and are very good at flying.
- Raptors mainly feed on frogs and lizards, as well as small birds, rodents and insects. Although their numbers are not large, they can effectively control the population of small animals in nature and occupy an important position in the ecosystem.
- It is worth mentioning that the protection level of all raptors is at or above the state second-level.

"Rules of bird watching"

- Bird watching is an important part of nature experience activities in the national park. By observing the shapes, movements, flying postures, feeding behaviors and unique voices of birds, we can more intuitively and truly understand the uniqueness and excitement of all things in nature. Bird watching should follow the basic principle of not disturbing the normal life of the birds, including not feeding, not getting too close, and not destroying the bird's habitat. When watching birds, our clothes and hats should not be too colorful to avoid disturbing the birds. We should watch the birds with a telescope and keep a record of what we have seen and learned. In the breeding season, you must be especially careful when watching birds to avoid disturbing so as so trigger abandoning behaviors of parent birds.

black eagle

crested serpent eagle

通用型解说 | 生物多样性解说　　解说编号：TY-WZ-15

15 暗夜猎手: 猫头鹰

解说要点

钱江源国家公园的猫头鹰：领角鸮、红角鸮、领鸺鹠

观察地点推荐

1. S02 古田山—古田庙：探秘常绿阔叶林自然体验路线
2. S03 瞭望台：国家公园科研体验路线
3. Q01 莲花塘：钱塘江寻源自然体验路线

解说方式 Interpretation Style

- ☑ 人员解说 Staff Interpretation
- ☐ 教育活动 Educational Activity
- ☑ 解说牌 Interpretation Signage
- ☐ 场馆解说 Hall Interpretation

解说季节 Interpretation Season

- ☑ 春季 Spring
- ☑ 夏季 Summer
- ☑ 秋季 Autumn
- ☑ 冬季 Winter

解说词

和白天活动的粗犷彪悍的鹰、雕相比，夜晚的猫头鹰因为昼伏夜出的特性和一双水汪汪的大眼睛，往往给人留下一种神秘又呆萌的感觉，让我们一时忘记它们也是一种食肉的猛禽。

如果大家有机会来参加钱江源国家公园里的夜观活动，可能会见到一些非常有趣的猫头鹰。比如，头上长“角”的领角鸮和红角鸮。

领角鸮：是一种在中国东部和南部地区广泛分布，比较常见的猫头鹰。领角鸮的脖子上有比较明显的浅沙色颈圈，像领子一样。领角鸮主要栖息于山地阔叶林和针阔叶混交林中，白天在树干上歇息，夜晚才出来捕食。它们的声音似柔和的“呜－”声，以大约 10～20s 的间隔重复，唱歌时间会持续 15min 或者更长。领角鸮不会自己筑巢，常营巢于天然树洞或者利用啄木鸟、喜鹊废弃的旧巢。

红角鸮：红角鸮是一种体形袖珍的猛禽，体长不过 20cm。眼睛为明亮的黄色，体色可分为灰色及棕色两种类型，身上羽毛多纵纹，乍看有些像树皮。红角鸮喜欢栖息在有树丛的开阔原野，是一种迁徙候鸟，常在 5 月下旬到来，9 月下旬离开。夏夜，在古田山庄附近就能听到它嘹亮的叫声，音似“chook”，节奏一长两短，约 2s 左右重复一次，几乎彻夜鸣叫，很容易辨识。

领鸺鹠：角鸮属的猫头鹰一般体形不大，是典型的夜行性猫头鹰，不具备在白天行动的能力，但是不是所有的猫头鹰都是昼伏夜出的，比如，领鸺（xiū）鹠（liú）就喜欢在白天出没。但通常领鸺鹠只在海拔较高的密林中活动，并不容易发现，偶尔能看见它们在阳光下飞翔，先急速地拍打翅膀作鼓翼飞翔，接着滑翔一段，如此交替。夜间，它的叫声音似短促的“呼－呼呼－呼”，重复频率快，细听有机会发现。

很多名字以“领”字开头的鸟，如领角鸮、领雀嘴鹎等，颈部都有明显的颈圈条纹，视觉上就像戴了个领子一样。此外，领鸺鹠头部后方具有一对“假眼”斑纹，酷似一对睁大的眼睛。领鸺鹠的体形很小，是浙江省体形最小的猫头鹰，只有拳头大小。它个头虽小，却能猎杀比它大型的鸟类，因此当领鸺鹠出现时总会引起小型鸟群的鼓噪骚动。因畏惧领鸺鹠捕捉幼鸟，雀鸟时常“组队”来驱逐巢区附近出没的领鸺鹠。有研究表明，领鸺鹠背后的大眼睛就是为了影响要骚扰它们的鸟群行为模式，从而降低它们受到威胁的风险。

注意事项

- 猫头鹰主要在夜间活动，如果选择白天介绍的话，注意搭配合适的图片、视频以及音频道具，解说才会更加生动。
- 成年人的手掌张开约有 20cm 左右宽，可以作为参考来比对鸟类的大小。

辅助道具推荐

- 各种猫头鹰照片（最好有站立和飞翔两种）。

General Interpretation | Biodiversity Section

Interpretation No. TY-WZ-15

15 Night hunter: owls

Key points

Owls in Qianjiangyuan National Park: collared scops owl, otus scops, collared owlets.

Observation places recommended

1. S02 Evergreen broad-leaved forest exploration route
2. S03 Scientific research experience route
3. Q01 Qiantang River source exploration route

Compared with the fierce eagles that are active in the daytime, the night owls often leave a mysterious and cute feeling because of their nocturnal characteristics and big adorable eyes, making us forget that they are also carnivorous raptors.

If you have the opportunity to participate in the night watching activities in Qianjiangyuan National Park, you may see some very interesting owls. For example, the collared scops owl and the otus scops with "horns" on their heads.

"Collared scops owls"

The collared scops owl is a relatively common owl. The neck of the collared scops owl has a relatively obvious light sand-colored collar. The collared scops owl mainly inhabits mountain broad-leaved forests and mixed coniferous and broad-leaved forest. It rests on tree trunks during the day and comes out to hunt at night. Their sound is like a soft "Woo—", which repeats in about 10－20 second, and the singing time will last for 15 minutes or more. Collared owls do not build their own nests. They often nest in natural tree holes or use old nests abandoned by woodpeckers and magpies.

"Otus scops"

The otus scops is a pocket-sized raptor, no more than 20cm in length. Its eyes are bright yellow, and its body color can be divided into two types: gray and brown. The feathers on the body have many vertical stripes, which look like bark at the first glance.

The otus scops like to live in open fields with bushes. It is a migratory bird that often arrives in late May and leaves in late September. In the summer night, you can hear its loud scream near Gutian Mountain Villa. The sound is like "chook", the rhythm is one long sound and two short sounds, and it repeats in about 2 seconds. It screams almost all night, which is easy to be recognized.

"Collared owlets"

Owls of the genus *Scops* are generally small. They are typical nocturnal owls and do not have the ability to prey during the day. However, not all owls appear at night, such as the collared owlets who like to hang out during the day. But usually collared owlets only move in dense forests at higher elevation, and it is not easy to find them. Occasionally, they can be seen flying in the sun. They first flap their wings sharply, and then glide for a while alternatively. At night, its call sounds like a short "whee – whee – whee", and the repetition rate is fast, and you will have a chance to find out if you listen carefully.

Many birds whose names begin with the word "collar", such as collared scops owl, collared finchbill, *etc*.. They have obvious stripes on their necks, like wearing a collar. In addition, the collared owlet has a pair of "fake eye" markings on the back of its head, which resembles a pair of wide-open eyes. The collared owlet is very small and also the smallest owl in Zhejiang Province, with the size of only a fist. Although it is small, it can hunt larger birds, so when the collared owlet appear, it will always cause small bird flocks to panic. Due to the fear of catching young birds, the birds often "group" to expel the collared owlets that appear near the nest area. Studies have shown that the big eyes behind the collared owlets are to interfere with the behavior patterns of the flock of birds that harass the collared owlets, thereby reducing their risk of being threatened.

Collared scops owl

Attention

- Owls are mainly active at night. If you choose to introduce them during the day, pay attention to prepare appropriate pictures, videos and audio tools to make the interpretation more vivid.
- The palm of an adult is about 20cm wide, which can be used as a reference to compare the size of birds.

Recommended auxiliary facilities and toolkits

- All kinds of owl photos [(preferably close view (standing) and far view (flying)]

通用型解说 | 生物多样性解说　　　解说编号：TY-WZ-16

16 命途多舛的两栖类

解说要点

古老的两栖类代表：中国大鲵和无斑肥螈的典型特征与生活环境

观察地点推荐

1. S01 古田飞瀑：丛林飞瀑自然体验路线
2. S02 古田山—古田庙：探秘常绿阔叶林自然体验路线
3. Q01 莲花塘：钱塘江寻源自然体验路线

解说方式 Interpretation Style

- ☑ 人员解说 Staff Interpretation
- ☐ 教育活动 Educational Activity
- ☑ 解说牌 Interpretation Signage
- ☐ 场馆解说 Hall Interpretation

解说季节 Interpretation Season

- ☑ 春季 Spring
- ☑ 夏季 Summer
- ☑ 秋季 Autumn
- ☑ 冬季 Winter

如果要大家选一种动物作为现在还活着的远古生物代表中国走向世界，你会选择谁呢？大熊猫、中华鲟、大鲵、鳄鱼……论萌，没有谁比得上大熊猫，但是论古老，就非大鲵莫属了。

「中国大鲵身上的一些原始特征」

中国大鲵，俗称“娃娃鱼”，是中国特有的一种两栖类，也是世界上现存体形最大的有尾两栖类，平均体长 1m 左右，最长可达 2m，即使放在人类中间，也是个高个子。但是只看长相，你是很难把它们和常见的青蛙、蟾蜍等两栖类联系在一起的，不是吗？

大家知道吗？大鲵的身体特征在 3 亿年前就出现了，而且从恐龙时代以来，基本没有改变过。这张照片上的大鲵看起来“又大又扁”，四肢短粗，尾巴长长的，眼睛小小的，有点像大一号的壁虎，但却不属于爬行类，湿黏的皮肤倒是和两栖类的青蛙有些相似。小时候的大鲵用鳃呼吸，长大后用肺呼吸，这才是它们被归为两栖类最重要的原因。

「在溪流洞穴中生活的大鲵」

大家知道大鲵生活在哪里吗？没错，在山间有水的洞穴里，幽暗的洞穴环境再加上昼伏夜出的特性，使得大鲵很难被捕食者发现。此外，大鲵也是一种寿命很长的物种，最长可达 100 岁。

20 世纪 50 年代前，大鲵曾在中国的黄河、长江、珠江等水域广泛分布，多到掀起一块石头就能见到；今天，我们却很难在野外见到野生大鲵的踪迹。这种在地球上生活了 3 亿年的古老生物最终却因人类而陷入灭绝的危机，已被 IUCN 列为极危物种。栖息地的破坏与人类对野味的贪婪欲望是大鲵面临的最主要的威胁。有文献记载，大鲵曾经也是钱江源国家公园的水下原住民，但如今已罕觅踪影。

「“小娃娃鱼”：无斑肥螈」

钱江源的溪流里还生活着一种“小娃娃鱼”——无斑肥螈，它们和“娃娃鱼”相比，虽然个头较小，外表简直就是一个模子刻出来的，但是这种“山寨”的特点却为它们带来了杀身之祸。市场上有许多不法分子将无斑肥螈冒充“娃娃鱼”幼苗兜售贩卖，而且要价不菲，这便无端加速了无斑肥螈的濒危速度。但其实无斑肥螈为了保持身体的湿润会分泌出一种粘液，这些黏液含有硫元素，人吃多了会中毒。

和大鲵一样，无斑肥螈也是一种古老的两栖动物，曾和恐龙一个年代，同时因为它们对于生活环境十分挑剔，只生活在水质清澈的高山溪流里，素有两栖“贵族”之称。而且无斑肥螈大部分时间都会待在水里，一旦水源被污染，它们的整个种群都会受到威胁。

和大鲵相比，无斑肥螈体形较小（体长 18～20cm），身体肥圆，皮肤裸露，看起来很像一条壁虎，当地人称“水虎”或“水和尚”。无斑肥螈遇到危险会有警戒反射：先伸直四肢，旋即翻过身来，露出腹部的橘红色斑块恫吓对方，1～2min 后感觉威胁消除后才恢复原状。

General Interpretation | Biodiversity Section

Interpretation No. TY-WZ-16

16 Amphibians

● Key points

Representatives of ancient amphibians: the typical characteristics and living environment of Chinese giant salamanders and spotless stout newt

● Observation places recommended

1. S01 Jungle and waterfall experience route
2. S02 Evergreen broad-leaved forest exploration route
3. Q01 Qiantang River source exploration route

If you were to choose an animal as the ancient creature that is still alive and represent China to the world, who would you choose, giant pandas, Chinese sturgeons, giant salamanders, crocodiles or others? No one can be comparable to giant pandas when it comes to cuteness, but when it comes to ancient times, it is the giant salamander.

"Some primitive characteristics of Chinese giant salamanders"

The Chinese giant salamander, commonly known as the baby fish. It is an amphibian peculiar to China and the largest existing tailed amphibian in the world, with an average body length of about 1 meter and a maximum length of 2 meters. Even if placed among humans, it is a tall guy. But just by looking at its appearance, it is difficult for us to associate it with common amphibians such as frogs and toads, right?

Do you know the physical characteristics of the giant salamander appeared 300 million years ago, and have basically not changed since the age of the dinosaurs? The giant salamander in this photo looks "big and flat" with chubby limbs, a long tail, and small eyes. It looks a bit like a big gecko, but it is not a reptile, and its wet and sticky skin is similar to frogs. When they are young, giant salamanders use their gills to breathe, and when they grow up, they breathe with their lungs. This is the most important reason why they are classified as amphibians.

"Giant salamanders living in a stream cave"

Do you know where the giant salamander lives? That's right, in the caves with water in the mountains, the dark cave environment coupled with the characteristics of nocturnal make the giant salamander difficult to be found by predators. In addition, the giant salamander is also a very long-lived species, with a lifespan of up to 100 years.

Before the 1950s, giant salamanders were widely distributed in China's Yellow River, Yangtze River, Pearl River and other waters. There were so many of them that you can find one just by lifting a stone in the stream; today, it is difficult to see giant salamanders in the wild. This ancient creature that has lived on the earth for 300 million years has finally fallen into a crisis of extinction because of humans, and has been listed as a critically endangered species by the International Union for Conservation of Nature (IUCN). The destruction of habitat and human greedy desire for eating wild animals are the main threats faced by giant salamanders. It is documented that the giant salamander was once an underwater native resident in Qianjiangyuan National Park, but now it is rarely found.

"'Little baby fish': the spotless stout newt"

There is also a kind of "little salamander" in the streams of Qianjiangyuan—spotless stout newt. Compared with the giant salamander, although it is smaller but they look exactly the same, which brought it a disaster. There are many lawbreakers who sell it as the giant salamander, which has accelerated the speed of endangerment of the spotless stout newt. But in fact, in order to keep the body moist, the spotless stout newt will secrete a kind of mucus, which contains sulfur, so people will be poisoned if they eat too much.

Like the giant salamander, the spotless stout newt is also an ancient amphibian as old as the dinosaur. At the same time, because it is very picky about its living environment to only live in high mountain streams with clear water, it is known as the amphibious "nobles". In addition, the spotless stout newt will stay in the water most of the time. Once the water source is contaminated, its entire population will be threatened.

Compared with the giant salamander, the spotless stout newt is smaller (about 18 – 20cm in length), has a plump body and bare skin. It looks like a gecko. The locals call it water tiger or water monk. The spotless stout newt will have a warning reflex when it encounters danger: it will straighten its limbs first, then turn over, revealing the orange-red patch on the abdomen to scare the opponent, and return to its original state after the threat is eliminated after 1 – 2 minutes.

Chinese giant salamanders

通用型解说 | 生物多样性解说 解说编号：TY-WZ-17

17 青蛙和蟾蜍

解说要点

钱江源国家公园多样的蛙类：泽陆蛙、华南湍蛙、大绿臭蛙、花臭蛙、中华大蟾蜍、弹琴蛙、镇海林蛙、棘胸蛙

观察地点推荐

1. S01 古田飞瀑：丛林飞瀑自然体验路线
2. S02 古田山—古田庙：探秘常绿阔叶林自然体验路线
3. Q01 莲花塘：钱塘江寻源自然体验路线
4. Q02 里秧田—大峡谷：遇见峡谷飞瀑自然体验路线

解说方式 Interpretation Style

☑ 人员解说 Staff Interpretation ☐ 教育活动 Educational Activity
☑ 解说牌 Interpretation Signage ☐ 场馆解说 Hall Interpretation

解说季节 Interpretation Season

☑ 春季 Spring ☑ 夏季 Summer
☑ 秋季 Autumn ☐ 冬季 Winter

解说词

如果大家在夏天的夜晚来到钱江源国家公园，一定可以在溪流、小河边听到那些此起彼伏、节奏明快、高低错落的蛙类大合唱。不过要想亲眼见到它们，可得好好寻觅一番。

它们有的隐藏在岸边草丛中静静观望，有的匍匐在静水中伺机而动，有的甚至不惧激流在浪花中“守株待兔”，上演着一场又一场捕食和求生的生存竞技。我们可以通过声音判断它们的方位，然后再通过身形、皮肤特征及行走的姿态的差异性来辨识是哪种蛙类。

「泽陆蛙」

这些青蛙里，分布最广泛的也是最常见的当属泽陆蛙。每个地方的泽陆蛙在花纹和色彩上会有些差异，但是它们的共同特点是背上皮肤比较粗糙，此外，它们背部的正中有一条竖向的浅色背脊线，这是辨识它们最明显的特征。如果实在拿不准的话，还可以观察它们的上下唇是否有一道道黑色的斑纹，也可以帮助大家识别泽陆蛙。一般我们可以在平原、丘陵和海拔 2000m 以下的山区稻田、水塘等静水区和附近的草丛观察到泽陆蛙。

「华南湍蛙」

钱江源国家公园山高流急，水流速度快，对于大多数喜欢静水的蛙类来说并不是理想的居住所。但是对于华南湍蛙来说，这是它们最喜爱的环境。华南湍蛙的脚趾上长着特大吸盘，可以帮助它们牢牢吸附于潮湿光滑的石面或瀑布里的石壁上，即使在接近垂直的崖壁上，它们也可以岿然不动。为了适应水流冲刷的环境，华南湍蛙的蝌蚪宝宝们也在嘴巴后方进化出一个大吸盘，可以帮助它们在激流涌动的环境中攀附在石块上逆流移动，不被冲走。

「大绿臭蛙和花臭蛙」

公园里蛙类众多，但是大都颜色低调。相比而言，个头较大、颜色鲜艳的大绿臭蛙和花臭蛙会更容易被发现。大绿臭蛙的背部为鲜艳的翠绿色，其他部位为浅棕色，四肢背面有深色横纹，与身边的苔藓、土壤和植物巧妙地融合为一个整体，是一种很好的保护色。

注意事项

- 围绕蛙类所在的不同生境、身上的纹理与色泽特征、叫声与习性的不同进行讲解。
- 可以结合野外实际情况，有选择地对蛙类进行解说。

辅助道具推荐

- 配合常见蛙类的照片，对特征进行解说

拓展活动推荐

- 夜观青蛙和蟾蜍。

雌雄大绿臭蛙体形差别很大，雄蛙体形只及雌蛙的一半。且雄蛙的两眼后各有一个褶皱状的声囊，鸣叫时会鼓起来。大绿臭蛙喜欢栖息在山区林间的溪流及其附近，和湍蛙一样，大绿臭蛙足底也自带吸盘，可以帮助它们牢牢地吸附在岩壁上，除此之外，大绿臭蛙的叫声也非常尖锐，在水流湍急的山涧中，这种声音无疑起到了高效交流的作用。

花臭蛙的皮肤也是绿色的，不过背部杂有很多深棕色或黑褐色近似圆形的大斑点，这即是花臭蛙“花”字的来源。和大绿臭蛙一样，花臭蛙也喜欢生活在环境潮湿、植被茂盛的山涧溪流边，常面朝溪水，蹲在溪边石头上耐心等候猎物光顾。一旦受到惊扰，还可以迅速躲进溪流中。但用肺呼吸的蛙类并不是潜水高手，一般在水内潜伏 10~20min 后，就必须游回岸边。

除了颜色鲜艳，大绿臭蛙和花臭蛙所属的臭蛙属，在遇到危险时，还会分泌出一种带有蒜味的滑腻粘液作为特殊的防御机制。如果手上有伤口，碰到这种粘液会有种刺痛感。但是大家不要因此对它们产生不必要的偏见哦。它们的这种味道并不显著，而且正是这些不太讨人喜欢的分泌物，才使得它们在残酷的生存斗争中安全地活了下来。

「中华大蟾蜍」

说起中华大蟾蜍，大家或许还有些陌生，但是如果提起它们的俗称“癞蛤蟆”，我想大概无人不知、无人不晓。蟾蜍常在傍晚或者夜间出没在路边的草丛、林下居民点周围的沟边、土穴等潮湿地带，行动迟缓。雄性蟾蜍背面为墨绿色或者褐绿色，雌性背面浅褐色，可以帮助它们和栖息的落叶层巧妙地融为一体。和青蛙相比，蟾蜍的长相并不讨喜。它们体形肥壮、皮肤粗糙、浑身长满各种大小不一的疣粒，总让人不敢靠近。蟾蜍是民间俗称的“五毒”之一，它们的身体在遇到危险时，会分泌出一种腺液，被人们误以为是具有剧毒的液体，实际上这只是蟾蜍进行自保的一种机制。一般那情况下，蟾蜍也不会轻易“放毒”的。

大绿臭蛙

中华大蟾蜍

「弹琴蛙」

“稻花香里说丰年，听取蛙声一片。”夏日的夜晚，我们经常可以在公园中听到各种鸣虫和蛙类此起彼伏的大合唱。平时大家听到的青蛙叫声大都是“呱—呱—呱”的声音，但是公园里却有一种蛙鸣是由2~3声低沉的“咕—咕—咕”组成的，颇似琴弦被弹奏的声音，因此得名弹琴蛙。弹琴蛙的体色浅灰，四肢有明显的横纹，趾端有个大吸盘。除此之外，雄弹琴蛙在鸣唱时，会有节奏地鼓起一对声囊，特别有趣。大家在夜观的时候不妨留心，看看能否通过声音或者外形认出弹琴蛙来。

弹琴蛙

「镇海林蛙」

镇海林蛙是中国特有的一种蛙类。通常生活在海平面至 1800m 范围内的山区。辨识它们的一个非常重要的特征是它们眼睛后侧颞部的黑色三角斑。此外，它们的皮肤也比较光滑，背部及体侧有少数小圆疣粒。镇海林蛙的体色会随着季节的不同而呈现出棕色、灰白色和暗棕色等不同的变化。镇海林蛙平时在繁茂的林间和杂草丛中活动，繁殖季节则会在水稻田、水塘等有草本植物的静水中繁殖。

镇海林蛙

「棘胸蛙」

如果大家在钱江源国家公园偶遇一种体形硕大、全身布满灰黑色斑纹的蛙类，说不定就是棘胸蛙。棘胸蛙是中国特有的大型野生蛙，俗称石鸡，因为个头较大、肉质细腻且富含营养元素而被称为“百蛙之王”，是过去当地人餐桌上的一道山珍野味，也因此遭遇了被捕杀的命运，同时，伴随着栖息地的破坏，棘胸蛙的野外种群数量迅速减少，已被 IUCN 列为近危物种。

棘胸蛙

General Interpretation | Biodiversity Section

Interpretation No. TY-WZ-17

17 Frog and toad

● Key points

Diverse frogs from Qianjiangyuan National Park: Asian grass frog, Chinese sucker frog, large odorous frog, Schmacker' s frog, Asiatic toad, China music frog, Zhenhai brown frog, Chinese spiny frog

● Observation places recommended

1. S01 Jungle and waterfall experience route
2. S02 Evergreen broad-leaved forest exploration route
3. Q01 Qiantang River source exploration route
4. Q02 Canyon and waterfall experience route

If you come to Qianjiangyuan National Park at a summer night, you can definitely hear those choruses of frogs that come and go. But if you want to see them with your own eyes, you have to look for them.

Some of them hide in the grass on the shore and look around silently, some are crawling in the still water waiting for opportunities, and some are not afraid of the rapids stream, staging a survival competition of predation and survival. We can locate them by sound, and then identify them based on the differences in body shape, skin characteristics and walking posture.

"Asian grass frog"

Among these frogs, the most widely distributed and the most common is the Asian grass frog. They have some differences in patterns and colors in different places, but their common feature is that the skin on their back is relatively rough, some resembling toads. In addition, there is a vertical light-colored back ridge in the center of their back, which is the most obvious characteristics to indentify them.You can also observe whether there are black markings on their upper and lower lips, which can also help you identify Asian grass frogs. In general, we can observe them in the plains, hills and mountainous areas below 2000 meters elevation, rice fields, ponds and other static water areas and nearby grasses.

"Chinese sucker frog"

Qianjiangyuan National Park has high mountains and fast currents, and it is not an ideal residence for most frogs who like still water. But for Chinese sucker frogs, this is their favorite

environment. The Chinese sucker frog's toes have large suction cups, which can help them to firmly adhere to the wet and smooth rock surface or the rock wall in the waterfall. Even on the closely vertical cliff wall, they can stay still. In order to adapt to the water flow, the tadpole babies of Chinese sucker frogs have also evolved a large sucker behind their mouths, which can help them cling to rocks and move against the current in a turbulent environment without being washed away.

"large odorous frog and Schmacker's frog"
There are many frogs in the park, but the colors are mostly low-key. In contrast, large odorous frogs and Schmacker's frogs have larger sizes and bright colors making them easier to be discovered. The back of the large odorous frog is bright emerald green, and the other parts are light brown. There are dark horizontal stripes on the back of the limbs. It is cleverly integrated with the moss, soil and plants around it, making it a good protective color.

Attentions

- Explain the different habitats of frogs, the texture and color characteristics of their bodies, as well as the differences in voices and habits;
- You can selectively explain the frogs based on the actual situation in the field.

Recommended auxiliary facilities and toolkits

- Conduct interpretation the characteristics with photos of common frogs

Recommended activities

- Night observation of frogs and toads

The size of male and female large odorous frogs is very different, and the size of male frogs is only half that of female frogs. And the male frog has a fold-like vocal sac behind each eye, which bulges when it screams. The large odorous frog likes to inhabit the stream by the mountain forest. Like the Chinese sucker frog, the large odorous frog also has its own suction cups on the bottom of the feet, which can help them firmly adhere to the rock wall. The voice of the large odorous frogs is also very sharp. In the turbulent mountain stream, this sound undoubtedly plays a role in efficient communication.

The skin of the Schmacker's frog is also green, but there are many dark brown or dark brown spots on the back. Like the large odorous frog, the Schmacker's frog also likes to live on the side of a mountain stream with a humid environment and lush vegetation. It often faces the stream and squats on the rocks, waiting patiently for its prey. It can quickly hide in the stream if it detects any danger. However, frogs that breathe with their lungs are not master divers. Generally, they must swim back to the shore after lurking in the water for 10 – 20 minutes.

In addition to its bright colors, the genus *Odorrana* that the large odorous frog and Schmacker's frog belongs to can secrete a slippery mucus with the smell of garlic as a special defense mechanism when in danger. If you have a wound on your hand, it will sting when you touch the mucus. However, please don't hold prejudice against them. The smell is not remarkable, and it is these unpleasant secretions that make them survive.

"Asiatic toad"

The Asiatic toad, commonly known as the toad. Toads often appear in wet areas such as the grass on the side of the road, the ditch and the dirt cave around the villages under the forest in the evening or at night, and they move slowly. The back of the male toad is dark green or brownish green, and the back of the female is light brown, which can help them to integrate into the deciduous layer. Compared with frogs, toads look unpleasant. They are fat, rough-skinned, and covered with warts of various sizes, making people afraid to approach them. Toads are one of the "five poisons" in the folk. When their bodies are in danger, they secrete a kind of glandular fluid, which is mistaken for highly toxic fluids. In fact, this is only a kind of toad's self-protection mechanism. Under normal circumstances, toads won't do it at ease.

"China music frog"

"The fragrance of rice blossoms indicating good harvest, and listen to the sound of frogs." On summer nights, we can often hear various chirping insects and frogs one after another in the national park. The sounds of frogs that you hear are mostly "quack – quack – quack", but there is a kind of frog croak in the park that is composed of 2 to 3 deep "coo – coo – coo" sounds, which is like a string being played.

The China music frog is light gray in color, with obvious horizontal stripes on the limbs and a large sucker at the end of the toe. In addition, when the male China music frog sings, it will rhythmically bulge a pair of vocal sacs, which is especially interesting. When you watch at night, you may wish to pay attention to see if you can recognize the frog by its sound or appearance.

"Zhenhai brown frog"

Zhenhai brown frog is a species of frog unique to China. They usually live in mountainous areas ranging from sea level to 1800 meters. One very important feature to identify them

Asiastic toad

large odorous frog

large odorous frog

Schmack[illegible]s frog

is the black triangular spots on the back of their eyes. In addition, their skin is relatively smooth, with a few small round warts on the back and sides. The body color of the Zhenhai brown frog will show brown, gray and dark brown with different seasons. Zhenhai brown frogs usually live in lush forests and weeds, but during the breeding season they will breed in still waters with herbaceous plants such as rice fields and ponds.

"Chinese spiny frog"

If you come across a large frog in Qianjiangyuan National Park with gray-black markings all over, it might be a Chinese spiny frog. The Chinese spiny frog is a large wild frog unique to China, commonly known as stone chicken. Known as the "king of hundred frogs" because of its large size, delicate meat and rich nutrients, it was a delicacy on the table of the locals in the past, and suffered the fate of being hunted and killed. At the same time, due to the destruction of the habitat, the wild population of Chinese spiny frog has rapidly decreased, and it has been listed as a near threatened species by IUCN.

通用型解说 | 生物多样性解说　　解说编号：TY-WZ-18

18 公园常见的爬行类

解说要点

1. 竹叶青、翠青蛇等蛇类的典型特征与生活习性
2. 蓝尾石龙子、平胸龟的典型特征与生活习性

观察地点推荐

1. S02 古田山—古田庙：探秘常绿阔叶林自然体验路线
2. S03 瞭望台：国家公园科研体验路线
3. Q01 莲花塘：钱塘江寻源自然体验路线

解说方式 Interpretation Style

☑ 人员解说 Staff Interpretation　☐ 教育活动 Educational Activity
☑ 解说牌 Interpretation Signage　☐ 场馆解说 Hall Interpretation

解说季节 Interpretation Season

☑ 春季 Spring　☑ 夏季 Summer
☑ 秋季 Autumn　☐ 冬季 Winter

「竹叶青与翠青蛇」

相比而言，蛇类是会令人有些恐惧的爬行类动物，但它们作为食物链中重要的捕食者，也发挥着不可或缺的作用。

竹叶青是令人闻之色变的一种有毒蛇类，它的成体长约1m，眼睛红色，背面绿色，腹部黄白色，体侧有红白各半或白色纵线纹，尾背及尾尖焦红色，因此在浙江又名焦尾青蛇。竹叶青栖息于溪沟边、灌丛中、岩壁或石上，以各种水域附近最常见，一般傍晚、夜间最活跃，主要捕食蛙类和蜥蜴，也吃鼠类。若在此时进入原生生境中，一定要注意观察树梢、路边的灌丛、落叶中是否有蛇出没。

无毒的翠青蛇眼大，背面纯绿，身上没有红色或橙色的纹路，因为外形和竹叶青很像，常常蒙受不白之冤，被误捕误杀。

「喜欢晒太阳的蓝尾石龙子」

在公园里，即使走在太阳照射下的水泥路上，我们留心观察路边向阳的山坡、排水沟可能会与一些小蜥蜴相遇，其中有一种身形修长的小蜥蜴尾巴是蓝色的，叫作蓝尾石龙子。和其他小蜥蜴相比，蓝尾石龙子最突出的应该就是它们宝蓝色的长尾巴以及背部5条明显的纵纹了，不过这只是幼年蓝尾石龙子的特征，成年后的蓝尾石龙子摇身一变，好像换了一件黄褐色的新衣，蓝色尾巴和纵纹也随着它们长大均逐渐消失不见。

蓝尾石龙子栖息于山区阳面的杂草灌丛处，喜欢趴在干燥且温度较高的石头上“晒太阳”，遇到惊扰时，它们会迅速钻入草丛或石缝中。此时，只要稍加耐心等待，它就可能会缓缓从石缝探出头来，你便可以对这种美丽的小蜥蜴端详一番啦。

「大头、鹰嘴的平胸龟」

平胸龟是世界上现存最古老的爬行动物之一，国家一级重点保护野生动物，被《IUCN红色物种名录》定为濒危物种。这是一种古老的龟类，和迄今为止发现的乌龟无论从外形还是习性上都很不同。除了“平胸龟”之外，它还有“大头龟”“鹰嘴龟”和“鹦鹉龟”等别名，每个名字都对应一个特征，比如，“平胸龟”得名于它们扁平的龟壳；“大头龟”对应它们和身子不相称的无法缩进龟壳的大头；“鹰嘴龟”则是因为它们嘴巴像鹰嘴一样呈钩曲状，看起来颇为锋利。

实际上它们的确“龟如其形”。它们钩曲状的鹰嘴不仅可以用来咬碎螃蟹、贝类等食物，在遇到危险时，还可以用来自我防卫。它们的爪子也很锐利，是捕食猎物的有力武器。此外，平胸龟不善游泳，却善攀爬，借助粗壮尾巴的支撑，它们能够在山涧中溯溪逆流而上，也能够攀爬树木吃鸟窝里面的鸟蛋和稚鸟，甚至能够攀爬呈直角的水泥墙……

平胸龟不喜欢炎热，一般只生活在山间清凉的溪流里，喜欢满是巨砾和碎石的水流湍急的山涧，具有昼伏夜出的习性。

General Interpretation | Biodiversity Section

Interpretation No. TY-WZ-18

18 Common reptiles

Key points

1. The typical characteristics and living habits of snakes such as Asian lanceheads and Chinese green snake
2. The typical characteristics and life habits of Shanghai elegant skink and big-headed turtle

Observation places recommended

1. S02 Evergreen broad-leaved forest exploration route
2. S03 Scientific research experience route
3. Q01 Qiantang River source exploration route

"Asian lancehead and Chinese green snake"
In contrast, snakes are somewhat scary reptiles, but as important predators in the food chain, they also play an indispensable role.

The Asian lancehead is a notorious venomous snake. Adult snakes can reach about 1m long with red eyes, green back, yellow and white belly, red and white or white vertical lines on the sides, and burnt red tail. So it is also known as the burnt tail green snake in Zhejiang Province. The Asian lancehead inhabits along streams, shrubs, rock walls or rocks, and is most common near various waters. Generally, it is most active in the evening and at night. It mainly preys on frogs and lizards, but also eats rodents. If you enter the native habitat at this time, you must pay attention to whether there are snakes in the treetops, roadside shrubs, and fallen leaves.

Non-poisonous Chinese green snakes have large eyes, pure green on the back, and no red or orange lines on their body. Because they look like the Asian lancehead, they often suffer injustice and are caught and killed by mistake.

"Shanghai elegant skink who likes to sunshower"
In the national park, even when walking on the concrete road under the sun, if we pay attention to the sunny hillsides and drains on the side of the road, you may see some small lizards. Among them, there is a slender lizard with a blue tail, called the Shanghai elegant skink. Compared with other small lizards, the most prominent thing about the Shanghai elegant skink is their royal blue tail and five obvious vertical stripes on the back, but this is only the characteristic of the young Shanghai elegant skink, the adult Shanghai elegant skink

change into yellowish-brown, and the blue tail and vertical stripes gradually disappear as they grow older.

The Shanghai elegant skink inhabits the weeds and shrubs on the sunny side of the mountain. They like to lie on dry and high-temperature stones to take a sun shower. When they are disturbed, they will quickly get into the grass or cracks in the rocks. At this time, just be patient until it slowly sticks out its head from the cracks in the rock, you can take a closer look at this beautiful little lizard.

“big-headed turtle”

The big-headed turtle is one of the oldest reptiles in the world. It is a state first-level key protected animal and is listed as an endangered species by the *IUCN Red Species List*. This is an ancient tortoise, which is quite different from the tortoises discovered so far in terms of appearance and habits. They also have different names based on their unique characteristics. For example, the “flat-breasted turtle” is named after their flat turtle shell; the “big-headed turtle” corresponds to its large head that is not commensurate with its body, and can't retract into the tortoise shell; the “hawk-billed turtle” is because its mouth is hooked like an eagle's beak, which looks quite sharp.

In fact, its hook-shaped mouth can not only be used to bite crabs, shellfish and other foods, but also can be used to defend itself when in danger. Its claw is also very sharp and a powerful weapon for prey. In addition, big-headed turtles are not good at swimming, but they are good at climbing. With the support of their thick tails, they can swim upstream in the mountain stream, also climb trees to eat the eggs and young birds in the nest, and even climb a vertical concrete wall…

The big-headed turtles don't like the heat, and generally only live in cool streams in the mountains. They like mountain streams full of boulders and gravels and turbulent water, with nocturnal behavior.

Shanghai elegant skink

通用型解说 | 生物多样性解说　　　　解说编号：TY-WZ-19

19 适应湍急溪流的鱼类

解说要点

1. 溪流生境中鱼类存在的重要性
2. 介绍常见溪流鱼种：光唇鱼、虾虎鱼

观察地点推荐

1. S02 古田山—古田庙：探秘常绿阔叶林自然体验路线
2. Q01 莲花塘：钱塘江寻源自然体验路线
3. Q02 里秧田—大峡谷：遇见峡谷飞瀑自然体验路线

光唇鱼

虾虎鱼

解说方式 Interpretation Style

- ☑ 人员解说 Staff Interpretation
- ☐ 教育活动 Educational Activity
- ☑ 解说牌 Interpretation Signage
- ☐ 场馆解说 Hall Interpretation

解说季节 Interpretation Season

- ☑ 春季 Spring
- ☑ 夏季 Summer
- ☑ 秋季 Autumn
- ☑ 冬季 Winter

解说词

和平原地区的河流相比，山间溪流不仅水量小、水温低、水流湍急，而且缺乏营养物质，环境变化大，这些条件共同塑造了溪流清澈的水质，同时也意味着它们不能为大多数鱼类提供非常合适的栖息地。因此，如果大家留心观察这里的溪流会发现，虽然这里的溪流水质很高，但却只有少数几种鱼类生活在这里。

「光唇鱼」

大家看，水里那一条条一指多长且背部有很清晰的横纹的鱼就是光唇鱼了。它们在水中游弋的时候，和水底的石头浑然一体，如同石头上的斑纹一样，所以当地人也将它称为“淡石斑”。我们很容易凭借这个特点将它们和溪流中的其他鱼类区分开。

石斑鱼一般生活在溪流中水流稍缓、水底较为平坦、多砾石的区域，以水中石头上的苔藓和藻类为食。长期的演化使得它们对溪流低温、急流、营养贫乏和水文动荡的环境适应良好，不过它们的繁殖能力较低且生长缓慢。在面对溪流环境的变化时，它们的抵抗力和恢复能力也都十分低下，人类活动对溪流的一点点干扰，都有可能威胁到它们整个种群的生存。

「虾虎鱼」

我们知道，鱼类大都是水中的游泳健将，但是钱江源的溪流中却有种鱼类为了适应湍急的溪流，干脆放弃了游泳技能的精进，将腹鳍进化成吸盘状，以便牢牢地贴在溪流底部的石头上，等待猎物靠近再将它们一举捕获。这种鱼叫虾虎鱼。它们大部分时候都趴在水底，静静地等待猎物上钩，偶尔才会“一蹿一蹿”地游泳行进，稍微挪动下身子。用“静若处子，动若脱兔”形容它们最合适不过了。

除了石斑鱼和虾虎鱼之外，溪流中还经常有一些其他的小鱼出没，石块是它们藏身的庇护所，和河床相近的保护色让它们在水中和我们捉迷藏，不容易被人发现，因此很多时候它们常被人们忽略，当地人称它们为“小杂鱼”。可别小看这些“小杂鱼”，它们是溪流生态系统中不可忽视的重要一环，同时它们的存在与否，也是钱塘江源头溪流水质优劣的重要指标。

注意事项

- 可以结合现场鱼类资源进行解说。
- 溪流环境路面较滑，注意安全，提醒访客不要戏水。

辅助道具推荐

- 石斑鱼、虾虎鱼照片

General Interpretation | Biodiversity Section

Interpretation No. TY-WZ-19

19 Fish that adapt to rapid streams

Key points

1. Importance of fish presence in stream habitats
2. Introduce common stream fish species, *Acrossocheilus fasciatus* and goby

Observation places recommended

1. S02 Evergreen broad-leaved forest exploration route
2. Q01 Qiantang River source exploration route
3. Q02 Canyon and waterfall experience route

Compared with rivers in the plains, mountain streams are not only small in water volume, low in water temperature, turbulent, but also lack in nutrients, and the environment changes greatly. These conditions together shape the clear water quality of the stream, and mean that they cannot provide a very suitable habitat for most fish. Therefore, if you pay attention to the streams here, you will find that although the water quality of the streams here is high, only a few species of fish live here.

"*Acrossocheilus fasciatus*"

Everyone, these finger-long fish in the water with very clear horizontal stripes on their backs are *Acrossocheilus fasciatus*. When they swim in the water, they are integrated with the rocks under the water, just like the markings on the rocks. Therefore, the locals also call it the grouper. We can easily distinguish the grouper from other fish in the stream through this feature.

Groupers generally live in streams where the water flow is slightly slower, the bottom is relatively flat with gravels. They feed on moss and algae on the rocks in the water. Long-term evolution has made them adapt well to the low temperature, rapids, and nutrient-poor and hydrologically turbulent environments of streams, but they have low reproductive capacity and slow growth. They are very fragile, any disturbance to the stream by human activities may threaten the survival of their entire population.

"Goby"

We know that most of the fish are good swimmers in the water, but in order to adapt to the turbulent stream, there is a kind of fish in the stream of the Qianjiang River.

They simply gave up their swimming skills

fish

shrimp

and evolved their pelvic fins into a sucker-like shape in order to stick firmly on the rocks at the bottom of the stream, waiting for the prey to approach. This kind of fish is called the goby. Most of the time, they lie on the bottom of the water, quietly waiting for their prey, and occasionally they swim by jumping and move their lower body slightly. It is most appropriate to describe them as "quiet as a lady, moving like a rabbit".

In addition to groupers and gobies, there are often other small fish in the stream. The rocks are their shelters. The protective color close to the river bed allows them to hide in the water and is not easy to be discovered, so they are often ignored by people. The locals call them "small fish". Don't underestimate these "small fish". They are an important part of the stream ecosystem that cannot be ignored. At the same time, their existence is also an important indicator of the quality of the streams at the source of the Qiantang River.

Attentions

- Conduct interpretation combined with on-site fish resources.
- The road in the stream environment is slippery, so pay attention to safety and remind visitors not to play in the water.

Recommended auxiliary facilities and toolkits

- Photos of the grouper and goby

通用型解说 | 生物多样性解说　　解说编号：TY-WZ-20

20 身怀绝技的鸣虫

● 解说要点

1. 常见鸣虫的分类
2. 常见鸣虫介绍：蟋蟀、纺织娘、蝉

● 观察地点推荐

1. S02 古田山—古田庙：探秘常绿阔叶林自然体验路线
2. Q01 莲花塘：钱塘江寻源自然体验路线
3. Q02 里秧田—大峡谷：遇见峡谷飞瀑自然体验路线

蝉
纺织娘
蝗虫

解说方式 Interpretation Style

☑ 人员解说 Staff Interpretation　☐ 教育活动 Educational Activity
☑ 解说牌 Interpretation Signage　☐ 场馆解说 Hall Interpretation

解说季节 Interpretation Season

☑ 春季 Spring　☑ 夏季 Summer
☑ 秋季 Autumn　☐ 冬季 Winter

解说词

大家听到过虫子的叫声吗？夏季是鸣虫大显身手的时节，在钱江源国家公园的夏夜，无论是高山峡谷或者溪流岸边，森林、灌丛和草丛中各种虫子常凑在一起，组成夏天的乐队，唱响国家公园的黑夜。

「常见鸣虫的分类」

通俗来讲，所有会叫的虫子都可以算做“鸣虫”，而其中最为大家熟知的是俗称蛐蛐和蝈蝈的两类鸣虫。蛐蛐是指蟋蟀，而蝈蝈的学名则叫螽（zhōng）斯。

大多数鸣虫都属于是生物学分类中直翅目下的昆虫，按照它们触角的长短，直翅目下又可以分为长角昆虫和短角昆虫两大类，长角的昆虫，比如蟋蟀和螽斯，主要在夜间活动，并通过摩擦翅膀来发声，声音相对比较响亮；短角的昆虫，比如蝗虫、蚂蚱和蚱蜢，大多在白天活动，它们会通过腿部和翅膀的摩擦来发声。

「常见鸣虫」

蟋蟀的样子我想大家一定非常熟悉了。但是蟋蟀有雌雄，我们怎么区分它们呢？大家注意观察它们的尾须，这种有两根尾须的，是雄蟋蟀，俗称二尾蛐蛐；这种不仅有两根尾须，还有一根长长的产卵器的，是雌蟋蟀，俗称三尾蛐蛐。除了外形，我们还可以通过它们的习性来判断雌雄。会鸣叫的蟋蟀为雄性，且好争斗，平时它们喜欢独居生活，只有在求偶季节，才会用叫声招揽异性。

螽斯是鸣虫中体形较大的一种，身体多为草绿色，触须长，但翅膀短。螽斯家族中最著名的歌唱家应该就是纺织娘了。夏日的夜晚，雄性纺织娘常常会连续“织—织—织”地叫个不停，叫声犹如织女在使用纺车，因此得名。

蝉虽不属于直翅目，但声音聒噪，尤其喜欢在炎热的午后躲在树荫里扯着嗓子喊叫，是大家最熟悉的夏日鸣虫，没有之一。蝉是夏季的象征，没有蝉的夏日是不完整的。蝉的一生几乎都在地下，它们能见到阳光的时间只有一个月左右。只有雄蝉才会鸣叫，它们为了找到心仪的对象，可谓是拼尽了一生的力气。此外，雄蝉的声音之所以这么大还得益于它们特殊的发声原理，像乐器中的鼓一样，是通过鼓膜的振动而发声的，也难怪声音如此之大。

注意事项

- 提前做好功课，帮助访客理清鸣虫的分类与常见鸣虫（蟋蟀等）的对应关系。
- 重点讲述比较特别的鸣虫，蝗虫因为声音不突出，不做主要介绍。
- 解说时配合鸣虫卡片，并为访客播放鸣虫音频。

辅助道具推荐

- 直翅目鸣虫分类卡片
- 蟋蟀、纺织娘、蝼蛄、蝗虫、蝉的照片
- 不同鸣虫声音音频

General Interpretation | Biodiversity Section

Interpretation No. TY-WZ-20

20 Chirping insects with talents

Key points

1. Classification of common chirping insects
2. Introduction to common chirping insects: crickets, katydids, cicadas

Observation places recommended

1. S02 Evergreen broad-leaved forest exploration route
2. Q01 Qiantang River source exploration route
3. Q02 Canyon and waterfall experience route

Have you heard the sound of insects? Summer is the time when the insects show their talents. On summer nights in Qianjiangyuan National Park, whether it is a mountain valley or a stream bank, various insects in the forest, bushes and grass often gather together to form a summer band singing all night long.

"Classification of common chirping insects"

In general, all the insects that can make sounds can be counted as "chirping insects", and the most well-known of them are the two types of chirping insects commonly known as crickets and grasshoppers.

If you ask a biologist what is a chirping insect, he will probably tell you this: most chirping insects belong to the Orthoptera in the biological classification. According to the length of their antennae, they can be under the Orthoptera with two categories: long-horned insects and short-horned insects. Long-horned insects, such as crickets and katydids, are mainly active at night and make sounds by rubbing their wings. The sound is relatively loud. Short-horned insects, such as locusts, and grasshoppers are mostly active during the day, and they make sounds by rubbing their legs and wings.

"Common chirping insects"

I think everyone is familiar with the look of crickets. But crickets have males and females, how do we distinguish them? Everyone pays attention to their tails. This one with two tails is a male cricket, commonly known as the two-tailed cricket; this one that has not only two tails, but also a long ovipositor, is a female cricket, commonly known as the three-tailed cricket.

In addition to appearance, we can also distinguish male and female by their habits. Chirping crickets are males and they are aggressive. They usually live alone. Only during the courting season will they use their calls to attract the female.

The katydid is the larger of the chirping insects. Its body is mostly grass green, with long tentacles but short wings. The most famous singer in the Katz family is called the weaving lady in Chinese. On summer nights, the male weaving lady often "zhi – zhi – zhi" chirps continuously, like a lady using a weaving wheel.

Cicadas do not belong to Orthoptera, but they have noisy voices. They especially like to hide in the shade of trees and shout in the hot afternoon. They are the most familiar summer bugs. Cicadas are a symbol of summer, and the summer without cicadas is incomplete. Cicadas spend almost all their lives underground. They can see the sun for only a month or so, and only male cicadas can scream. It can be said that they have tried their best to find their favorite partner. In addition, the sound of male cicadas is so loud thanks to their special sounding mechanism. Like drums in musical instruments, they sound through the vibration of the tympanic membrane. No wonder the sound is so loud.

katydid

katydid

Attentions

- Do your homework in advance to help visitors clarify the correspondence between the classification of chirping insects and common chirping insects (crickets, *etc.*).
- Focus on the more special chirping insects instead of the locusts.
- Combine with chirping insect introduction cards and play audio from chirping insects.

Recommended auxiliary facilities and toolkits

- Orthoptera classification card
- Photos of crickets, katydids, mole crickets, locusts, cicadas
- Audio of different sounds of insects

通用型解说 | 生物多样性解说　　　　解说编号：TY-WZ-21

21 蜻蜓？还是豆娘？

● 解说要点

1. 蜻蜓与豆娘的外形特征、生活史与生活习性
2. 蜻蜓、豆娘与水环境的关系

● 观察地点推荐

1. S02 古田山—古田庙：探秘常绿阔叶林自然体验路线
2. Q01 莲花塘：钱塘江寻源自然体验路线
3. Q02 里秧田—大峡谷：遇见峡谷飞瀑自然体验路线

正在交配的赤基色蟌

巨齿尾溪蟌

解说方式 Interpretation Style

☑ 人员解说 Staff Interpretation　☐ 教育活动 Educational Activity

☑ 解说牌 Interpretation Signage　☐ 场馆解说 Hall Interpretation

解说季节 Interpretation Season

☑ 春季 Spring　☑ 夏季 Summer

☑ 秋季 Autumn　☐ 冬季 Winter

解说词

初夏时节，公园的水边常出现一种大家非常熟悉的昆虫——蜻蜓，还有一种常被误以为是蜻蜓的昆虫——豆娘（又叫螅，cōng）。

「蜻蜓与豆娘的外形特征」

豆娘和蜻蜓的外表十分相似，它们都拥有修长的身形、两对轻盈的翅膀和一双炯炯有神的大眼睛。但是要想识别它们并不难。首先，从身材上来看，蜻蜓体形更加粗壮挺拔，而豆娘则娇小纤细；其次，蜻蜓的大眼睛一般是贴在一起的，豆娘的双眼则明显分开；另外，蜻蜓休息时翅膀是水平打开的，而豆娘休息时双翅会合拢，竖直地立在背上。

「蜻蜓与豆娘的生活史与生活习性」

蜻蜓和豆娘一生会经历卵、稚虫和成虫三个阶段，几乎每个阶段都离不开水。它们习惯将卵产在水面或者水生植物上，由卵孵化而成的稚虫叫作“水虿”，生活在水中，靠腮呼吸，会以水蚤、孑孓（jié jué，蚊子幼虫）为食，大的水虿甚至可以捕食小鱼和蝌蚪。不同蜻蜓和豆娘的水虿期长短不一，短则两三个月，长则七八年才会成熟，羽化成成虫。蜕变为成虫的蜻蜓和豆娘一般会在水虿栖息的水域附近活动。

我们常说蜻蜓是益虫，这是因为它们以捕食苍蝇、蚊子等昆虫为生，这种食性特点对控制蚊虫传播的疾病（如疟疾等）有相当的作用。蜻蜓头上那对巨大的复眼是它们捕食猎物的最佳武器，它们是自然界中捕捉目标猎物成功率最高的物种之一。此外，作为食物链中的一环，蜻蜓和豆娘以及它们的稚虫也是两栖动物、鱼类和雀鸟的重要食物来源。

「蜻蜓、豆娘与水环境的关系」

蜻蜓和豆娘对水体的变化也十分敏感。以前，人们会通过蜻蜓低飞来判断要下雨了，但是如今我们在城市已经很难见到蜻蜓的身影。水质的恶化或者水体的消失，是蜻蜓和豆娘生存最主要的威胁。它们是水质环境优劣的检察官，一般蜻蜓和豆娘出现较多的地方，水质一定不会太差。

拓展知识点

「蜻蜓的古老历史」

- 蜻蜓是一种古老的生物，最早的化石记录表明3亿年前它们的祖先就存在于地球上了，比我们熟悉的恐龙出现的年代还要早一些。那时候的蜻蜓个头超大，身长近1m。随着捕食者的增多和生存环境的改变，蜻蜓才逐渐演化成今天的样子。目前，世界上约有5000种蜻蜓，它们除了外形差异之外，在翅膀的煽动方式等方面也不尽相同。

「蜻蜓的复眼」

- 拥有复眼结构的昆虫有：苍蝇、蜜蜂、蟋蟀、蝴蝶、蟑螂等。其中，蜻蜓的复眼是所有昆虫中最大的，它们鼓鼓地突出在头部的两侧，占头部总面积的2/3以上，由28000个小眼面组成。复眼的体积越大，小眼面越多，昆虫的视力也越强。

General Interpretation | Biodiversity Section

Interpretation No. TY-WZ-21

21 Dragonfly vs damselfly

● Key points

1. The appearance characteristics, life history and habits of dragonflies and damselflies
2. The relationship between dragonflies, damselflies and water environment

● Observation places recommended

1. S02 Evergreen broad-leaved forest exploration route
2. Q01 Qiantang River source exploration route
3. Q02 Canyon and waterfall experience route

In early summer, a familiar insect often appears on the waterside of the park—the dragonfly, and another insect that is often mistaken for the dragonfly—the damselfly.

"The appearance of dragonfly and damselflies"
Damselflies and dragonflies are very similar in appearance, they both have a slender body, two pairs of light wings and a pair of big piercing eyes. But it is not difficult to identify them. First of all, from the body, the dragonfly is more sturdy and taller, while the damselfly is petite and slender; secondly, the big eyes of the dragonfly are generally close together, and the eyes of the damselfly are clearly separated; in addition, the wings of the dragonfly are open horizontally when resting, and when the damselfly is resting, her wings will close together and stand upright on her back.

"The life history and habits of dragonflies and damselflies"
Dragonfly and damselfly will go through three stages of egg, larva and adult in their lives, and almost every stage cannot be separated from water. They are used to lay their eggs on the water or on aquatic plants. The larvae hatched from the eggs are called "naiad". They live in the water and breathe on their gills. They feed on water fleas and larvae (jié jué, mosquito larvae). The big naiad can even prey on small fish and tadpoles. Different dragonflies and damselflies have different hydrostatic periods, ranging from as short as two to three months or seven or eight years before they mature and emerge into adults. Dragonflies and damselflies, which metamorphose into adults, generally move around the waters where the naiad inhabits.

We often say that dragonflies are beneficial insects. This is because they feed on flies, mosquitoes and other insects for their livelihoods. This feeding characteristic has a considerable effect on controlling mosquito-borne diseases (such as malaria, *etc*.). The giant compound eyes on dragonflies' heads are their best weapons for prey. They are one of the species with the highest success rate in capturing target prey in nature. In addition, as a link in the food chain, dragonflies and damselflies and their larvae are also important food sources for amphibians, fish and birds.

"The relationship between dragonflies, damselflies and water environment"

Dragonflies and damselflies are also very sensitive to changes in water bodies. In the past, people used dragonflies flying low to determine whether it was going to rain, but now we can hardly see dragonflies in cities. The deterioration of water quality or the disappearance of water bodies is the main threat to the survival of dragonflies and damselflies. They are judges of the water quality. Generally, where dragonflies and damselflies appear more often, the water quality can't be too bad.

dragonfly

damselfly

damselfly

Extra information

"The ancient history of dragonflies"

- Dragonflies are an ancient creature. The earliest fossil records indicate that their ancestors existed on the earth 300 million years ago, which is earlier than the time when the familiar dinosaurs appeared. At that time, the dragonfly was very large and nearly one meter long. With the increase of predators and the change of living environment, the dragonfly gradually evolved into what it is today. There are currently about 5,000 species of dragonflies in the world. In addition to the differences in appearance, they are also different in the way of inciting wings.

"Dragonflies's compound eyes"

- Insects with compound eyes are: flies, bees, crickets, butterflies, cockroaches, *etc*.. Among them, the compound eyes of dragonflies are the largest of all insects. They bulge out on both sides of the head, occupy more than two-thirds of the total area of the head, and are composed of 28,000 small eye faces. The larger the volume of the compound eye, the more small eye faces, and the stronger the vision of the insect.

通用型解说 | 生物多样性解说　　　　解说编号：TY-WZ-22

22 点亮夏夜的精灵：萤火虫

● 解说要点

1. 萤火虫对原真自然环境的重要指示意义
2. 钱江源国家公园不同萤火虫的特点与习性差异

● 观察地点推荐

1. S01 古田飞瀑：丛林飞瀑自然体验路线
2. Q02 里秧田—大峡谷：遇见峡谷飞瀑自然体验路线

锯角萤　萤火虫　穹宇萤
窗萤　锯角萤　萤火虫

解说方式 Interpretation Style

☑ 人员解说 Staff Interpretation　☐ 教育活动 Educational Activity
☑ 解说牌 Interpretation Signage　☐ 场馆解说 Hall Interpretation

解说季节 Interpretation Season

☑ 春季 Spring　☑ 夏季 Summer
☑ 秋季 Autumn　☐ 冬季 Winter

「萤火虫对原真自然环境的重要指示意义」

各位朋友，说起萤火虫，大家脑袋里会浮现出哪些关键词呢？浪漫、梦幻、童话、回忆或者传说？也有可能有很多人，特别是年轻人从来没有在野外见到过萤火虫。但如果我们询问年长的爷爷奶奶，可能很多人会说：小时候见的可多了，是童年很重要的玩伴和回忆。

大家知道吗？全世界的萤火虫一共1000多种，目前在中国生活着的萤火虫有200～300种。过去，它们曾是夏夜必不可少的一道风景，但如今，无论城市还是乡村，见到萤火虫都变得极为奢侈。环境的变化，是萤火虫消失的最主要的一个原因。萤火虫对栖息地的要求极高，不仅要水质清澈，还要植被丰富，同时也不能有光污染，它们可不想在找对象的时候有那么多竞争对手。

许多萤火虫爱好者不惜跋山涉水前往人迹罕至的荒野，只为见一见这种可爱的昆虫。钱江源国家公园就是长江三角洲地区萤火虫的最佳观赏地之一。环境教育调研团队仅在2019年夏季的一次现场调查中就发现，这里至少生活有6种萤火虫，其中还包括一些中国特有的种类。

「钱江源国家公园不同萤火虫的特点与习性差异」

按照萤火虫幼虫生活的环境，可将萤火虫分为陆生、水生和半水生三种类型。全世界的水生萤火虫只有7种，其中只在我国分布的有2种，生活在钱江源国家公园里的雷氏萤就是这两种水生萤火虫之一。雷氏萤的幼虫主食为水里的小型软体动物，比如螺蛳。羽化后的雷氏萤成虫体形较大，鞘翅呈黄色，会发出闪烁的黄绿色光。和“重口味”的幼虫不同，雷氏萤长大后只喝露水，生命周期也只有1～2周。

还有一种会集群同步发光的萤火虫：穹宇萤。和大多数萤火虫一样，穹宇萤也是雄虫发光，但令人惊讶的是它们会像约定好了一样，成群同时发出相同频率的闪光信号。如果野外种群数量够大，它们同时发光的壮观程度，堪比满天星斗，这也是穹宇萤得名的原因。穹宇萤的幼虫是半水栖的代表，主要生活在河流、小溪、瀑布等非常潮湿的地方。

不是所有的萤火虫长大后都会飞行，窗萤就只有雄虫飞行，雌虫则不会飞，雌虫和幼虫形态相似，像穿着厚厚盔甲的战士，有很强的辨识度。

也不是所有的萤火虫都是在夜晚活动的，国家公园里的锯角萤是就是一种在白天求偶的萤火虫，不过作为萤火虫家族的成员，它们的雄萤在夜晚也会发出微弱的光。

General Interpretation | Biodiversity Section

Interpretation No. TY-WZ-22

22 Fireflies that light up the summer night

Key points

1. Significance of fireflies to the original natural environment
2. The characteristics and habits of different fireflies in Qianjiangyuan National Park

Observation places recommended

1. S01 Jungle and waterfall experience route
2. Q02 Canyon and waterfall experience route

"Important significance of fireflies to the original natural environment"

Dear friends, when talking about fireflies, what keywords will pop up in your heads, romantic, dream, fairy tale, Memories or legend? There may also be many people, especially young people, who have never seen fireflies in the wild. But if we ask older grandparents, many people may say: we saw a lot when we were young, and they were very important playmates and memories in childhood.

Do you know there are more than 1,000 species of fireflies in the world, and there are currently two to three hundred species of fireflies living in China? In the past, they were an indispensable scenery in summer nights, but nowadays, seeing fireflies in cities and villages has become extremely luxurious. Environmental changes are one of the main reasons for the disappearance of fireflies. Fireflies have extremely high requirements for their habitat. They must not only have clear water, but also rich vegetation, and they must also be free from light pollution. They don't want to have so many competitors when they are looking for partners.

Many lovers of fireflies do not hesitate to trek through the mountains and rivers to the wilderness off the beaten track, just to see this cute insect. Qianjiangyuan National Park is one of the best places to watch fireflies in the Yangtze River Delta. The environmental education research team found that at least 6 species of fireflies live here only in a field survey in the summer of 2019, including some species unique to China.

"The characteristics and habits of different fireflies in Qianjiangyuan National Park"

According to the living environment of firefly larvae, fireflies can be divided into three types: terrestrial, aquatic and semi-aquatic. There are only seven species of aquatic fireflies in the world. Among them, there are only two species distributed in our country. The Lei's firefly living in the national park is one of these two species. The staple food of the larvae of the Lei's firefly is small mollusks in the water, such as snails. After emergence, the adult Lei's fireflies are larger, the elytron are yellow, and emit a flickering yellow-green light. Unlike the "heavy taste" larvae, Lei's firefly only drinks dew when it grows up, and its life cycle is only one to two weeks.

There is also a kind of fireflies that glow together in groups: Qiongyu firefly. Like most fireflies, the Qiongyu fireflies also emit light from males, but it is surprising that they will emit flash signals of the same frequency at the same time in groups. If the population in the wild is large enough, they can glow spectacularly at the same time, which is comparable to a starry sky, which is why Qiongyu firefly gets its name. The larvae of the fireflies are semi-aquatic representatives, mainly living in rivers, streams, waterfalls and other very humid places.

Not all fireflies can fly when they grow up. For window fireflies , only males can fly but females cannot fly. Females and larvae are similar in shape, like warriors wearing thick armor, you can easily recognize them in the wild.

Not all fireflies are active at night. Sawhorn fireflies in the park are a kind of fireflies that court during the day, but as members of the fireflies family, their male fireflies also emit weak light at night.

Attentions

- Fireflies were a very common insect in the past, but because of the destruction of their habitat, they can only exist in the imagination or memories of most people.
- The original ecological environment of Qianjiangyuan provides a habitat for fireflies.

Recommended auxiliary facilities and toolkits

- Photos of different fireflies
- Illustration of firefly pulse signal

Extra information

"Fireflies of Qianjiangyuan"

- The environmental education research team found at least 6 species of fireflies in an on-site survey in the summer of 2019 alone, namely, Lei's firefly, *Luciola anceyi* Olivier, *Luciola satoi*, Qingyu firefly, window firefly and sawhorn firefly. Each firefly has its own unique characteristics.

"How to protect fireflies"

- The main reason for the disappearance of fireflies is the destruction of nature by humans: destruction of habitats, massive exploitation of groundwater, soil pollution caused by pesticides and fertilizers, light pollution in cities, and the massive introduction of alien organisms that have led to loss of biodiversity, wetlands, *etc*..

通用型解说 | 生物多样性解说 解说编号：TY-WZ-23

23 神奇动物在钱江源：蚁墙蜂

解说要点

蚁墙蜂独特的巢穴防御策略

观察地点推荐

1. S02 古田山—古田庙：探秘常绿阔叶林自然体验路线
2. S03 瞭望台：国家公园科研体验路线

解说方式 Interpretation Style

☑ 人员解说 Staff Interpretation ☐ 教育活动 Educational Activity
☑ 解说牌 Interpretation Signage ☐ 场馆解说 Hall Interpretation

解说季节 Interpretation Season

☑ 春季 Spring ☑ 夏季 Summer
☑ 秋季 Autumn ☑ 冬季 Winter

解说词

各位朋友，请猜一猜这张照片上的动物是什么？蚂蚁？蜜蜂？实际上这是科学家 2014 年在古田山新发现的物种，名字叫作蚁墙蜂。有朋友可能就要问了，那到底是蚂蚁还是蜜蜂呢？

其实，蚁墙蜂和蜜蜂关系更近一些，属于蛛蜂家族。蛛蜂家族世世代代以捕食蜘蛛为生，因此得名。蚁墙蜂也不例外。蛛蜂妈妈们个个都是捕食蜘蛛的好手：它们会用毒刺将蜘蛛麻痹，然后拖进提前准备好的巢穴里。巢穴里有很多房间，每个房间里放一只蜘蛛，作为小蛛蜂的食物。新发现的“蚁墙蜂”也会通过这样的方式为后代储存食粮，但是和已知蛛蜂筑巢行为不同的是，巢穴最外部的小室不是空置的，而是填满了蚂蚁的尸体。这种用蚂蚁尸体来制造蚁墙，封住巢穴来保护它的后代的策略及行为十分独特，在整个动物界中还是首次发现。

为什么蚁墙蜂要这么做呢？科学家的进一步研究发现，原来蚂蚁尸体释放的化学物质能够掩盖幼虫的气味，保护幼虫免遭捕食者的攻击。这也是科学家首次发现自然界中会用“化学武器”保护幼仔的动物。和同类黄蜂的蜂巢相比，这种雌蜂所建蜂巢的天敌寄生率明显低得多。

研究蚁墙蜂的德国科学家泰伯博士说，他第一次看到填满蚂蚁尸体的巢穴时，联想起中国古老的长城。如同长城保护了中华民族一样，蚁墙也保护了这个新物种的后代免受天敌的攻击。

泰伯博士还表示，古田山自然生态系统应该包含了蛛蜂、蜘蛛、蚂蚁、胡蜂和寄生蜂之间的相互作用网络，因此，一定还有很多其他令人惊叹的现象有待发掘。

注意事项

- 蚁墙蜂的发现，极大地丰富了钱江源国家公园的生物多样性，也为钱江源国家公园的保护多提供了一种理由。

辅助道具推荐

- 蚁墙蜂及其巢穴照片
  蚁墙蜂的图片故事
  蚁墙蜂在植物的中空部位筑巢。它们会在巢穴中用泥土隔出一个个独立的小空间，在内部的空间中产卵，并将捕捉蜘蛛或其他节肢动物麻醉后存储在其中，作为幼虫孵化后的食物。更奇特的是，它们会在外部的空间中堆放满满的蚂蚁尸体，用死亡蚂蚁的味道掩盖幼虫的气味，防止被捕食者发现。这是人类首次在蜂类中发现这种恐怖的修筑“蚁墙”进行防御的行为，也是它们中文名“蚁墙蜂”的由来。

拓展活动

- 蚁墙蜂巢穴模型搭建。

拓展知识点

- 全世界科学家已经发现并且命名的物种有 130 多万个，据估计地球上可能还有数以百万的物种未被发现。每年科学家都会发现新的物种，2014 年，全球科学家一共新发现了 18000 个物种，其中，蚁墙蜂就位列前十。

General Interpretation | Biodiversity Section

Interpretation No. TY-WZ-23

23 Bone-house wasp

Key points

The bone-house wasp unique nest defense strategy

Observation places recommended

1. S02 Evergreen broad-leaved forest exploration route
2. S03 Scientific research experience route

Dear friends, please guess what is in this photo, ant or bee? In fact, this is a new species discovered by scientists in Gutian Mountain in 2014, named the bone-house wasp. Some friends may ask, is it an ant or a bee?

In fact, the bone-house wasp has a closer relationship with bees, belonging to the spider wasp family. The spider wasp family has been born to prey on spiders for generations, hence the name. The bone-house wasp is no exception. Bone-house wasps are excellent hunters: they paralyze the spider with a poisonous stinger, and then drag it into the nest prepared in advance. There are many rooms in the nest, and a spider is placed in each room as food for the spider wasp. The newly discovered bone-house wasp will also store food for offspring in this way, but unlike the known nesting behavior of spider wasps, its last chamber is not empty, but filled with ants' bodies. This strategy and behavior of using ants body to create an ant wall and seal the nest to protect its offspring is very unique, and it is the first time it has been discovered in the entire animal kingdom.

Why does the bone-house wasp do this? Further research by scientists found that the chemicals released by the ant's bodies can mask the smell of the larvae and protect the larvae from predators. This is also the first time scientists have discovered animals in nature that use "chemical weapons" to protect their babies. Compared with the hive of similar wasps, the parasitic rate of natural enemies of this hive is significantly lower.

Dr. Taber, a German scientist who studies this bone-house wasp said that when he first saw the ant wall filled with ant corpses, he thought of the ancient Great Wall of China. Just as the Great Wall protects the Chinese nation, the dead-ants' wall also protects the descendants of this new species from natural enemies.

Dr. Taber also said that the natural ecosystem of Gutian Mountain should include the interaction network between spider wasps, spiders, ants, wasps and parasitic wasps. Therefore, there must be many other amazing phenomena to be discovered.

Attention

- The discovery of the bone-house wasp has greatly enriched the biodiversity of Qianjiangyuan National Park and provided another reason for the protection of it.

Recommended auxiliary facilities and toolkits

- Photos of the bone-house wasp and its nest
- Photo story of the bone-house wasp
  The wasps build their nests in hollow parts of plants. They construct their nest into a collection of small rooms by mud. Then they lay their eggs in the inner rooms and store the captured spiders or other arthropods in them after anesthesia, as food for the larvae after hatching. Even more bizarrely, they fill the outer space rooms with dead ants, masking the smell of the dead ants to protect their babies from predators. This is the first time humans have observed the dreaded defensive tactic of building "ant walls" in bees. Hence they get their Chinese name as "ant-wall bee".

Recommended Activities

- The bone-house wasp's nest model building

Extra information

- Scientists around the world have discovered and named more than 1.3 million species, and it is estimated that there may be millions of species on the earth that have not been discovered. Scientists discover new species every year. In 2014, scientists around the world discovered a total of 18,000 new species, of which the bone-house wasp was among the top ten.

通用型解说 | 生物多样性解说　　解说编号：TY-WZ-24

24 小微“湿地”：滴水崖壁

● 解说要点

1. 滴水崖壁上常见的物种观察：苔藓、滴水珠、秋海棠、圆叶挖耳草、小沼兰……
2. 滴水崖壁的形成条件

● 观察地点推荐

1. Q01 莲花塘：钱塘江寻源自然体验路线
2. Q02 里秧田—大峡谷：遇见峡谷飞瀑自然体验路线

圆叶挖耳草　秋海棠　小沼兰　滴水珠

解说方式 Interpretation Style

☑ 人员解说 Staff Interpretation　☐ 教育活动 Educational Activity
☑ 解说牌 Interpretation Signage　☐ 场馆解说 Hall Interpretation

解说季节 Interpretation Season

☑ 春季 Spring　☑ 夏季 Summer
☑ 秋季 Autumn　☑ 冬季 Winter

解说词

各位朋友，我们通常把眼前这块大石头叫作滴水崖壁，因为溪流的缘故，这里常年都能得到水分的补充，因此一直可以保持湿润的状态。这样的环境为一些偏好湿润环境的生物提供了非常适合的栖息地。大家可以凑近来观察一下，看看这上面有什么。

「苔藓」

滴水崖壁上肉眼可见最多的植物就是苔藓了。这是一种对大家来说既熟悉又陌生的生物。熟悉是因为它们随处可见，而陌生则是我们常常对它们视而不见，因此对它们的了解认识十分有限。苔藓是个大家族，全世界有两万多种苔藓，大家可以数一数我们面前的这处滴水崖壁上有多少种苔藓吗？[可结合前面关于苔藓的介绍（解说编号：TY-WZ-06）]

苔藓是滴水崖壁上最早的定居者，也是这片岩石荒地的开拓者。它们会分泌出一种特殊的有机酸，可以加速岩石的风化，日积月累，原来裸露的崖壁上逐渐形成最初的一层薄薄的土壤，为其他植物提供落脚之地。

「滴水珠」

这种植物叫作滴水珠，是我国特有的一种植物。大家可以通过观察告诉我它为什么叫这个名字吗？没错，在它的叶片和叶柄连接处有个绿色的圆球，就像一滴水珠一样，因此才有了这个名字。那大家再猜一下，这滴“水珠”有什么作用呢？生物学家把它叫作“珠芽”，里面其实孕育着小滴水珠宝宝。当珠芽成熟脱落后，会“滚落”到滴水崖壁的其他地方，长成新的植株。这种看似随意却又独特的繁殖方式，是滴水珠对滴水崖壁荫蔽且湿润环境适应的体现，可以保证它们子孙后代的延续繁衍。

「秋海棠」

这棵开着粉红色小花的植物叫秋海棠。它是一种典型的具有“绿叶红花”特征的草本植物。请大家仔细观察下它们有几片花瓣。是的，秋海棠的花瓣有三片和四片两种，其中，三片的是雌花，四片的是雄花。除了花朵独特，秋海棠的果实也非常别致，像一架微型的纸飞机一样，十分惹人喜欢。

注意事项

- 解说员可以根据情况，选择2~3个物种进行解说，最好选择可以实际观察到的物种。
- 滴水崖壁环境比较湿滑，注意提醒访客注意安全。
- 滴水崖壁上的植物十分脆弱，注意不要破坏它们的生境。

辅助道具推荐

- 花岗岩岩石
- 圆叶挖耳草（带叶片、捕虫囊）照片

「圆叶挖耳草」

在滴水崖壁上的众多植物中间，有一种叶片特别小巧的食虫植物：圆叶挖耳草。它们的叶片小巧玲珑，形似纽扣，花瓣则为淡紫色，由纤细的花梗托起，形似“挖耳勺”，因而得名。圆叶挖耳草体形迷你，常紧贴在潮湿岩壁上生长，如果不仔细看，很容易被大家误以为是苔藓一类的植物。那么小棵的植物是怎么捕虫的呢？请大家看向我手中的照片。这根茎上生有一些半透明空心小球，我们把它叫作捕虫囊，上有盖子，平时紧闭，内部气压较低，当小昆虫触及时会自动弹开并将昆虫吸入。

「小沼兰」

滴水崖壁上还有一种迷你的兰花：小沼兰。每一棵植株上只有一片心形叶，约拇指指甲盖大小。小沼兰常在四五月份开花，和大家常见的兰花一样，小沼兰也会从叶片基部伸出一根直立纤细的花莛，上满缀满芝麻粒般大小的黄色小花。

「滴水崖壁的形成」

行走在国家公园里，像这样的滴水崖壁随处可见。但是出了公园，就不一定能再见到这样的生境了。对于滴水崖壁来说，石头和水二者缺一不可。一般来说，形成滴水崖壁的石头大都为花岗岩，质地坚硬，不容易被风化，所以才会形成崖壁这样的立面；此外，公园湿润多雨的气候使得这里常年水流不断，不断浸润着崖壁，才形成滴水崖壁这样特别的环境：土层稀薄、潮湿清凉且空气中富含负氧离子。

小沼兰

General Interpretation | Biodiversity Section

Interpretation No. TY-WZ-24

24 Micro wetland: dripping cliffs

● Key points

1. Observation of common species on dripping cliffs: moss, miniature green dragon, hardy begonia, striped bladderwort, small bogorchis (*Oberonioides pusillus*)···

2. The formation conditions of the dripping cliff

● Observation places recommended

1. Q01 Qiantang River source exploration route

2. Q02 Canyon and waterfall experience route

Dear friends, we usually call the big rock in front of us the dripping cliff. Because of the stream, it can be replenished with water all year round, so it can always be moist. Such environment provides a very suitable habitat for some organisms that prefer a humid environment. You can take a closer look and see what's on it?

"Moss"

The most visible plant on the dripping cliff is moss. This is a creature that is both familiar and unfamiliar to everyone. Familiarity is because they can be seen everywhere, while unfamiliarity is because we often turn a blind eye to them, our understanding of them is very limited. Moss is a big family. There are more than 20,000 kinds of mosses in the world. Can you count how many kinds of moss there are on this dripping cliff in front of us?

The moss is the earliest settler of the dripping cliff and the pioneer of this rocky wasteland. They secrete a special organic acid, which can accelerate the weathering of the rock. Over time, the original thin layer of soil gradually forms on the exposed cliff wall, which provides a place for other plants to settle.

"Dripping drops"

This kind of plant is called the miniature green dragon, also known as the dripping drops in Chinese, which is a kind of plant unique to our country. Can you tell me why it is called this name through observation? That's right. There is a green ball at the junction of its leaf and petiole, like a drop of water, so it got this name. Then guess again, what does this drop of water do? Biologists call it "pearl bud", which actually breeds its baby with small drops of water. When the pearl buds mature and fall off, they will "roll down" to other places on the dripping cliff and grow into new plants. This seemingly random but

Oberonioides pusillus

unique method of reproduction is a manifestation of the adaptation of dripping drops to the sheltered and humid environment of dripping cliffs, which can ensure the continued reproduction of their offspring.

"Hardy begonia"

This plant with small pink flowers is named hardy begonia. It is a typical herb with the characteristics of "green leaf and red flower". Please carefully observe how many petals they have? Yes, there are two kinds, three and four petals. Three-petal flowers are female and four-petal flowers are male. In addition to the unique flowers, the fruit of begonia is also very unique, like a miniature paper airplane, very attractive.

"Striped bladderwort"

Among the many plants on the dripping cliffs, there is a carnivorous plant with very small leaves: the striped bladderwort. Their leaves are small and exquisite, shaped like buttons, and their petals are lavender. They are held up by slender pedicels and shaped like "ear scoops". The striped bladderwort is tiny and often grows close to wet rock walls. If you don't look closely, it is easy to be mistaken for moss and other plants. So how do small plants

Attentions

- The interpreter can choose 2～3 species to conduct interpretation according to the situation. It is best to choose a species that can actually be observed.
- The environment of the dripping cliff wall is relatively slippery, please remind visitors to pay attention to safety.
- The plants on the dripping cliff are very fragile, so be careful not to damage their habitat.

Recommended auxiliary facilities and toolkits

- Granite rocks
- Photos of striped bladderwort (with leaves, insect traps)

catch insects? Please look at the photo in my hand. There are some translucent hollow spheres growing on this rhizome, we call it the insect trap. There is a cover, which is normally closed tightly, and the internal air pressure is low. When small insects touch it, they will automatically open and suck in the insects.

"*Oberonioides pusillus*"

There is also a kind of mini orchid on the dripping cliff: *Oberonioides pusillus*. There is only one heart-shaped leaf on each plant, about the size of a thumb fingernail. They often bloom in April and May. Like most common orchids, *Oberonioides pusillus* also protrude from the base of the leaves with an upright slender scape, covered with small yellow flowers in the size of sesame seeds.

"The formation of dripping cliffs"

Walking in the park, dripping cliffs like this can be seen everywhere. But once you leave the national park, you may not be able to see such a habitat again. For dripping cliffs, both stone and water are indispensable. Generally speaking, most of the stones that form the dripping cliffs are granite, which are hard in texture and not easy to be weathered, so the cliffs are formed. In addition, the wet and rainy climate of the national park makes the water flow continuously throughout the year, constantly infiltrating the cliffs. Therefore, the special environment of dripping cliff wall is formed: water flows all the year round, the soil is thin, moist and cool, and the air is rich in negative oxygen ions.

Striped bladderwort

通用型解说 | 生物多样性解说　　　　解说编号：TY-WZ-25

25 荫蔽的林下生境

● 解说要点

1. 林下代表性植物：阔叶箬竹、蕨类等
2. 林下生境与森林生态系统的关系

● 观察地点推荐

1. S02 古田山—古田庙：探秘常绿阔叶林自然体验路线
2. S03 瞭望台：国家公园科研体验路线
3. Q01 莲花塘：钱塘江寻源自然体验路线

阔叶箬竹

林下的蕨类

林下的蕨类

解说方式 Interpretation Style

☑ 人员解说 Staff Interpretation　☐ 教育活动 Educational Activity
☑ 解说牌 Interpretation Signage　☐ 场馆解说 Hall Interpretation

解说季节 Interpretation Season

☑ 春季 Spring　☑ 夏季 Summer
☑ 秋季 Autumn　☑ 冬季 Winter

在来到森林之前，我想大多数人心里对森林植物最大的想象一定是高大的乔木。但是当我们来到森林才发现，森林里的生物种类之多远超出我们的想象。

无论是高大的乔木，还是攀缘的藤本，无论是林下的灌木，还是紧贴地面生长的苔藓，每一种生物仿佛都能在森林占据一席之地。

「阔叶箬竹」

对于林下植物来说，最稀缺的资源就是阳光了。为了获取更多的阳光，它们也是使出了各种招数。我们眼前这种植物叫阔叶箬竹。大家注意看，这种竹子和平时我们见过的竹子有什么不一样吗？首先，阔叶箬竹不像毛竹一样拥有高大笔直的竹秆，最高也不过 2m。此外，阔叶箬竹的叶子比一般竹叶都要宽大，可以帮助它们最大限度地接收阳光。阔叶箬竹叶是南方人常用来包粽子的一种粽叶。

「蕨类」

早在恐龙出现之前，蕨类植物就已经在地球上生活了，我国有 2600 多种蕨类植物，大多分布在南方地区。蕨类植物依靠孢子繁殖。掀开叶片背面，如果看到许多棕色的似虫卵状的结构，有可能就是蕨类植物的孢子囊。蕨类植物是有性繁殖的，换句话说这些孢子囊中的小孢子是有性别的，它们必须要借助水才能完成不同性别的孢子传播和受精。此外，孢子成熟后也需要依靠风或者水传播出去，才能在适宜的环境下萌发生长。这也是为什么蕨类植物虽然生活在陆地，却总生长在阴湿环境下的原因。

「密度效应、林窗和林下生物」

森林的树冠像是一把巨大的遮阳伞，森林越茂密，林下光照越少。但是，大家有没有想过，随着森林不断生长，为什么这些树冠没有把森林上层的光全部遮住呢？这就不得不提到森林自我更新机制——林窗。森林中的植物似乎约定好了似的，会让群落保持在合适的密度，到了一定阶段就会自然死亡，以便为林下的幼苗提供生长的空间。此外，这些喜阳光植物的幼苗通常比阴生植物生长速度快得多，以便接收更多的阳光。

注意事项

- 走到林下，结合具体物种进行解说；
- 林下阴生世界是森林生态系统必不可少的一个重要组成部分，是一个相对的环境，会随着森林的演替发生动态变化。

辅助道具推荐

- 蕨类植物标本
- 林窗照片

General Interpretation | Biodiversity Section

Interpretation No. TY-WZ-25

25 Shade habitat under forest

● Key points

1. Representative understory plants: *Indocalamus latifolius*, ferns, *etc.*
2. The relationship between understory habitat and forest ecosystem

● Observation places recommended

1. S02 Evergreen broad-leaved forest exploration route
2. S03 Scientific research experience route
3. Q01 Qiantang River source exploration route

Before coming to the forest, I think the biggest imagination for most people of forest plants must be tall trees. But when we came to the forest, we discovered that the number of species in the forest was far beyond our imagination.

Whether it is a tall tree, a climbing vine, a shrub under the forest, or a moss growing close to the ground, every creature seems to have a place in the forest. As the builder of the forest, with its tall trunks and luxuriant branches and leaves, the arbor almost monopolizes the most sunny part of the upper forest.

"*Indocalamus latifolius*"

For understory plants, the most scarce resource is sunlight. In order to get more sunlight, they also tried various tricks. The plant in front of us is called *Indocalamus latifolius*. Please pay attention, this kind of bamboo is different from the bamboo we have seen before. First of all, it does not have tall and straight bamboo poles like moso bamboo, and the height is only two meters. In addition, its leaves are wider than ordinary bamboo leaves, which can help it receive the maximum sunlight. Their leaves are also used by southerners to make *zongzi*, a kind of rice dumplings.

"Fern"

The fern is probably the most common plant in the shaded environment under the forest, and it is also one of the most difficult to be recognized in the wild.

Ferns have already lived on the earth long before the appearance of dinosaurs. There are more than 2,600 kinds of ferns in our country, most of which are distributed in the south. Ferns rely on spores to multiply. Open the back of the leaves, if you see many brown insect egg-like structures, it may be the sporangia of ferns. What's interesting is that ferns

ferns under the forest

reproduce sexually. In other words, the microspores in these sporangia have gender. They must use water to complete the transmission and fertilization of spores of different sexes. In addition, after the spores mature, they need to be spread out by wind or water to germinate and grow in a suitable environment. This is also the reason why although ferns live on land, they always grow in a humid environment.

"Density effect, forest gaps and understory creatures"

The canopy of the forest is like a huge parasol. The denser the forest, the less sunlight under the forest. But have you ever wondered, as the forest continues to grow, why don't these canopies cover all the light from the upper layers of the forest? This has to mention the forest self-renewal mechanism—forest gap. The plants in the forest seem to have been agreed, and the community will be kept at an appropriate density, and will die naturally at a certain stage in order to provide space for the growth of seedlings under the forest. In addition, the seedlings of these sun-loving plants usually grow much faster than shady plants in order to receive more sunlight.

Attentions

- Go to the forest and explain in combination with specific species.
- The world of shaded forests is an essential part of the forest ecosystem. It is a relative environment that will dynamically change with the succession of forest.

Recommended auxiliary facilities and toolkits

- Fern specimens
- Photos of the forest gap

通用型解说 | 生物多样性解说 解说编号：TY-WZ-26

26 枯枝落叶下的隐秘世界

解说要点

1. 枯枝落叶下的代表生物：真菌、假水晶兰、蚯蚓、锹甲
2. 枯枝落叶是土壤有机质的重要来源，也是森林生态系统物质循环必不可少的一个环节

观察地点推荐

1. S02 古田山—古田庙：探秘常绿阔叶林自然体验路线
2. Q01 莲花塘：钱塘江寻源自然体验路线

解说方式 Interpretation Style

☑ 人员解说 Staff Interpretation ☐ 教育活动 Educational Activity
☑ 解说牌 Interpretation Signage ☐ 场馆解说 Hall Interpretation

解说季节 Interpretation Season

☑ 春季 Spring ☑ 夏季 Summer
☑ 秋季 Autumn ☑ 冬季 Winter

解说词

钱江源国家公园的原始森林里，每年树木都会落叶，假设每年的落叶只有 1cm，那么 100 年后，枯枝落叶就会堆成 1m 高，但是我们脚下的落叶肉眼可见的部分却远低于这个数字，这些落叶都去哪儿了呢？

答案就在这些枯枝落叶的下方。科学家把森林中由枯枝落叶组成的这一层叫作凋落物层。别看它们表面一片枯寂，轻轻扒开一片叶子，就会发现一个我们经常视而不见的隐秘世界。

这个世界里最常见的成员就是各种各样的真菌和细菌，比如蘑菇、木耳等。林下光照少、湿度大的环境为它们的生长提供了十分有利的条件。在古田山的森林中，分布有 200 多种大型真菌，其中大部分都是依靠枯枝落叶、倒木等为食物生长，比如大家熟悉的灵芝、猴头菇、银耳等。这些真菌就是落叶消失的秘密中转站，它们会把枯枝落叶分解成土壤中的有机质，对生态系统物质循环起着关键作用。

除了真菌之外，在适宜的季节和环境下，我们有可能观察到一种神奇而美丽的腐生植物——球朵假水晶兰。它们和真菌一样，也是生长于潮湿且富含有机质的林下，依靠与真菌合作分解各种有机物获取能量，完全不需要阳光和叶绿素也能生长，它们完全抛弃了叶绿素，活成了一株玲珑剔透的水晶一般的植物。

假水晶兰虽然叫兰但并不是真正意义上的兰花，通常兰花下方有一片比较大的花瓣，整朵花呈两侧对称，而假水晶兰的花呈辐射对称。假水晶兰对生境依赖性较高、难以栽培，一年中大部分时间都埋在地下，仅在花期和果期才会露出地面。

除了植物，林下布满落叶、朽木的土壤层内也有许多隐秘的生物，比如蚯蚓，它是土壤中常见的动物分解者。而许多锹甲、金龟的幼虫也是分解枯枝落叶的主要选手。

这是调研团队在现场一段朽木中发现的许多小型锹甲的幼虫和成虫的照片，保守估计至少有十几只，因为有了它们的存在，朽木、落叶中的养分才能迅速回到土壤，完成土壤中碳和养分的循环。

注意事项

- 找一处有枯枝落叶层的地方，轻轻扒开，观察枯枝落叶层的厚度、组成等特点，及其与下层土壤的关系；
- 一般枯枝落叶下都会有真菌（注意不一定是蘑菇），但能结合有蘑菇的地点解说更好。
- 注意提醒访客不要对枯枝落叶层、蘑菇等进行过度干扰。

辅助道具推荐

- 土壤分层剖面图
- 假水晶兰的照片
- 锹甲照片

General Interpretation | Biodiversity Section

Interpretation No. TY-WZ-26

26 The hidden world under fallen leaves

● Key points

1. Representative creatures under fallen leaves: fungi, *Monotropastrum*, earthworms, stag beetles

2. The fallen leave is an important source of soil organic matter and an indispensable link in the material cycle of forest ecosystems

● Observation places recommended

1. S02 Evergreen broad-leaved forest exploration route

2. Q01 Qiantang River source exploration route

In the virgin forest of Qianjiangyuan National Park, leaves will fall every year. Assuming that the annual deciduous leaves thickness are only 1 centimeter, then 100 years later, the fallen leaves will be piled up to a meter high, but the visible part of the fallen leaves under our feet is much lower than this number, where do all these fallen leaves go?

The answer lies under these dead branches. Scientists call the layer of fallen leaves in the forest the fallen leaveslayer. Although the surface looks deadly quiet, if you gently pull away some leaves there is a hidden world that we often ignore.

The most common members of this world are all kinds of fungi and bacteria, such as mushrooms and *Auricularia auricula*. The environment of low sunlight and high humidity under the forest provides very favorable conditions for their growth. In the forest of Gutian Mountain, there are more than 200 kinds of large fungi, most of which rely on fallen leaves and fallen wood for nutrients, such as the familiar *Ganoderma lucidum*, *Herician erinaceus*, and snow fungus. These fungi are secret transfer stations for the disappearance of fallen leaves. They decompose the fallen leaves into organic matter in the soil and play a key role in the material cycle of the ecosystem.

In addition to fungi, under the right season and environment, we may observe a magical and beautiful saprophytic plant——*Monotropastrum.*

Like fungi, they also grow in the moist and organic-rich forest. They rely on the cooperation with fungi to decompose various organic matter to obtain nutrients. They can grow without sunlight and chlorophyll so as to evolve into a delicate and transparent plant.

Although the *Monotropastrum* is called orchid in Chinese, it is not an orchid. Usually, there is a relatively large petal below the orchid. The whole flower is bilaterally symmetrical, while the *Monotropastrum* flower is radially symmetrical. *Monotropastrum* plants are highly dependent on the habitat and difficult to cultivate. They are buried in the ground most of the year, and only appear on the ground during the flowering and fruiting periods.

In addition to plants, there are also many hidden creatures in the soil layer covered with fallen leaves and dead wood. For example, earthworms are common animal decomposers in the soil. Larvae of many stag beetles and scarabs larvae are also the main players in breaking down litters.

This is a photo of larvae and adults of many small stag beetles found by the research team in a piece of dead wood. It is estimated that there are at least a dozen of them. Because of their existence, nutrients in dead wood and fallen leaves can quickly return to the soil, completing the cycle of carbon and nutrients in the soil.

Attentions

- Find a place where there is a fallen leaves layer, gently pull it, and observe the thickness and composition of the fallen leaves layer and its relationship with the underlying soil.
- Generally, there will be fungi under the fallen leaves (note that it is not necessarily a mushroom), but it is better to conduct interpretation in a place that has mushrooms.
- Pay attention to remind visitors not to disturb the fallen leaves layer, mushrooms, *etc*..

Recommended auxiliary facilities and toolkits

- Profile of soil layers
- Photos of *Monotropastrum*
- Photos of stag beetles

通用型解说 | 生物多样性解说　　　　解说编号：TY-WZ-27

27 溪流王国

● 解说要点

1. 清澈溪流生境的成因与特点
2. 在溪流里生活的代表性物种：虾虎鱼、光唇鱼、青蛳、狡蛛、东方巨齿蛉

● 观察地点推荐

1. S02 古田山—古田庙：探秘常绿阔叶林自然体验路线
2. Q01 莲花塘：钱塘江寻源自然体验路线
3. Q02 里秧田—大峡谷：遇见峡谷飞瀑自然体验路线

解说方式 Interpretation Style

- ☑ 人员解说 Staff Interpretation
- ☐ 教育活动 Educational Activity
- ☑ 解说牌 Interpretation Signage
- ☐ 场馆解说 Hall Interpretation

解说季节 Interpretation Season

- ☑ 春季 Spring
- ☑ 夏季 Summer
- ☑ 秋季 Autumn
- ☑ 冬季 Winter

「溪流生境的特点」

钱江源国家公园内有十多条大大小小的溪流，它们大都形成于山间裂缝或者凹陷处，河道狭窄，水流湍急，与平原河流形态有很大不同。溪流从高处一路蜿蜒向下，每隔数十米就有一个跌水瀑布或平流浅滩，枯水季节，常有大石头裸露，在丰水期则又被淹没。在地势高差和四季的更迭中，溪流生境变化多端，为多种生物的生存提供了丰富的选择。

「在溪流里生活的代表性物种」

在溪流中，生活着虾虎鱼、华南湍蛙等生物。它们就像水中的“蜘蛛侠”一样，拥有特殊的吸盘，可以帮助它们牢牢地吸附在石头上，任凭激流冲刷，还能稳若泰山。

在水流稍缓处，生活着虾虎鱼的邻居——光唇鱼。溪流水一般比较浅，且水质清澈，一般的鱼类很容易暴露自己。光唇鱼与溪流环境适应的结果就是它们那一身横条纹的衣裳，可以帮助它们模拟林间斑驳的光影，和水底的石头巧妙地融合在一起，很难被发现，因此也被称为“石斑鱼”。

流动的溪流浅水区中还有一些颜色青黑、体形修长的螺蛳贴附在水底的大石头上，这就是《舌尖上的中国 2》里提到的开化青蛳。和常见的螺蛳不同，青蛳是优质水源的质量检测官，主要生长在没有污染的活水中，以水生植物和有机碎屑物为食。青蛳还是溪流的净化官，它们分泌的黏液有助于水中的残渣沉降，对水质有一定的净化作用。

在小溪内的石头上还能见到一类长着“大长腿”的蜘蛛——狡蛛。它们依靠长长的足可以牢牢抓住石头，即使不慎失足落水也不怕，狡蛛还可以在水面行走和潜水，有些种类潜水时间可达35min。狡蛛主要生活在落叶层、溪流边的草丛灌木和石壁上，以“守株待兔”的方式捕捉小鱼、水生昆虫等，因此又被称为捕鱼蛛。

在雨后的夜晚，在溪流边，我们有机会见到一种体形巨大且长着“长牙”的昆虫，它是世界上现存最大的昆虫——越中巨齿蛉的亲戚：东方巨齿蛉。东方巨齿蛉体长可达 10cm 以上，雄虫的“长牙”尤其威武。齿蛉科昆虫多生活在生态环境优越、植被茂密的山间溪流、大河中，以各种水生昆虫为食。幼虫生活在干净的中，成虫生活在水边的灌丛，对水质变化敏感，常用作指示生物进行水质监测。

溪流还为蜻蜓、豆娘、石蝇、蜉蝣及其幼虫等水生昆虫提供了栖居的场所，这些昆虫则是依靠溪流生活的一些鸟类的食物，如燕尾、红尾水鸲、白鹡鸰等。溪流的开阔环境为我们近距离观察这些鸟类提供了很好的条件。

General Interpretation | Biodiversity Section

Interpretation No. TY-WZ-27

27 The Stream Ecosystem

Key points

1. Formation and characteristics of the clear stream habitat

2. Representative species that live in streams: gobies, groupers, green freshwater snails, *Dolomedes*, *Acanthacorydalis orientalis*

Observation places recommended

1. S02 Evergreen broad-leaved forest exploration route

2. Q01 Qiantang River source exploration route

3. Q02 Canyon and waterfall experience route

"Characteristics of the stream habitat"

There are more than a dozen large and small streams in Qianjiangyuan National Park. Most of them are formed in mountain cracks or depressions. The rivers are narrow and the currents are turbulent, which are very different from plain rivers. The stream winds all the way down from a high place, and there is a waterfall or shallow ponds every tens of meters. In the dry season, large rocks are often exposed, and they are submerged in the water during the wet season. In the changing of the four seasons, the stream habitats change a lot, providing wealthy choices for a variety of organisms.

"Representative species that live in a stream"

In the rush of water, there are aquatic creatures such as gobies and Chinese sucker frogs. They are like the "spider-man" in the water, with special suction cups that can help them firmly adhere to the stone, allowing them to keep still in the rapid streams.

In the place where the current is slightly slower, lives the neighbour of the goby, the grouper. The stream water is generally shallow and the water quality is clear, and fish can easily expose themselves. The way grouper fish adapt to the stream environment is their horizontal striped clothes, which can help them to subtly blend with the underwater stones and are difficult to be discovered, so they are also called "freshwater groupers".

In the shallow water area of the flowing stream, there are some greenish-black slender snails attached to the big rocks under the water. This is the Kaihua green freshwater snail mentioned in *Bite of China 2*.

Unlike common snails, green freshwater snails

are quality inspectors that mainly grow in unpolluted living water and feed on aquatic plants and organic debris. Green freshwater snails are also the purifiers of streams. The mucus they secrete helps the residues in the water settle and has a certain purification effect on water quality.

On the rocks in the creek, you can also see a kind of spider with "big long legs"—*Dolomedes*. They rely on their long feet to firmly grasp the stone, even if they accidentally fall into the water, they are not afraid at all. *Dolomedes* can also walk and dive on the surface of the water. Some species can dive up to 35 minutes. *Dolomedes* mainly live on deciduous layers, bushes and shrubs beside streams, and on stone walls. They catch small fish and aquatic insects, so they are also called fishing spiders.

On the night after the rain, we will see a huge and long-fanged insect, sometimes by the stream, especially in the evening after rain, which is the relative of the largest surviving insect in the world *Acanthacorydalis fruhstorferi*.That is *Acanthacorydalis orientalis*. It can reach more than 10 cm in length. Their males have larger "long teeth", and their bodies tend to be darker and thicker. Insects of the Corydalidae family mostly live in mountain streams and rivers with superior ecological environment and dense vegetation. They feed on various aquatic insects and are sensitive to the changes in water quality. They are often used as indicator to monitor water quality.

The stream also provides home for aquatic insects such as dragonflies, damselflies, stoneflies, mayflies and their larvae, and these insects are food for some birds that depend on the stream, such as forktails, red-tailed water thrushes, and white wagtails. The open environment of the stream provides good conditions for us to observe these birds up close.

通用型解说 | 生物多样性解说　　解说编号：TY-WZ-28

28 植物的明争暗斗：种间竞争

● 解说要点

1. 种间竞争的概念
2. 乔木、附生植物、寄生植物和林下阴生植物的竞争手段

● 观察地点推荐

1. S02 古田山—古田庙：探秘常绿阔叶林自然体验路线
2. S03 瞭望台：国家公园科研体验路线
3. Q01 莲花塘：钱塘江寻源自然体验路线

解说方式 Interpretation Style

- ☑ 人员解说 Staff Interpretation
- ☐ 教育活动 Educational Activity
- ☑ 解说牌 Interpretation Signage
- ☐ 场馆解说 Hall Interpretation

解说季节 Interpretation Season

- ☑ 春季 Spring
- ☑ 夏季 Summer
- ☑ 秋季 Autumn
- ☑ 冬季 Winter

「什么是种间竞争」

自然界的每一片土地都是一个大舞台，这个舞台上每时每刻都有好戏上演。看似寂静无声的森林，除了有动物之间的厮杀争斗之外，植物们也丝毫不相让，暗自进行着争夺地盘的较量。生物学家把物种之间为了争夺阳光、养分和生存空间等的竞争称为“种间竞争”。

「生物的各种“竞争”手段」

大家看这片林子，其实就是一处没有硝烟的战场。高大的乔木虽然拥有占据上层的绝对优势，但是它们哪一个不是从一颗种子逐渐“升级打怪”，一路成长起来的呢？林火、淹水、低温、少光……每时每刻都会威胁着它们的生存。为了高处的阳光，乔木们从生来就只有一个目标：向上！向上生长！

乔木凭借自身实力占据高位，有些植物则通过曲线救国的方式，站在“巨人”的肩膀上获取更多的阳光。公园里的藤本植物，比如香花崖豆藤、中华猕猴桃等，常常借助高大的乔木树干盘旋而上，以便获取更多的阳光。它们有的只是借乔木这个“梯子”往上爬，有的则会努力用藤蔓将乔木“绞”死。而短茎萼脊兰等附生植物，则会将家安在林冠层上，从而获取更多的光照。

另外，还有一种寄生植物，不仅将家安在别人身上，还要把根扎进别人的树木里，吸取别人的养分，比如桑寄生、槲寄生（寄生在松树上）。近年来在古田山阔叶林中记录到的花似蘑菇的穗花蛇菰就是一种全寄生植物，需要从其他植物的根上获得营养才能生存。

森林越往下层，阳光越稀缺，竞争也越激烈。八角莲和箬竹等草本植物，为了扩大光合作用的面积，进化出了大而薄的叶片；蕨类植物还会通过“下毒”的方式，抑制竞争者生长。

竞争关系看似残酷激烈，但不同物种正是通过竞争找到了自己合适的生态位，各取所需，同时提高了整个系统对资源的利用效率，对维持生物多样性有重要作用。

注意事项

- 种间竞争是指不同物种之间的竞争，是相对于种内竞争而言的，后者是指同一物种之间的竞争。

辅助道具推荐

- 附生植物、寄生植物、阴生植物照片

General Interpretation | Biodiversity Section

Interpretation No. TY-WZ-28

28 Battle between plants: interspecies competition

Key points

1. Concept of interspecies competition
2. Competitive methods of trees, epiphytes, parasitic plants and understory plants

Observation places recommended

1. S02 Evergreen broad-leaved forest exploration route
2. S03 Scientific research experience route
3. Q01 Qiantang River source exploration route

"What is interspecies competition"

Every piece of land in the natural world is a big stage, and there are stories staging at all times. In the seemingly silent forest, in addition to fightings between animals, the plants also do not give in at all, secretly engaged in a contest for territory. Biologists call the competition among species for sunlight, nutrients, and living space as "interspecies competition".

"Various 'competitive' methods of biology"

Everyone, this forest is actually a battlefield without guns. Although tall trees have the absolute advantage of occupying the upper level, all of them start from a seed and grow up all the way? Forest fire, flooding, low temperature, lack of light··· There are threats throughout their lives. For the sunshine, all the trees have only one goal in common from birth: upward! grow up!

Arbors occupy a high position by themselves, while some plants stand on the shoulders of "giants" to obtain more sunshine. The vines in the park, such as *Millettia dielsiana*, Chinese actinidia, *etc*., often use the tall tree trunks to hover up to get more sunlight. Some of them just borrow the "ladder" of trees to climb, while others try to "hang" the trees to death with vines. However, epiphytes such as *Sedirea subparishii* will place their homes on the forest canopy to get more light.

There is also a parasitic plant that not only sets the home on others, but also pierces the roots into other trees and absorbs nutrients from them, such as mulberry mistletoe, *European mistletoe* (parasites on pine trees). In recent years, the mushroom-like flower *Balanophora spicata*, which has been recorded in the broad-leaved forest of Gutian Mountain in recent years, is a fully parasitic plant that needs nutrients from the roots of other plants to survive.

The lower the forest, the scarcer the sunlight and the fiercer competition. Herbs such as *Dysosma* and *Indocalamus* have evolved large and thin leaves in order to expand the area of photosynthesis; ferns will also "poison" to inhibit the growth of competitors.

The competitive relationship seems cruel and fierce, but it is through competition that different species find their own suitable niches, which improves the efficiency of resource utilization and plays an important role in maintaining biodiversity.

Attentions

- Interspecies competition refers to competition between different species; it is relative to intraspecies competition, which refers to competition between the same species.

Recommended auxiliary facilities and toolkits

- Photos of epiphytes, parasitic plants, and shade plants

通用型解说 | 生物多样性解说　　　　解说编号：TY-WZ-29

29 大自然的“伪装者”

解说要点

1. 拟态的概念
2. 代表性的拟态物种举例：竹节虫、蝴蝶、食芽蝇、兰花

观察地点推荐

1. S02 古田山—古田庙：探秘常绿阔叶林自然体验路线
2. Q01 莲花塘：钱塘江寻源自然体验路线
3. Q02 里秧田—大峡谷：遇见峡谷飞瀑自然体验路线

竹节虫

枯叶蛾

拟态胡蜂的舟山黎柄角蚜蝇

枯叶蛾的幼虫

拟态鸟类的蝴蝶幼虫

解说方式 Interpretation Style

☑ 人员解说 Staff Interpretation　☐ 教育活动 Educational Activity

☑ 解说牌 Interpretation Signage　☐ 场馆解说 Hall Interpretation

解说季节 Interpretation Season

☑ 春季 Spring　☑ 夏季 Summer

☑ 秋季 Autumn　☐ 冬季 Winter

「拟态的概念」

角色扮演、模仿游戏是我们平时生活中的娱乐方式。自然界中，形形色色的生命也在上演着“模仿游戏”，这是它们在大自然的演化中形成的各自独特的生存策略。其中最为大家熟悉的“模仿游戏”就是拟态。它是指一种生物模拟另一种生物或环境的现象。这些物种是大自然中最神奇的魔术师，要认出它们，可得练就一双火眼金睛。

「代表性的拟态物种举例」

如果大家看过电影《神奇动物在哪里》，一定对其中的“护树罗锅”印象深刻，自然界中最接近这种神奇动物的，竹节虫算一个。竹节虫生活在热带或者亚热带的丛林中，以树叶等植物为食，为了躲避天敌的追捕，它们会把自己伪装成树木或草叶的样子，这样鸟儿就不容易发现它们了。竹节虫的身体一般都非常细长，一动不动时，真的像是植物的一部分呢。即使移动时，它们还会模仿树枝随风晃动的样子，因此很难被发现。

说起拟态昆虫，大家比较熟悉的可能还有蝴蝶和飞蛾家族等。尤其是其中的枯叶蝶 / 蛾，对枯叶的模仿简直惟妙惟肖。但不是每只毛毛虫都有机会蜕变成蝴蝶，为了躲避天敌的追捕，许多蝴蝶的幼虫也都是拟态界的高手。比如，柑橘凤蝶初龄的幼虫还会把自己拟态成鸟粪的样子，以躲避鸟类的捕食。最后一个阶段的幼虫身体为绿色，头顶长有一对黄色的丫形腺，受到惊吓时会弹出，加上头顶的眼斑，看起来特别像一条蛇。

食蚜蝇是“攻击型”的拟态代表，它们会将自己“打扮”成可怕的杀手——有螫刺的蜜蜂或胡蜂的样子。要识别它们可以在它们停下时观察一下它们有几对翅膀：蜜蜂或者胡蜂有前后 2 对翅膀，而食蚜蝇与苍蝇是近亲，虽有 2 对翅膀，但因为后面的翅膀退化，看起来只有 1 对前翅。此外，它们的眼睛也略有不同，食蚜蝇的复眼几乎相连，而蜜蜂、胡蜂的复眼则相距较远。

自然界的模仿游戏屡见不鲜，就连植物也要参与，而植物中最杰出的拟态高手非兰花莫属。比如，蜂兰为了骗取传粉昆虫的信任，不仅会把自己拟态成雌蜂的样子，还会通过模仿雌蜂性外激素的方式，吸引雄蜂来和自己“交配”，最终达传粉的目的。

注意事项

- 拟态现象是自然界生物与环境协同进化的结果，产生了很多有趣的现象，这部分讲解如果能结合实物或者图片等效果会更好。

辅助道具推荐

- 拟态动、植物卡片

拓展活动推荐

- 准备一些拟态动植物卡片，与访客玩互动小游戏“大家来找碴”或“谁是模仿大王”，寻找那些了不起的拟态生物。

[参考动植物]

- 竹节虫
- 变色龙
- 枯叶蝶蛾
- 叶虫
- 蜂兰
- 兰花螳螂
- 乌林鸮
- 章鱼
- 叶尾守宫（难度极高）

General Interpretation | Biodiversity Section

Interpretation No. TY-WZ-29

29 Mimetic species in Qianjiangyuan National park

Key points

1. The concept of mimicry
2. Examples of representative mimetic species: stick insects, butterflies, flower flies, orchids

Observation places recommended

1. S02 Evergreen broad-leaved forest exploration route
2. Q01 Qiantang River source exploration route
3. Q02 Canyon and waterfall experience route

"The concept of mimicry"

Role-playing and imitation games are the way we entertain in our daily lives. In nature, all kinds of life are also playing "imitation games", which are their unique survival strategies formed in the evolution of nature. One of the most familiar "imitation games" is mimicry. It refers to the phenomenon in which one organism imitates another organism or the environment. These species are the most amazing magicians in nature. To recognize them, you have to look carefully.

"Examples of representative mimetic species"

If you have seen the movie *Fantastic Beasts and Where to Find Them*, you must be impressed by the "bowtruckle". The closest thing to this kind of magical animal in nature is the stick insect. Stick insects live in tropical or subtropical jungles and feed on plants such as leaves. In order to avoid being hunted by natural enemies, they will disguise themselves as trees or blades of grass, so that birds cannot easily spot them. The body of stick insects is generally very slender, and when they are not moving, it really looks like a part of a plant. Even when they move, they imitate the way branches sway in the wind, so they are hard to be spotted.

Speaking of mimetic insects, the butterfly and moth families may be familiar to everyone. In particular, the orange oakleaf butterfly/moth mimics the dead leaf perfectly. But not every caterpillar has the opportunity to transform into a butterfly. In order to avoid hunting by natural enemies, many butterfly larvae are also masters in the mimicry world. For example, the larva of the Asian swallowtail butterfly needs to undergo many metamorphosis before it grows up. The larva of the last stage is green with a pair of yellow Y-shaped glands on the top of the head, and the eye spots on the top of the head look particularly like a snake. Besides this, the larvae will mimic themselves like bird

droppings to avoid predation by birds.

Hoverflies are mimicry representatives of the "aggressive" type. They "dress up" themselves as terrible killers-stinging bees or wasps. To identify them, you can observe how many pairs of wings they have when they stop: bees or wasps have two pairs of wings before and after, while hoverflies and flies are close relatives, although there are two pairs of wings, because the rear wings are degraded, it looks like there is only one pair of forewings. In addition, their eyes are slightly different. The compound eyes of hoverflies are almost connected, while the compound eyes of bees and wasps are far apart.

Imitation games in nature are not uncommon, even plants have participated, and the most outstanding mimicry master among plants is the orchid. For example, in order to deceive the trust of pollinators, the bee orchid not only mimics itself as a female bee, but also attracts male bees to "mate" with itself by mimicking the female bee pheromone, and ultimately achieves the purpose of pollination.

Attentions

- Mimicry is the result of the co-evolution of natural organisms and the environment, which has produced many interesting phenomena. This part of the interpretation will be better if it can be combined with real objects or pictures.

Recommended auxiliary facilities and toolkits

- Cards of mimetic animals and plants

Recommended activities

- Prepare some cards of mimetic animals and plants, and play interactive games with visitors "Let's Find the Difference" or "Who is the Imitation King" to find those amazing mimics.

"Reference animals and plants"

- stick insect
- chameleon
- orange oakleaf butterfly/moth
- *Phyllium*
- bee orchid
- orchid mantis
- great grey owl
- octopus
- *Uroplatus* (very difficult)

3.2 通用型村庄与传统文化解说方案

General interpretation of village and traditional culture

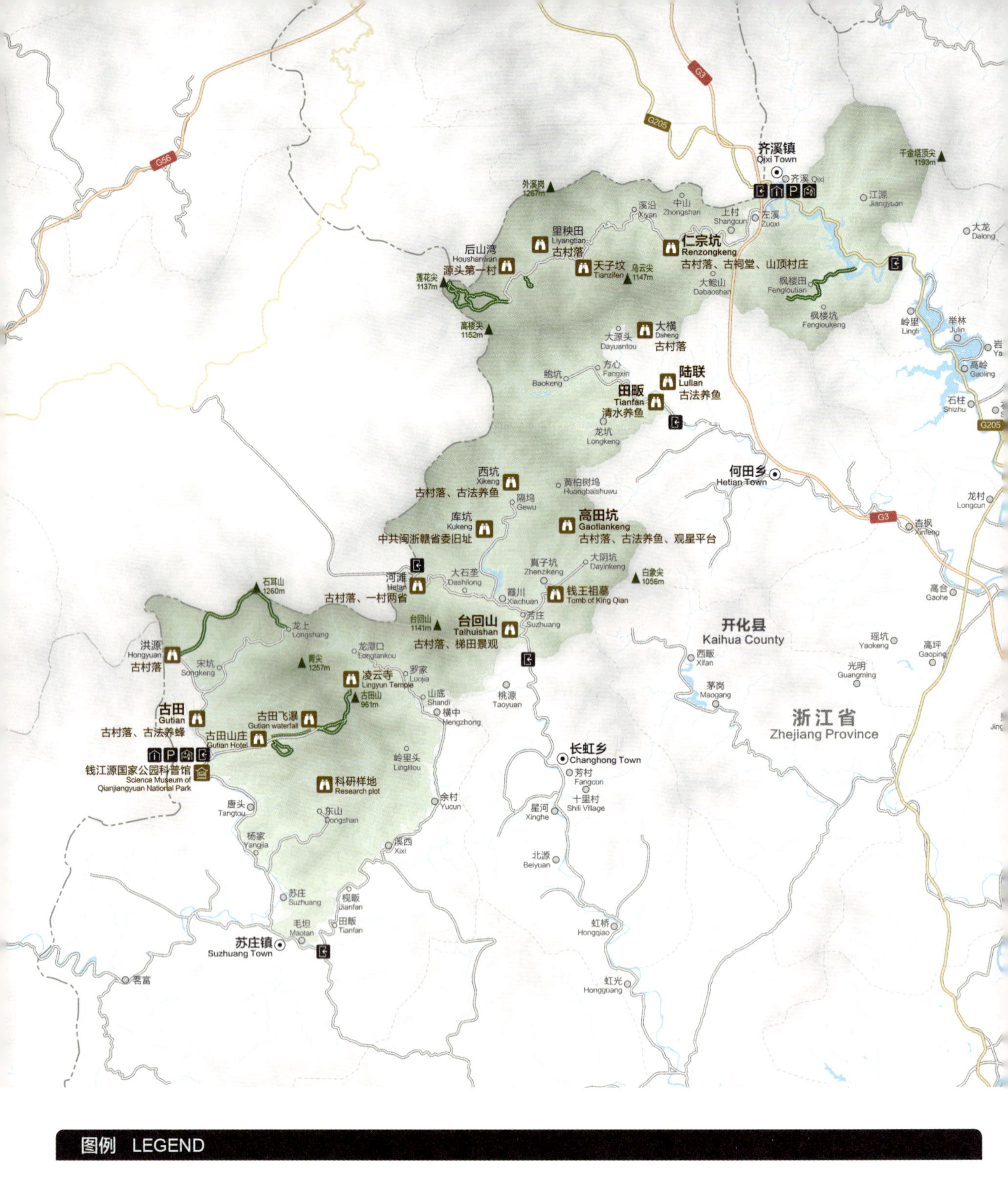

图例 LEGEND

图例	名称	图例	名称	图例	名称	图例	名称
G3	高速公路 Freeway		乡镇界线 Town Boundary		行政村 Administrative Village		救护室 Ambulance room
G205	国道 National Highway		县道 County Road		自然村 Natural Village		景点 Scenic Spot
	河流 River		步道 Trail		出入口 Gateway		科普馆 Science Museum
	省界线 Province Boundary		山峰 Mountain Peak		游客中心 Visitor Center		
	县界线 County Boundary		乡镇 Town	P	停车场 Parking Lot		

0km 4km 8km

认识钱江源国家公园的传统村落

探访钱江源国家公园的过程中，你会发现这里不仅拥有旖旎的自然风光，同时还有诸多从历史悠久的传统村落。公园所在的开化县是浙江、安徽、江西 3 省 7 县交界之地，历史上有大量人口因为官宦隐逸、躲避战乱等原因而迁居到此，唐宋时期尤盛。相应地，各种生产生活方式与文化习俗也在这里汇聚、发展，昔日荒芜面貌得以改变。

开化建县至今已 1000 多年，据不完全统计，全县有 340 多个姓氏，多数为外地迁入。钱江源地区同样如此。外地人来到这里，按宗族姓氏建成村庄，在历史中形成了“亲疏居有族，少长游有群”的同姓聚族而居的特点，以血缘关系为纽带的宗族在社会关系的组织、协调以及乡民教化等方面发挥着重要作用，很多人文传统都是借助宗族文化得以传承和发展。由于钱江源地处偏僻、交通不便以及山居生活自给自足的生产方式，这些来自不同地方迁移而成的村庄，直到今天，仍有一定的文化差异性保留下来。

长期生活在钱江源地区的人们，在悠久的历史中逐渐经历了从认识自然、依靠自然、理解自然再到尊重自然、顺应自然和守护自然的过程，“因地制宜”“道法自然”等朴素的自然观念也慢慢深入人心，并且代代传承。钱江源地区可耕地资源非常稀少，人们常用“九山半水半分田”来形容这里的自然地理格局。但自然却以另外一种形式给予生活在钱江源的山民以丰富的馈赠。

在这个部分，笔者为解说员提供了关于钱江源国家公园传统村庄的解说内容。以人与自然的共生为核心，提取出 17 个具有代表性的解说点，旨在帮助访客认识这些古老村庄背后依山傍水的生态智慧、传承古法的超卓技艺，以及千百年来当地人与自然和谐共生过程中所凝聚的朴素自然观。

解说员可根据实际情况灵活调用本手册的解说点及具体内容。值得注意的是，不同于自然资源主题的步道或场地解说，带领访客参观古村落的过程中难免会遇到正在此地生活的村民，对在地居民的生活方式和日常习惯请予以尊重；同时，如能适当将在地居民作为角色引入到解说活动中来，这将会是一种更具互动和亲切感的方式。比如，安排一顿具有地方风味的饭菜，在餐桌上为访客介绍与当地饮食文化有关的内容；在特定的时节带领访客亲自体验当地的风俗节日；或者提前与养蜂人、竹编匠人、制茶手艺人等共同筹备，为访客进行现场技艺展示与解说。如果能引发面前受众的好奇心与参与感，解说就已经成功一半了！

Meet the traditional villages in Qianjiangyuan National Park

During the visit to Qianjiangyuan National Park, you will find that it not only has beautiful scenery but also has many traditional villages since ancient times. Kaihua County, where the park is located, is the junction of seven counties in the three provinces of Zhejiang, Anhui, and Jiangxi. A large number of people have moved here in history because of retirement, avoiding wars and other reasons, especially during the Tang and Song Dynasties. Correspondingly, various production, lifestyles and cultural customs have also gathered and developed here, changing this former barren land.

During the past around 1,000 years since Kaihua County was established, there are more than 340 surnames in the county, most of which are moved in from other places according to incomplete statistics. The same thing applies to the Qianjiangyuan area. Outsiders came here and built villages based on the clan's surnames. In history, they formed the characteristics of living together with the same surnames. The clan based blood relationship plays an important role in the organization and coordination of social relations and the education of villagers. Many traditions are inherited and developed with the help of clan's culture. Due to the remoteness of Qianjiangyuan, the inconvenient transportation, and the self-sufficient production methods of mountain life, these villages migrated from different places still preserve some cultural differences to this day.

People who have lived in the Qianjiangyuan area for a long time have gradually experienced the process from meeting nature, relying on nature, understanding nature to respecting nature, adapting to nature and protecting nature. Simple natural concepts such as "adjusting measures to local conditions" "Following the natural world" "the unity of man and nature" have gradually penetrated into the hearts of local people and have been passed down from generation to generation. The arable land resources in the Qianjiangyuan area are very scarce, and people often use "90% mountains, 5% water and 5% farmland" to describe the natural geographical pattern here. But nature has given rich gifts to the mountain people living in Qianjiangyuan in another form. In this section, we will provide interpretations about traditional villages in Qianjiangyuan National Park. From the symbiosis between people and nature in different villages, we have extracted 17 typical and representative interpretation stops, designed to help visitors understand the ancient villages of Qianjiangyuan from different perspectives from the continuation and inheritance of ancient techniques, the ecological wisdom of living by the mountains and rivers revealed in daily life, as well as the harmonious coexistence between man and nature. The interpreter can conduct interpretation flexibly based on the actual situation.

In addition, interpreters can use various means when conducting traditional village interpretation. Different from the trails or site interpretations of natural resources, you will inevitably meet the villagers who live here

while leading visitors to the ancient villages. Please respect the lifestyle and daily habits of local residents; in the meanwhile, it will be interactive and intimate if you can include the villagers into the interpretation. For example, arrange a meal with local flavors, and introduce local food culture; take visitors to experience the local customs during certain seasons; or contact beekeepers, bamboo craftsmen, tea making craftsmen, masters of ancient inkstone production in advance and conduct on-site technique demonstrations and interpretations for visitors. And It will be the best if you can conduct on-site interpretation when introducing the stories of steles, ancient trees and starry sky.

通用解说 | 村落与传统文化解说　　解说编号：TY-CL-01

01 人口汇聚之地，多元文化之乡

● 解说要点

1. 公园的自然地理条件与人口迁移的历史
2. 人口迁移对当地语言、饮食、宗族文化等的影响

● 观察地点推荐

1. 苏庄镇古田村、唐头村
2. 长虹乡高田坑自然村、库坑村、河滩自然村
3. 齐溪镇仁宗坑村

解说方式 Interpretation Style

- ☑ 人员解说 Staff Interpretation
- ☐ 教育活动 Educational Activity
- ☑ 解说牌 Interpretation Signage
- ☐ 场馆解说 Hall Interpretation

解说季节 Interpretation Season

- ☑ 春季 Spring
- ☑ 夏季 Summer
- ☑ 秋季 Autumn
- ☑ 冬季 Winter

解说词

钱江源地区位于浙江、安徽、江西三省交界处，与福建靠近，地理位置偏僻、地形多山且交通不便。独特的地缘关系使得这里接纳了大量因官宦隐逸、战乱或灾害等原因而迁移至此的移民。

第一批迁移至此的人口可以追溯至唐代，至今已有一千多年的历史。在这一千多年里，“迁移”二字始终与这里的居民相伴随。据不完全统计，如今开化全县共有340多个姓氏，几乎都是由外地迁入的。他们有的来自皖南，属于徽州文化圈，因此随他们迁来的有徽派建筑和徽州方言等；有的来自江西，饮食偏辛辣，对本地的食物产生了深远的影响；这里还有来自福建的客家人，他们带来了闽南语和客家文化。

有趣的是，虽然不同人群在此汇聚，但却没有互相同化，保留了各自的差异性。一方面是由于地理上的天然阻隔，使得这里交通十分不便，沟通成本比较高，另一方面，则是受到小农经济的影响，人们定居于此后，便过上了自给自足的小农生活，不需要进行太多的交流，因此才使得这里呈现出多种文化共存的现象。

以方言为例，开化的三大方言为开化话（属于吴语系）、华埠话、马金话（属于徽州语系），除此之外，这里还有淳安话、闽南语……一共十几种方言。独特的语言文化使得开化许多地方出现了“一村不同语”“一家多语”和“一人多语”的奇妙现象。

语言是人们沟通的桥梁，而血缘则是人们聚居成村落的基础。外地人来到这里，常按宗族姓氏建成村庄，比如古田村的居民多姓赖，高田坑村的居民多姓余，中山村的居民多姓汪。这样的聚居特点使得以血缘关系为纽带的宗族文化得以在钱江源得到传承和发展。宗族内部长幼有序，通常由族内辈分最高、年龄最大、威望最重的人担任族长，主持处理宗族的大小事务。村中一般都建有祠堂，平时用来议事社交，节日则用来进行祭祀活动和聚会。如今，很多村落的祠堂不仅延续了其宗族祭祀的功能，同时也成为村民们日常社交和文化娱乐的去处，在新时代仍具有凝聚人心的力量。

注意事项

- 可以结合具体村落解说，但要注意提前了解该村落人口迁移自哪里，姓什么，有什么特色文化。

General Interpretation | Village and Traditional Culture Section

Interpretation No. TY-CL-01

01 A land of immigrants and diverse cultures

● Key point

1. The physical geography of the national park and the history of population migration
2. The influence of population migration on local language, diet, clan culture, *etc*.

● Observation places recommended

1. Gutian Village and Tangtou Village of Suzhuang Town
2. Gaotian Keng Natural Village, Kukeng Village and Hetan Natural Village of Changhong Town
3. Renzong keng Village of Qixi Town

Qianjiangyuan National Park is located at the junction of Zhejiang Province, Jiangxi Province and Anhui Province, near Fujian Province. It's a place with remote geographical location, mountainous terrain and inconvenient transportation. Because of its unique geographical relationship, it has accepted a large number of immigrants because of the seclusion of officials, wars or disasters.

The first people to move here can be traced back to the Tang Dynasty, which has a history of more than 1,000 years. During this time, the word "migration" has accompanied the inhabitants here. According to incomplete statistics, there are more than 340 surnames in Kaihua County, almost all of which were moved in from other places. Some of them came from southern Anhui Province which belonged to Huizhou cultural circle, they have brought with them Huizhou architecture and Huizhou dialect, and *etc*. Some came from Jiangxi Province and had a spicy diet, which had a Profound influence on local food. There are also Hakka people from Fujian who brought with them the Hokkien language and hakka culture.

What's interesting is that although different groups of people meet here, they don't assimilate with each other, and retain their differences. On the one hand, the natural geographical conditions have made communication inconvenient and communication cost high. On the other hand, under the influence of small-scale peasant economy, the people settled here have lived a self-sufficient smallholder life for a long time, and don't need too much communication. As a result, a multicultural coexistence phenomenon was presented here.

Take dialects as an example. There are three major dialects in Kaihua, which are Kaihua dialect, Huabu dialect, and Majin dialect

respectively. Besides, there are Chun’an dialect, Minnan dialect… Altogether there are over ten dialects. The unique language culture has led to the emergence of “different languages in a village” “multiple languages in a family” and “multiple languages in a person” in many civilized places.

Language is a bridge for people to communicate, and blood is the basis for people to live together in a village. Outsiders come here and often build villages according to clan surnames. For example, most residents in Gutian Village have the surname Lai, most residents in Gaotian Pit Village have the surname Yu, and most residents in Zhongshan Village have the surname Wang. This kind of settlement makes the clan culture, which is bound by blood relationship, inherit and develop in Qianjiangyuan. The elders and children of the clan are in order, and the patriarch is usually the person with the highest rank, the oldest age and the most prestige in the clan, who is in charge of the affairs of the clan. Ancestral temples are generally built in villages, usually used for social meetings, and used for sacrificial activities and gatherings during festivals. Nowadays, ancestral temples in many villages not only continue their function of clan sacrifice, but also become the daily social and cultural entertainment destinations for villagers, which still have the power to unite people in the new era.

Attentions

- The explanation can be combined with the specific village, but it should be noted in advance to know where the population of the village migrates from, what is its surname and what characteristic culture it has.

通用解说 | 村落与传统文化解说　　　　解说编号：TY-CL-02

02 九山半水半分田的农业坚守

● 解说要点

1. 山多地少的山地格局与梯田的关系
2. 公园内梯田上经常种植的农作物：水稻、龙顶茶、油茶、油菜

● 观察地点推荐

1. 长虹乡台回山自然村、高田坑自然村
2. 齐溪镇仁宗坑村
3. 何田乡陆联村、田畈村
4. 苏庄镇古田村、唐头村

解说方式 Interpretation Style

☑ 人员解说 Staff Interpretation　☐ 教育活动 Educational Activity
☑ 解说牌 Interpretation Signage　☐ 场馆解说 Hall Interpretation

解说季节 Interpretation Season

☑ 春季 Spring　☑ 夏季 Summer
☑ 秋季 Autumn　☑ 冬季 Winter

解说词

大家来到钱江源国家公园，觉得这里土地上生长出来的最丰富的资源是什么？水、森林……农田呢？特别少，是吗？公园内几乎 93% 的土地都被森林覆盖，可耕地资源十分稀缺。为了获得日常生活所需的食物，人们便在坡度较缓的地方修筑梯田，并在梯田之上种植水稻、茶叶、油菜等农作物。

水稻是钱江源地区的主要粮食作物，这里气候湿润、热量充沛，为水稻的生长提供了天然的水热条件。水稻种植讲究时节，钱江源地区的农民常在清明和谷雨前后将稻谷撒向田间。农用的肥料也是就地取材，人们会将山间随处可见的野草收割后做成肥料投入田间，或者将自家的厨余垃圾、大粪灰、鸡鸭毛拌成的肥料块投入田间，这些都是天然的有机肥料。秋天稻谷收割以后，人们会利用自家房屋二楼窗户外几根长杆搭建的晒场晾晒稻谷，极大地利用了空间。

大家有种植过水稻吗？水稻生产程序十分复杂，需要经播种、育苗、插秧、管理和收割等诸多环节，几乎是单位面积投入劳动力最高的一种农作物。但是大家走访村落的时候会发现，村子里现在留守的大都是老年人和小孩，他们劳动力十分有限，因此许多曾经的水田都已荒废或者种上了玉米等农作物。

相比于水稻的精耕细作，茶树的种植就省事许多。在公园内山地海拔较高、坡度较陡的位置，经常可见一亩亩油绿齐整的茶园，随着梯田层叠起伏。钱江源地区地处浙江、安徽、江西三省交界的“茶叶金三角”地区，这里山高林茂、云雾环绕，气候温和多雨且坡度较陡，不易积水，是高山龙顶茶的重要产地。

开化素有“中国油茶之乡”之称。得益于本地独特的山区小气候，这里出产的油茶含油率高、品质佳，是钱江源地区重要的油料作物和本地的特色产业之一。除了普通油茶之外，这里还是我国特有树种浙江红花油茶的重要产地。每年初春，苏庄镇的千亩红花油茶都会形成一道亮丽的风景。

钱江源地区还有一种非常重要的油料作物：油菜。每到春天来临，长虹乡台回山自然村的油菜花便争相绽放，漫山遍野布满梯田，与梯田、白云、村庄一起组成一幅美丽的画卷。除了吸引游人之外，油菜花田每年这个时候还会有一批特殊的客人准时光顾，这就是中华蜜蜂。它们在采蜜的同时也帮助油菜花完成了传宗接代的任务，夏初时节，农民们便可以收割油菜籽了。

拓展知识点

- 梯田是在丘陵山坡地上沿等高线方向修筑的条状阶台式或波浪式断面的田地，不仅可以增大田地的使用面积，同时还可以是治理坡耕地水土流失的有效措施，具有蓄水、保土、保肥等功效。

General Interpretation | Village and Traditional Culture Section

Interpretation No. TY-CL-02

02 Agricultural development in the mountainous area

● Key points

1. The relationship between mountains and terraced fields
2. Crops on the terraces of the national park: rice, Longding tea, oil tea, rape

● Observation places recommended

1. Gaotiankeng Natural Village and Taihuishan Village of Changhong Town
2. Renzongkeng Village of Qixi Town
3. Lulian Village and Tianfan Village of Hetian Town
4. Gutian Village and Tangtou Village of Suzhuang Town

When you come to Qianjiangyuan National Park, what are the most abundant resources on the land? Water, forest··· How about Farmland? Very little, isn't it? Almost 93% of the national park is forested and arable land is scarce. In order to obtain food for daily life, people built terraces on the slope of the land and planted rice, tea, rape and other crops on the terraces.

Rice is the main food crop in Qianjiangyuan area, where the climate is humid and the heat is abundant, which provides the natural hydrothermal conditions for the growth of rice. Farmers in Qianjiangyuan often scatter rice to the fields around Qingming Festival and grain rain. The fertilizer used for farming is also local. People harvest the grass that can be seen everywhere in the mountains and put it into the fields, or they put their kitchen waste, dung ash, feathers of chickens and ducks into the fields, which are natural organic fertilizers. When the rice is harvested in the autumn, people dry it in a drying yard built with a few long poles outside the windows of their houses on the second floor. In this way, they make full use of the space.

Have you ever grown rice? The rice production process is very complex, requiring many links, such as sowing, seedling raising, transplanting, management and harvesting. It is almost one of the crops with the highest labor input per unit area. However, when you visit the villages, you will find that most of the people left behind in the villages are old people and children. Their labor force is very limited, so many of the former paddy fields have been abandoned or planted with crops such as corn.

Compared with the intensive cultivation of rice, the cultivation of tea trees is much easier.

In the national park, where the mountain is at a higher altitude and the slope is steeper, you can often see a neat and green tea garden, rising and falling with the terraced fields. Qianjiangyuan area is located in the "tea golden triangle" area at the junction of Zhejiang, Anhui and Jiangxi provinces. Surrounded by mountains, luxuriant forests and clouds, and with a steep slope, it is not easy to collect water. It is an important producing area for the Gaoshan Longding tea.

Kaihua is known as "the hometown of Chinese camellia". Thanks to the local unique mountain microclimate, the camellia produced here has high oil content and good quality, and is an important oil crop and one of the local characteristic industries in Qianjiangyuan area. In addition to the ordinary camellia, here is the important area of producing Zhejiang safflower camellia that is unique to China. Every year in early spring, thousands of acres of safflower camellia will form a beautiful scenery.

There is also a very important oil crop in Qianjiangyuan area: rape. Every spring, the rape flowers in the Taihui Mountain Village of Chanhong Township compete to bloom all over mountains, fields and terraced fields. Terraced fields, white clouds and the village together constitute a beautiful picture. In addition to attracting tourists, rape flower field has a special group of visitors on time at this time of year: the Chinese bee. In addition to collecting honey, they also help the rape plants to carry on the family line. In early summer, farmers can harvest the rapeseed.

Extra information

- Terraces are strips of terraces or wavy sections built along the contour lines on the ground of hills and slopes. It can not only increase the usable area of farmland, but also be an effective solution to control soil and water loss in sloping farmland. It has the effects of water storage, soil preservation and fertilizer preservation.

通用解说 | 村落与传统文化解说　　解说编号：TY-CL-03

03 好山好水出好茶

解说要点

1. 开化有两种茶：龙顶茶与黄金茶
2. 两种茶叶与当地环境、当地人的关系

观察地点推荐

1. 苏庄镇古田村、横中村
2. 长虹乡高田坑自然村

解说方式 Interpretation Style

☑ 人员解说 Staff Interpretation
☐ 教育活动 Educational Activity
☑ 解说牌 Interpretation Signage
☐ 场馆解说 Hall Interpretation

解说季节 Interpretation Season

☑ 春季 Spring
☑ 夏季 Summer
☑ 秋季 Autumn
☐ 冬季 Winter

解说词

中国人喝茶的习惯由来已久，无论是经典的采自茶树的茶叶，还是用其他植物叶片、花朵，亦或果实、种子制成的各色茶饮，缔造了我国茶产品的多样性。在物产丰富的钱江源地区，就有两种富有代表性的茶：一种是经典的绿茶——开化龙顶，另一种则是由山间的灌木柳叶蜡梅的嫩叶风干制成的黄金茶。

「龙顶茶」

我们先来说说龙顶茶。注意不是著名的西湖龙井哦。不过龙顶茶和龙井茶，一个是在钱塘江源头，一个是在钱塘江尾部，水土不同，茶叶的口感自然也不尽相同。龙顶茶是一种典型的高山云雾茶，钱江源地区群山环绕，林茂水丰，光照充足却不会有阳光曝晒，常年云雾不散，加上没有污染和太多人类活动干扰，开化龙顶的品质和营养价值绝对不亚于龙井，只是深居简出，天性低调不善营销而已。

据《开化县志》记载，明崇祯年间开化芽茶已被列为贡品。光绪年间，茶叶开始作为外贸商品对外出口，开化绿茶产业也随之快速发展。史料记载最盛时期为民国年间，年产量达到数千担（1 担 =50kg），许多历史悠久的茶园至今仍在种植生产。

「黄金茶」

如果说龙顶茶是养在深山的大家闺秀的话，那么黄金茶应该称得上是人见人爱的小家碧玉了。在开化，几乎家家户户都会饮黄金茶，尤其是在夏季，黄金茶更是解暑生津必不可少的一道饮品。

黄金茶的原材料来自柳叶蜡梅。这是一种半常绿灌木，生长在钱江源国家公园的山坡沟谷。这种看似不起眼的野生植物，却在当地人们的日常生活中发挥着重要的作用。

好山好水才能出好茶，无论是龙顶茶的清冽馨香，还是黄金茶的清甜口感，都离不开钱江源国家公园千百年来保留的原生自然环境，以及人与自然长期和谐共生的过程中积淀的智慧结晶。作为国家公园周边地区的特色生态产品，它既是当地自然保护历史的见证，也承载着区域未来绿色发展的希望。

注意事项

- 重点介绍龙顶茶、黄金茶与当地人生活密不可分的关系，背后折射出人与自然的相伴相生的关系。注意不要变成推销茶叶的。

辅助道具推荐

- 龙顶茶茶园
- 柳叶蜡梅野外植株
- 龙顶茶与龙井茶干茶叶，黄金茶茶包

拓展活动推荐

- 参观茶园，进行茶叶的采摘与制作。

General Interpretation | Village and Traditional Culture Section

Interpretation No. TY-CL-03

03 Good environment makes perfect tea

● Key points

1. There are two kinds of tea in Kaihua: Longding tea and golden tea
2. The relationship between the two kinds of tea and the local environment and local people

● Observation places recommended

1. Gutian Village and Henghzong Village of Suzhuang Town
2. Gaotiankeng Natural Village of Changhong Town

Chinese people have a long habit of drinking tea, whether it is a classic tea leaf from a tea tree, or a variety of teas made from other plant leaves, flowers or fruits, seeds, creating the diversity of tea products in China. There are two types of representative tea in the rich Qianjiangyuan area: one is the classic green tea—Kaihua Longding, and the other is the golden tea made from the tender leaves of the mountain shrub—*Chimonanthus salicifolius*.

"Longding tea"

Let's talk about Longding tea first. Similar to the famous West Lake Longjing, Longding tea also belongs to the green tea. However, one of them is at the source of the Qiantang River and the other is at the end of the Qiantang River. The water and soil are different, and naturally the taste of tea is different. Longding tea is a typical high-altitude cloud tea. The Qianjiangyuan area is surrounded by mountains with abundant forests and water, with sufficient light but not too much, perennial clouds and mist, plus no pollution and interference from human activities. As a result, the quality and nutritional value of Kaihua Longding tea is absolutely no less than that of Longjing tea.

According to the *Records of Kaihua County*, the Kaihua bud tea was listed as a tribute during the Ming Chongzhen period. During the Guangxu period, tea was exported as a foreign trade commodity, and the Kaihua green tea industry also developed rapidly. Historical materials record the most prosperous period as the Republic of China, with annual production reaching thousands of dans (1dan=50kg), and many historic tea gardens are still being operated and producing tea.

"Golden tea"

If Longding tea is a lady hidden in the deep mountains, then the golden tea should be regarded as a lovely girl. In Kaihua, almost every household will drink the gold tea, especially in summer, gold tea is an

indispensable drink to relieve summer heat.

The raw material of golden tea comes from *Chimonanthus salicifolius*. This is a semi-evergreen shrub that grows in hillside valleys in Qianjiangyuan National Park. This seemingly unremarkable wild plant plays an important role in the daily lives of local people.

Only good mountains and water can produce good tea. Whether it is the clear fragrance of Longding tea or the sweet taste of gold tea, it is inseparable from the original natural environment that Qianjiangyuan National Park has retained for thousands of years, and the long-term relationship between people and nature. The wisdom crystal accumulated in the process of harmonious coexistence. As a special ecological product in the surrounding areas of the national park, it is not only a witness of the history of local nature conservation, but also carries the hope of the future green development of the region.

Attentions

- Please focus on the inseparable relationship between Longding tea and golden tea and the lives of local people. The relationship between man and nature is reflected behind it. Please pay attention, don't conduct your interpretation as a salesperson of tea.

Recommended auxiliary facilities and toolkits

- Longding Tea Garden
- Wild plants of *Chimonanthus salicifolius*
- Dry Longding tea and Longjing tea, golden tea bags

Recommended activities

- Visit the tea garden to pick and make tea

通用解说 | 村落与传统文化解说　　解说编号：TY-CL-04

04 传统手榨茶籽油

● 解说要点

1. 茶籽油作为一种健康的食用油而广受欢迎
2. 油茶是自然对开化山民的慷慨馈赠

● 观察地点推荐

苏庄镇古田村

解说方式 Interpretation Style

- ☑ 人员解说 Staff Interpretation
- ☐ 教育活动 Educational Activity
- ☑ 解说牌 Interpretation Signage
- ☐ 场馆解说 Hall Interpretation

解说季节 Interpretation Season

- ☑ 春季 Spring
- ☑ 夏季 Summer
- ☑ 秋季 Autumn
- ☐ 冬季 Winter

解说词

大家平时进厨房吗？如果有的话，你们家做饭用什么油呢？有的人家用大豆油、花生油，有的人家用橄榄油，还有的人家用芝麻油和猪油调味……我们平时吃的油按照来源的不同可以分为植物油和动物油两大类。动物油直接在家用于烹饪已经较少，更多的是品种丰富的植物油。除了刚才大家说的，今天我为大家介绍一种当地特产——茶籽油。

茶籽油产自油茶的种子。油茶又叫茶子树、茶油树等，泛指山茶属中种子含油率较高、生产效益好的可供产油的物种。被称作“九山半水半分田”的钱江源地区，由于地势起伏，地形坡度大，不适合传统的农耕活动，但却为油茶的生长提供了极佳的条件。这里出产的油茶含油率高、品质好，营养价值很高。

开化县素有“中国油茶之乡”的美称，主要种植普通油茶，也有少量小叶油茶和浙江红花油茶。其中，浙江红花油茶因为对海拔、光照、气温、降水、土壤性质等自然条件要求更严苛，主要分布在高海拔地区阳光充足、降水丰沛的地方，营养价值更高，但数量少，成品油产量极低，仅为白花茶油的1%～2%，稀少而珍贵。

中国利用油茶榨油的历史悠久。据古籍记载，中国自元代开始就开始取油茶果榨油，明清时期油茶作为油料作物广泛种植，如今油茶已经成为开化县苏庄镇农户增收致富的主导产业之一，当地每年还会举办山茶油开榨节，祈愿丰产。

油茶所产的茶籽油又称茶油，其中的不饱和脂肪酸含量高达90%以上，对血脂的代谢、降低胆固醇都有积极的作用，联合国粮食及农业组织（FAO）也将其列为推荐的健康型食用油，不仅在我国，在欧美、东南亚等地也很受欢迎。

拓展知识点

「茶籽油的生产工艺」

- 手工榨油是开化当地的传统技艺，所得茶籽油颜色金黄、气味清香、沉淀物和杂质少，烹饪时烟点高，适宜久贮藏。具体的制作流程包括如下7个步骤：烘炒—碾粉—蒸粉—做饼—入榨—出榨—入缸。

「入榨所用的传统木榨」

- 传统手工榨油机的核心是一根长5m以上、直径不小于1m的粗硕“油槽木”，其中心凿出一个长2m，宽40cm的“油槽”，用于装填油胚饼。开榨时，利用重约15kg重的油锤的自重，压榨油槽中的胚饼，挤压出其中的茶油。

General Interpretation | Village and Traditional Culture Section

Interpretation No. TY-CL-04

04 Traditional hand-made camellia oil

● Key points

1. The camellia oil is popular as a healthy cooking oil
2. Camellia is a generous gift of the nature to the people of Kaihua

● Observation places recommended

Gutian Village of Suzhuang Town

Do you usually go to the kitchen? If so, what oil does your family use for cooking? Some people will use soybean oil and peanut oil; some people will use olive oil, and some people will use sesame oil and lard for seasoning··· The oil we usually have can be divided into two categories based on different sources, vegetable oil and animal oil. Animal oil is rarely used in cooking at home directly. What we usually use is more of a rich variety of vegetable oil. In addition to what we just mentioned, today I will introduce a local specialty—camellia oil.

The camellia oil is produced from the seeds of camellia. Camellia is also called tea seed tree, tea oil tree, *etc.*. It refers to the species with high oil content and good production efficiency of seeds of *Camellia* genus for oil production. The Qianjiangyuan area, known as "90% mountains, 5% water and 5% farmland", is not suitable for traditional farming activities due to the undulating terrain and big terrain slope, but it provides excellent conditions for the growth of camellia. The camellia produced here has high oil content, good quality and high nutritional value.

Kaihua County is known as the "hometown of Chinese camellia", mainly cultivating common camellia. There are also a small amount of small leaf camellia and Zhejiang safflower camellia. Among them, Zhejiang safflower camellia has stricter requirements on natural conditions such as altitude, light, temperature, precipitation, and soil properties. It is mainly distributed in high-altitude areas with plenty of sunshine and abundant precipitation. Its nutritional value is higher, but its quantity is small, and its refined oil output is extremely low, only 1%–2% of white flower camellia oil. So Zhejiang safflower camellia is rare and precious.

China has a long history of using camellia to squeeze oil. According to ancient records, China has been extracting oil from camellia since the Yuan Dynasty. Camellia was widely

planted as an oil crop during the Ming and Qing dynasties. Today, camellia has become the one of the leading sources of income for farmers in Suzhuang Town, Kaihua County, the local will also hold the Camellia Oil Ectraction Opening Festival every year to pray for harvest.

The tea seed oil produced by camellia is also called the tea oil, and its unsaturated fatty acid content is as high as 90% or more, which has a positive effect on the metabolism of blood lipids and lowering cholesterol. The Food and Agriculture Organization of the United Nations also lists it as a recommended healthy cooking oil. It is very popular only in China, but also in Europe, America, Southeast Asia and other places.

Extra information

"The production process of camelia oil"

- Squeezing oil by hand is a traditional local culture. The resulting camellia oil has a golden color, a fragrant smell, few deposits and impurities, and a high smoke point during cooking. It is suitable for long-term storage. The specific production process includes the following seven steps: roasting – milling powder – steaming powder – making a dough – putting it in – starting squeezing – taking it out.

"Traditional wood squeezer used for squeezing"

- The core of the traditional manual oil squeezer is a thick oil wood with a length of more than 5 meters and a diameter of not less than 1 meter. A 2 meter long and 40 centimeter wide "oil tank" is cut in the center for filling oil seed dough when squeezing, use the weight of an oil hammer weighing about 15kg to squeeze the camellia oil out of the dough.

通用解说 | 村落与传统文化解说　　　　解说编号：TY-CL-05

05 无辣不欢与风干腌制的智慧

解说要点

1. 喜食辣椒与钱江源地理环境的密切关系
2. 风干腌制与钱江源四季、生活的关系
3. 白腊肉背后的环境与文化意义

观察地点推荐

1. 齐溪镇里秧田村、仁宗坑村
2. 长虹乡高田坑自然村

解说方式 Interpretation Style

- ☑ 人员解说 Staff Interpretation
- ☐ 教育活动 Educational Activity
- ☑ 解说牌 Interpretation Signage
- ☐ 场馆解说 Hall Interpretation

解说季节 Interpretation Season

- ☑ 春季 Spring
- ☑ 夏季 Summer
- ☑ 秋季 Autumn
- ☑ 冬季 Winter

我想问问大家分别来自哪里。能不能介绍自己家乡饮食在口味上的特色？比如，四川人喜欢麻辣，贵州则偏爱酸辣，上海人喜欢浓油赤酱而口味偏甜，而广东人则喜欢白灼和清炖以保留食材的天然口感。如果说到浙江，大家觉得应该用哪种味道来形容最为合适呢？江浙菜给人的印象一般都口味清淡而偏甜。但是位于浙江西部山地的钱江源地区，却与大家熟悉的江浙菜风味大不相同。

「喜食辣椒与钱江源地理环境的密切关系」

如果大家已经品尝过当地美食，一定感受到了当地菜色中不同于江浙地区的辣味，而且口味偏重（调味略咸）。为什么地处江南的开化地区却有这样的饮食偏好呢？这还要从本地的自然环境开始讲起。

我们常说一方山水养一方人。由于钱江源地处山区，地形复杂多样，且常年云雾环绕，温润潮湿，为了祛湿除寒，食用香辛料自然成了生活在山间村落的人们的不二之选。除此之外，这种习俗的形成也与钱江源的地理位置有关。由于位于三省交界之地，当地有大量从江西、安徽等迁入的外来人口，同时带来了他们的饮食习惯。因此，开化虽地处浙江西部，但饮食习惯却和地理环境相似度更高的江西、安徽更为接近。

「风干腌制与钱江源四季、生活的关系」

风干和腌制食物是开化当地餐桌上另一类常见的食材。

偏僻的地理位置和不太便利的交通，让钱江源的人们必须在食物稀缺的冬季来临之前储备好过冬的食粮。先民们在食物制作上发挥了充分的想象力和卓越的智慧，试图与时间赛跑，在食物丰盛的季节将其风干褪去水分，并用盐腌制，大大延长了食物的存放时间；等到食物短缺的季节，再用泡水等方法让其“恢复”水分，依然可以享用新鲜的美味。

在长期的实践中，人们不断调整着风干和腌制食材的加工方法，这些原材料经历了时间的魔法，在当地人的巧手侍弄下形成各种独特的风味，白腊肉、腊猪肝、各色干菜、葛粉等都是当地的特色风味。

「白腊肉背后的环境与文化意义」

大家肯定都吃过腊肉，但开化有一道名菜——白腊肉，却不是人人都有福气品尝。

和四川和重庆地区的腌制腊肉不同，开化的白腊肉，特指在过年前后最冷的几天，取新鲜的前胸连着肋条部位的带肥精肉，不经腌制，在天然寒冷的环境中挂在通风的木梁下风干晾晒而成。白腊肉的口感鲜美温和，经过风干和微发酵的自然熟成过程，更增添了独特的醇厚口感，或炒，或炖或慢火焖透，味道俱佳。

由于不用盐腌制，吃白腊肉十分讲究时节，每年春节和三四月份最佳，之后就很难吃到正宗的白腊肉了。如果遇到暖冬，当年白腊肉的口感也会受到影响。随着气候变化和传统文化逐渐淡出人们的生活，未来这道食材能否在国家公园的餐桌上被保留，与我们每一个人都息息相关。

General Interpretation | Village and Traditional Culture Section

Interpretation No. TY-CL-05

05 Spicy, air-dried and preserved food

Key points

1. The close relationship between the chili-loving dietary habitat and the geographical environment of Qianjiangyuan

2. The relationship between air-dried and marinated food and Qianjiangyuan's four seasons as well as life

3. The environmental and cultural significance behind white bacon

Observation places recommended

1. Liyangtian Village and Renzongkeng Pit Village of Qixi Town

2. Gaotiankeng Natural Village of Changhong Town

I would like to ask where do you come from. Can you introduce the taste of your hometown diet? For example, Sichuan people like spicy food; Guizhou people prefer hot and sour food; Shanghai people like thick red sauce and sweet taste; while Guangdong people like the natural taste of the ingredients preserved by boiling and stewing. When it comes to Zhejiang, what kind of flavor do you think is the most appropriate? The impression of Jiangzhe cuisine is generally light and sweet. However, the food in the Qianjiangyuan area which is located in the mountainous region of western Zhejiang is very different from that of Jiangsu and Zhejiang that we are all familiar with.

"The close relationship between the chili toving dietary habitat and the geographical environment of Qianjiangyuan"

If you have already tasted the local food, you must have felt it: spicy (slightly salty). Why does the Kaihua region in the southern Yangtze River have such a dietary preference? This also starts from the local natural environment.

We often say that each place has its own way of supporting its own inhabitants. Because Qianjiangyuan is located in a mountainous area, the terrain is complex and diverse, and it is surrounded by clouds and mist all year round, moist and humid. In addition, the formation of this custom is also related to the geographical location of Qianjiangyuan. Since it is located at the border of the three provinces, a large number of immigrants from Jiangxi, Anhui, *etc.* have brought their dietary habits. Therefore, although Kaihua is located in the west of Zhejiang, its dietary habit is closer to that in Jiangxi and Anhui, where the geographical similarity is higher.

"The relationship between air-dried and marinated food and Qianjiangyuan's four seasons and life"

Air-dried and marinated food is another common type of food on local tables.

The remote location and inconvenient transportation make people in Qianjiangyuan have to store food before winter when food is scarce. The ancestors have exerted their full imagination and excellent wisdom in food production, trying to race against time.They dry off and marinate food with salt when harvesting, which greatly extends the storage time of food. Wait until the season of food shortage, people can still enjoy the fresh and delicious by soaking dry food into the water.

In the long-term practice, people continue to adjust the processing methods of air-dried and preserved ingredients. These raw materials have experienced the magic of time and formed various unique flavors under the ingenuity of local people. White bacon, marinated pork liver, and various colors of dried vegetables and kudzu powder are all local specialties.

"Environment and cultural meaning behind white bacon"

Everyone must have eaten bacon, but Kaihua has a famous dish – white bacon, but not everyone has the luck to try it.

Unlike the marinated bacon in Sichuan and Chongqing, the Kaihua white bacon refers to the fat-rich meat with fresh ribs on the front chest. It is hung in the coldest days before and after the Chinese New Year, without marination, air-dried in a natural cold environment. The taste of white bacon is delicious and mild. The natural aging process of air-drying and micro-fermentation adds a unique mellow taste. It tastes the best when being fried, stewed or simmered.

Because it is not salted, white bacon is a seasonal food. It tastes the best during Spring Festival, March and April. After that, it is difficult to eat authentic white bacon. If there is a warm winter, the taste of white bacon will also be affected. As climate change and traditional culture gradually fade out of people's lives, whether or not this ingredient will be kept on the table in the national park in the future is closely related to each of us.

通用解说 | 村落与传统文化解说 解说编号：TY-CL-06

06 古法养蜂

解说要点

1. 什么是古法养蜂
2. 古法养蜂技艺的传承、改良与山林的保护

观察地点推荐

1. 苏庄镇古田村
2. 齐溪镇里秧田村

解说方式 Interpretation Style

- ☑ 人员解说 Staff Interpretation
- ☐ 教育活动 Educational Activity
- ☑ 解说牌 Interpretation Signage
- ☐ 场馆解说 Hall Interpretation

解说季节 Interpretation Season

- ☑ 春季 Spring
- ☑ 夏季 Summer
- ☑ 秋季 Autumn
- ☑ 冬季 Winter

“九山半水半分田”的自然地理格局，决定了钱江源地区很难发展大型规模性农业，但是大自然却赋予了这里一年四季交替开放的各种野花，比如，春天的油菜花、柃木和紫云英，夏天的甜槠、拐枣，秋天的五倍子和冬天的枇杷等。这些丰富的蜜源植物使这里生产的蜂蜜品质格外的高。

大家这几天在国家公园参访的过程中不知道有没有在路边、树下、农家院落等地观察到沿路布设这样的圆形的木桶（展示蜂桶照片）。这就是传统养蜂工具：蜂桶。和标准化、规模化的蜂蜜生产方式不同，古法养蜂是我们祖辈智慧的结晶。大家想想，为什么蜂桶是圆柱形而不是我们以前看过的方形的呢？有朋友说，这是用大树的树干做的。没错，传统的蜂桶确实是将自然圆木的树心挖空后制成的，但原因并不仅仅是为了充分利用原材料的自然形状。其实，当地人饲养的中华蜂有喜欢在蛀空的树干中筑巢的自然习性，蜂桶的设计是最大程度上还原它们喜欢的自然生存空间的形态和结构。

蜂桶建造好以后，一旦有蜂群入住，人们便在上下两侧加上木质挡板，再用泥土或牛粪将缝隙封好，仅保留几个进出口，确保桶中通风又防雨，安全舒适。接下来，就把酿造的工作交给蜜蜂了。当地人一般一年只在秋末冬初采集一次蜂蜜，此后便由蜂群自然越冬，有时还会适当补喂一些糖水帮助其过冬。

当然，随着自然保护区和国家公园的建设，使用原木制作蜂桶的方法早已经停止，今天我们看到的其实大部分是人工制作的拼接的木桶，但也还有极少数沿用了几代人的老式蜂桶依然在被使用。

这里养殖的中华蜂是我国传统的蜜蜂品种。别看它们个头小，但是每天采的花粉和花蜜重量几乎和它们自身的体重相当。遗憾的是，这种古老的本土物种，近年来却面临着严峻的生存危机。

早在清代，就有人为了产量高和便于驯化等特点，引入俄罗斯的黑蜂和欧洲的意大利蜂。这些外来物种的出现、定居和繁衍对中蜂种群产生了严重的威胁。研究发现，凶悍又狡猾的意大利蜂会伪装成中华蜂的雄蜂，躲过“守卫”混入蜂巢，吃饱喝足后还会杀死蜂王，导致中华蜂种群的分崩离析。不仅如此，还有研究发现，同一区域内意大利蜂种群大于中华蜂时，中华蜂蜂王的交配成功率会明显下降。

但是，意大利蜂虽然霸道，但它们并不能完全取代中华蜂的生态功能，特别在高海拔地区，因为中华蜂的耐寒性要明显优于意大利蜂。可想而知，如果中蜂的野生种群消失，很多高海拔地区的植物可能会面临缺乏授粉昆虫而无法繁殖的威胁，进而引发一系列的物种连锁反应。

如今，我们市场上出售的蜂蜜大多数都产自意大利蜂。据研究，自中国引进西方蜜蜂一百多年来，我国本土蜜蜂——中华蜂的分布区域缩小 75% 以上，种群数量减少 80% 以上。我国东北平原和华北平原几乎已经很难看到有人在养殖中华蜂。因此，及时设立相关的保护区，恢复我国的中华蜂野生种群刻不容缓。

General Interpretation | Village and Traditional Culture Section

Interpretation No. TY-CL-06

06 Traditional bee-keeping

● Key points

1. What is traditional bee-keeping
2. The inheritance and improvement of traditional bee-keeping skills and the protection of mountains and forests

● Observation places recommended

1. Gutian Village of Suzhuang Town
2. Liyangtian village of Qixi Town

The natural geographical pattern of "90% mountains, 5% water, 5% farm" determines that it is difficult for Qianjiangyuan to develop large-scale agriculture industry. However, nature has given it a variety of wildflowers that bloom alternately throughout the year. For example, the rape flowers in spring, cypress wood and Chinese milkvetch, the sweet oachestnut and *Hovenia* during summer, the nutgall tree in fall and the loquat in winter.

During your visit to the national park these days, did you notice such round wooden barrels (showing photos of bee barrels) along the roadside, under trees or in farmyards, *etc*.? This is the traditional bee-keeping tool: bee bucket. Unlike standardized and large-scale honey production methods, traditional bee-keeping is the wisdom of our ancestors. Think about it, why is the bee bucket cylindrical rather than the square we have seen before? A friend said that this was made with the trunk of a big tree. Yes, traditional bee buckets are indeed made by hollowing out the heart of natural logs. But the reason is not just to make full use of the natural shape of raw materials. In fact, the Chinese bees raised by locals have the natural habit of building nests in hollow tree trunks. The design of the bee bucket is to restore the shape and structure of the natural living space they love.

Once a bee colony moves in, people will add wooden baffles on the upper and lower sides, and then seal the gap with mud or cow drops, leaving only a few entrances and exits to ensure that the bucket is ventilated, rainproof, safe and comfortable. Next, everything is left to the bees. Local people generally only collect honey once a year in the late autumn and early winter, after which the bee swarm will take a break, and sometimes beekeepers will supplement some sugar water to help them through winter.

Of course, with the construction of nature reserves and national parks, the method of using logs to make bee buckets has ceased for a long time. Most of which we see today are artificially made spliced wooden barrels, but there are also some that have been used for generations. Old-fashioned bee buckets are still being used.

The Chinese bees raised here are traditional bee species in China. Although they seem small, they can carry pollen and nectar almost the same as their own weight every day. Unfortunately, this ancient native species has faced a severe survival crisis in recent years.

As early as the Qing Dynasty, some people introduced Russian black bees and European Italian bees for the high yield and easy domestication. The emergence, settlement and reproduction of these alien species posed a serious threat to the Chinese bee population. The study found that the fierce and sly Italian bees will disguise as the male of the Chinese bee, avoid the "guard" and sneak into the hive, and kill the queen bee, resulting in the collapse of the Chinese bee population. Not only that, but also studies have found that when the population of Italian bees in the same area is larger than that of Chinese bees, the mating success rate of Chinese queen bees will decrease significantly.

However, although Italian bees are overbearing, they cannot completely replace the ecological functions of Chinese bees, especially at high altitudes, because the cold resistance of Chinese bees is significantly better than that of Italian bees. It is conceivable that if the wild population of the Chinese bee disappears, many plants at high altitudes may face the threat of lack of pollinating insects and cannot reproduce, which may trigger a series of species chain reactions.

Today, most of the honey sold on our market is produced by Italian bees. According to the study, since the introduction of Western honey bees for more than 100 years, the distribution area of native bees in China has shrunk by more than 75%, and the population has decreased by more than 80%. It's almost hard to see anyone breeding Chinese bees on the Northeast Plain and North Plain of China. Therefore, it is imperative to set up relevant protected areas in time and restore the wild population of Chinese bees in China.

通用解说 | 村落与传统文化解说　　解说编号：TY-CL-07

07 古法养鱼

解说要点

1. 古法养鱼的原理
2. 古法养鱼对原生鱼种的保护

观察地点推荐

1. 长虹乡高田坑自然村、台回山自然村
2. 何田乡陆联村、田畈村
3. 齐溪镇仁宗坑村

解说方式 Interpretation Style

☑ 人员解说 Staff Interpretation　☐ 教育活动 Educational Activity
☑ 解说牌 Interpretation Signage　☐ 场馆解说 Hall Interpretation

解说季节 Interpretation Season

☑ 春季 Spring　☑ 夏季 Summer
☑ 秋季 Autumn　☑ 冬季 Winter

解说词

在钱江源地区，很多依傍溪流而居的农户，都会紧邻溪流一侧河岸开挖鱼池，顺水流方向前后分别开进出水孔，并在中间修筑跌水坝或在进水口一侧沿河岸修筑略高于自然水位的导水渠，利用水位高低差援引溪涧山泉，一进一出，常年活水不断。

鱼池建好以后，当地人会从溪流中捕获一些草鱼投放其中，以黑麦草或厨余果蔬长期投喂。这些食谱健康清淡，又终年在低温冷水的水流中“锻炼身体”的鱼生长缓慢，肉质紧实而鲜美，尤以炖汤风味最佳。

「古法养鱼对原生鱼种的保护」

古法养鱼始自唐代，至今仍保留着这种传统。清水鱼的平均养殖时间至少都有 3～5 年，而每家每户的鱼池里通常都还有至少一条数年甚至数十年的“鱼王”。很多当地老人介绍，自家鱼池里的大鱼多是年少时捕获并养殖于此，“鱼王”与他们一同见证了光阴流转、四季更迭。正是这些不经意间的守护和传承，为保护本土鱼类的种苗资源作出了积极的贡献。

注意事项

- 古法养鱼的关键在于通过鱼池构造的巧妙设计，利用水位差援引活水进行养殖。
- 古法养鱼的传统，体现了人根据自然规律，对其加以适当的改造和利用的生态智慧。

拓展活动推荐

- 可以带着访客去品尝当地的清水鱼，请当地人讲解清水鱼从鱼池到餐桌的整个过程，探寻背后人与自然和谐共生的淳朴自然观。

拓展知识点

「自给自足的生态庭院」

- 古法养鱼的鱼池上方一般都搭建遮阴竹架，因为鱼类一般不喜欢在强光环境下长时间活动，因此鱼池不宜长期暴晒。竹架的两边种植豆角、黄瓜、丝瓜、香瓜等爬藤蔬菜，一边提供遮阳功能，一边也兼顾日常的生产供给。庭院中空地上种植的常用的蔬菜和香草的干枯老叶，以及庭院中的杂草都可以直接投喂池鱼，而鱼池也方便了庭院的就近灌溉，如此形成一个小而有序的生态系统，也是当地人自给自足的生态庭院。

General Interpretation | Village and Traditional Culture Section

Interpretation No. TY-CL-07

07 Ancient fish farming

Key points

1. Principles of ancient fish farming
2. Protection of native fish species by ancient fish farming

Observation places recommended

1. Gaotiankeng Natural Village and Taihuishan Natural Village of Changhong Town
2. Lulian Village and Tianfan Village of Hetian Town
3. RenzongkengVillage of Qixi Town

In the Qianjiangyuan area, many farmers living near the stream will dig fish ponds next to the bank of the stream side, open inlet and outlet holes along the flow, and build a falling dam in the middle using the difference in water level to invoke the mountain spring with continuous water flowing all year round. The types of fish they raised are generally grass carp. The local people feed them with ryegrass or healthy kitchen waste like fruits and vegetables. The fish that "exercise" in the cold flowing water all year round grow slowly, and its meat is firm and delicious, especially perfect for making fish stew.

"Protection of native fish species by ancient fish farming"

The ancient method of fish farming started in the Tang Dynasty and still retains till this day. The average breeding time of freshwater fish is at least three to five years, and there is usually at least one "fish king" in the pond of each household for several years or even decades. Many local elders said that most of the big fish in their own fish ponds were caught and farmed here when they were young, and the "fish king" witnessed the time and the change of seasons with them. It is the inadvertent guarding and inheritance that have made a positive contribution to the protection of the seedling resources of native fish.

Attentions

- The key to ancient fish farming is to use fresh water with ingenious fishpond structure design.
- The tradition of ancient fish farming embodies the eco-intelligence of people to properly transform and use nature according to the laws of nature.

Recommended activities

- You can bring the visitors to taste the local freshwater fish, ask the locals to explain the whole process of the freshwater fish from the fish pond to the table, and explore the simple nature philosophy behind the harmony between man and nature.

Extra information

"Self-sufficient ecological courtyard"

- In the ancient method of fish farming, a shaded bamboo frame is usually built above the fishpond. Since fish generally do not like to be active for a long time under strong light, the fishpond should not be exposed to sunlight for too long. Both sides of the bamboo frame are planted with vines, beans, cucumbers, loofahs, muskmelons and other climbing vegetables. While providing shade, they also take into account the daily production supply.
- The dried old leaves as well as the weeds in the courtyard can be directly fed to the pond fish, and the fishpond also facilitates the nearby irrigation of the courtyard, thus forming a small and orderly ecosystem. It is also a self-sufficient ecological courtyard for local people.

通用解说 | 村落与传统文化解说　　解说编号：TY-CL-08

08 舌尖上的开化青蛳

解说要点

1. 开化青蛳与钱江源水质的关系
2. 关于青蛳美味与物种保护之间的平衡的思考

观察地点推荐

1. 苏庄镇横中村
2. 齐溪镇里秧田村

解说方式 Interpretation Style

☑ 人员解说 Staff Interpretation　☐ 教育活动 Educational Activity
☑ 解说牌 Interpretation Signage　☐ 场馆解说 Hall Interpretation

解说季节 Interpretation Season

☑ 春季 Spring　☑ 夏季 Summer
☑ 秋季 Autumn　☐ 冬季 Winter

解说词

螺蛳是中国人餐桌上非常常见的一味时令食材。在中国，只要有淡水的地方，几乎就有螺蛳的身影。而各位来钱江源之前，可能已经听说这里有一种中国特有种：青蛳。美食纪录片《舌尖上的中国 2》里说，“最好的螺蛳是藏在开化山里的青蛳……”这对美食家极富诱惑的宣传，对于青蛳，却是不幸来敲门的信号。

「开化青蛳与钱江源水质的关系」

开化青蛳生活在钱江源头的清澈溪流中，这里泥少水净，因此长出来的青蛳也与众不同。和一般螺蛳相比，青蛳的外形更加修长，颜色也更深，吃起来有淡淡的苦味。其肉质是青绿色的，可能这也是青蛳的由来吧。

青蛳前面加上“开化”二字，意味着这是当地特有的一种物产，出了开化，就很难找到开化青蛳了。这是因为开化青蛳对水温、水质和周围环境的要求很高，水一定是要流动的，而且不能太深，水中要有大大小小的碎石，且不能有污染，钱江源优质的山涧溪流恰好为开化青蛳的生长提供了得天独厚的条件。

「关于青蛳美味与物种保护之间的平衡的思考」

青蛳一年四季都有，尤其是在炎热的夏季产量最高。据说，过去当地人采集青蛳特别容易，只要在夏天中午太阳最大的时候，在小溪水浅的地方扔一把柳枝，傍晚再去会发现枝条上爬着不少青蛳。

但是随着旅游业的发展和《舌尖上的中国 2》的热播，各地的食客纷纷慕名而来，青蛳作为一道地方特色菜其野外资源被大量消耗，现在的溪流里已经很难有当年青蛳生长的盛况了。餐桌上的青蛳个头也小了很多，若无合理的控制，会不会有一天青蛳也成了童年的回忆，只留在纪录片中呢？

注意事项

- 讲青蛳的目的不是为了宣传美食，而是为了让访客意识到青蛳与钱江源环境的关系，进而选择保护这个物种。
- 解说青蛳可以在两种情境下：一种是在带领访客餐桌上品尝青蛳的时候，一种是带领访客观察溪水里的青蛳的时候。

辅助道具推荐

- 青蛳照片
- 餐桌上的青蛳

General Interpretation | Village and Traditional Culture Section

Interpretation No. TY-CL-08

08 Taste of Kaihua green freshwater snail

● Key points

1. The relationship between Kaihua green freshwater snail and Qianjiangyuan water quality
2. Thinking about the balance between the deliciousness of green freshwater snails and the protection of species

● Observation places recommended

1. Hengzhong Village of Suzhuang Town
2. Liyangtian Village of Qixi Town

Snails are very common seasonal ingredients on the Chinese table. In China, as long as there is fresh water, there are snails. Before you came to Qianjiangyuan, you may have heard that there is a kind of Chinese specialty: green freshwater snails. The food documentary *A Bite of China 2* said, "The best snail is the green freshwater snail hidden in the Kaihua Mountain···" However, this tempting publicity for foodies isn't the best news for snails.

"Relationship between Kaihua green freshwater snail and Qianjiangyuan water quality"

Kaihua green freshwater snails live in a clear stream at the source of the Qiantang River, where there is little mud, so the green freshwater snails here are unique as well. Compared with the common snail, the green snail has a more slender shell and a darker color, and it tastes slightly bitter. Its meat is kind of greenish, maybe that's how it gets its name.

The word Kaihua is added in front of the green freshwater snail, which means that this is a local specialty. When you leave Kaihua, it is difficult to find the green freshwater snail. This is because the green freshwater snails have high requirements on water temperature, water quality and surrounding environment. The water must be flowing but not too deep. There must be gravels in the water without any pollution. The Qianjiangyuan is a high-quality mountain stream that happens to provide unique conditions for the growth of green freshwater snails.

"Think about the balance between the deliciousness of green freshwater snails and the protection of species"

Green freshwater snails are available all year round, especially in hot summer seasons. It is said that it is very easy for the locals to collect green freshwater snails in the past. During the middle of the day, if you throw a handful of

willow branches in the shallow water of the stream when you go back in the evening, you will find many green freshwater snails crawling on the branches.

However, with the development of tourism and the popularity of *A Bite of China 2*, diners from all over the country have come here. As a local specialty, the wild resources of green freshwater snails have been consumed in a lot. You can barely see a huge amount of green freshwater snails growing in the stream. The green freshwater snails served on the dinner table are also much smaller. Without proper control, will the green freshwater snails one day become memories of childhood to remain in the documentary?

Attentions

- The purpose of talking about freshwater snails is not to promote food, but to make visitors be aware of the relationship between green freshwater snails and the environment of Qianjiangyuan, so as to protect this species.
- Interpreting the green freshwater snail can be conducted in two situations: one is to take the visitor to taste the green freshwater snail on the table, and the other is to bring the visitor to observe the green freshwater snail in the stream.

Recommended auxiliary facilities and toolkits

- Photos of green freshwater snails
- Green freshwater snails on the dining table

通用解说 | 村落与传统文化解说　　解说编号：TY-CL-09

09 乡愁记忆之开化气糕

解说要点

1. 典型开化地方美食气糕
2. 独特口感与水质、自然环境背后的关系

观察地点推荐

苏庄镇古田村、唐头村、横中村

解说方式 Interpretation Style

- ☑ 人员解说 Staff Interpretation
- ☐ 教育活动 Educational Activity
- ☑ 解说牌 Interpretation Signage
- ☐ 场馆解说 Hall Interpretation

解说季节 Interpretation Season

- ☑ 春季 Spring
- ☑ 夏季 Summer
- ☑ 秋季 Autumn
- ☑ 冬季 Winter

「气糕作为典型开化地方美食」

大家这两天来到开化，有没有尝过什么地道的开化美食？早餐有没有吃到著名的“气糕”呢？如果说兰州人的一天是从一碗兰州拉面开始的，开化人的一天是则是从一块热气腾腾的气糕开始的。

对于每个开化人来说，早晨如果能来一份香糯适中、洁白晶莹、松软有弹性的气糕，一整天可都会元气满满。街头巷尾的早餐店，如果缺了气糕都不能算完整。除了作为日常早餐，每年农历七月半和重阳节的时候，开化家家户户也都会蒸气糕，就像端午节吃粽子和中秋节吃月饼一样，这一习俗一直延续至今。

作为开化人，无论走到哪里，都惦记着家乡的这道美食。有漂泊异乡的开化游子曾经试图按照同样的原料和工艺，在其他地方制作气糕，但是总做不出那种熟悉的松软微酸的口感。气糕难道也会“水土不服”吗？

「独特口感与水质、自然环境背后的关系 」

当地人相信，气糕的独特口感，得益于当地洁净的水质为米浆的发酵提供了最适宜的条件。离开开化，水质和其中微生物组成一旦变化，气糕的口感可能也就改变了。

就这样，气糕成了开化游子日日思念却始终解不开的一种乡愁。所以说开化气糕是乡愁记忆的符号，真是再合适不过。

注意事项

- 气糕作为开化人早餐必不可少的一道食物，是开化人心中特殊的开化符号。
- 离了开化的气糕，会因为“水土不服”而味道不再那么正宗，背后其实是世世代代人与自然相处过程中所生成的独特味道。

辅助道具推荐

- 气糕照片
- 气糕，可以给访客直接品尝

拓展活动推荐

- 品尝气糕美味，请当地厨师或者早餐店老板为大家示范讲解气糕制作过程。

拓展知识点

- 气糕又称“东方比萨”，由当地产的早稻米经过泡米、磨浆、发酵、蒸熟四道工序制作而成，其表面一般撒有香干、肉丝、辣椒、虾皮、香菇等材料，蒸熟出锅后再撒上翠绿的葱花，一道美味可口的气糕就做成了。
- 受中亚热带湿润季风气候的影响，钱江源地区四季分明、雨水丰沛、光照适宜，适宜许多农作物的生长，而喜欢温暖湿润环境，尤其可以利用山间沼泽湿地或坡地梯田种植的水稻，则理所当然成为当地最重要的主食。而磨米成浆发酵后蒸制的开化气糕，则是其中最有特色的一道。

General Interpretation | Village and Traditional Culture Section

Interpretation No. TY-CL-09

09 Taste of steamed rice cake

● Key points

1. Steamed rice cake as a classic Kaihua specialty
2. The relationship between the unique taste and the water quality as well as natural environment

● Observation places recommended

Gutian Village, Tangtou Village and Hengzhong Village of Suzhuang Town

"Steamed rice cake as a classic Kaihua specialty"
We have been here in Kaihua for two days. Have you tasted any authentic Kaihua cuisine? Did you eat the famous "Steamed rice cake" for breakfast? If the people in Lanzhou start the day with a bowl of Lanzhou ramen, the people in Kaihua start the day with a hot steamed rice cake.

For everyone in Kaihua, if they can get a mild, white, crystal, soft and elastic Kaihua steamed rice cake in the morning, they will be full of energy all day long. A breakfast shop on the street can't be complete without a Kaihua steamed rice cake. In addition to being a daily breakfast, on the 15th day of the 7th month of the lunar calendar and the Chung Yeung Festival, every family in Kaihua will also steam rice cakes, just like eating Zong Zi on the Dragon Boat Festival and eating moon cakes on the Mid-autumn Festival. This custom has continued to this day.

As a Kaihua person, no matter where he goes, he always remembers this food in his hometown. A Kaihua's man, wandering in foreign countries, once tried to use the same ingredients and craftsmanship to make Kaihua steamed rice cakes elsewhere, but he couldn't always get the familiar soft and slightly sour taste. Can Kaihua steamed rice cakes also be "unaccustomed"?

"Relationship between the unique taste and the water quality as well as natural environment"
The locals believe that the unique taste of the Kaihua steamed rice cakes is due to the clean local water quality that provides the most suitable conditions for the fermentation of rice milk. Once you leave Kaihua, the water quality and microbial composition in it will change, resulting in the change of the taste of the Kaihua steamed rice cake.

In this way, the Kaihua steamed rice cake has become a kind of homesickness that the wanderer misses every day but is unable to solve. Therefore, you can say Kaihua steamed rice cake is a symbol of nostalgic memory.

Attentions

- Kaihua rice cake is an essential food for the Kaihua people during breakfast, as well as a special culture symbol.
- The Kaihua steamed rice cake without Kaihua will not taste so authentic because of the unaccustomed water and soil once it is made outside Kaihua. In fact, what behind it is the unique taste generated by generations of people getting along with nature.

Recommended auxiliary facilities and toolkits

- Photos of Kaihua steamed rice cakes
- Steamed rice cakes for tasting

Recommended activities

- To taste the delicious Kaihua steamed rice cake, please ask the local chef or breakfast shop owner to demonstrate the Kaihua steamed rice cake making process for everyone.

Extra information

- The Kaihua steamed rice cake, also known as "oriental pizza", is made from locally produced early rice through four processes: rice rinsing, refining, fermentation and steaming. The surface is generally sprinkled with dried tofu, shredded pork, pepper, dried shrimp, mushrooms and other ingredients, steamed and then sprinkled with green onions.Till then, a delicious Kaihua rice cake is made.
- Affected by the humid monsoon climate in the mid-subtropics, the Qianjiangyuan area has four distinct seasons, abundant rain, and suitable light so that it is suitable for the growth of many crops. Rice, which likes a warm and humid environment, especially can be planted in mountain swamps or sloping terraces, of course becomes the most important local staple food. Kaihua rice cake steamed after milled rice pulp is the most distinctive one.

通用解说 | 村落与传统文化解说　　解说编号：TY-CL-10

10 竹编技艺的传承

● 解说要点

1. 竹编技艺与传统朴素的生活智慧
2. 将竹编技艺作为一种生态产品去传承和保护

● 观察地点推荐

长虹乡台回山自然村

解说方式 Interpretation Style

- ☑ 人员解说 Staff Interpretation
- ☐ 教育活动 Educational Activity
- ☑ 解说牌 Interpretation Signage
- ☐ 场馆解说 Hall Interpretation

解说季节 Interpretation Season

- ☑ 春季 Spring
- ☑ 夏季 Summer
- ☑ 秋季 Autumn
- ☑ 冬季 Winter

「竹编技艺与传统朴素的生活智慧」

钱江源国家公园山里盛产毛竹等各种竹子，其中最常见的毛竹直径达碗口粗细。竹子全身都是宝。除了采集新鲜的竹笋食用之外，竹子也是建造房屋的重要建材，还是制作家具和各种日用品必不可少的原材料。

当地居民的院子里，随处可见由竹子制成的各种家用器具，比如，晾晒干菜用的竹匾，可背可提的竹篮、竹筐，斗笠，簸箕，扫把，盛水的杯子，捕鱼捕虾的渔猎工具……竹子在当地人的巧手匠心之下，变换成各种各样实用的物件，丰富而便捷了当地人的日常生活，简单之中有一种自然朴素的美。

大家看看这个竹筒，实际上是一种腌菜的工具，粗壮的竹筒经过打磨，不仅方便易用，而且腌菜后还带着淡淡的竹子清香。而这个挎篮，看起来平凡无奇，实际上编织起来是十分讲究的。这个孔隙的大小，既要方便盛装蔬菜时抖落泥沙，又不能漏掉小棵的菜苗果实。此外，这个篮子上还有很多细节，比如，花纹的样式，篮身和提手的用材差异，以及包边的工艺，处处都是学问。

竹子并不能直接用来编织，要经过劈、削、磨、刮等各种工艺，制成宽度和厚度都一样的竹篾作为原材料。

仅仅有原材料还不够，竹编的技巧也是繁复多样的。除了传统的经纬编织方法，当地人还发展出许多全新的技法，比如，疏编、插、穿、削、锁、钉、扎、套等，这样编出来的花样图案更多，造型也更丰富、更美观。

「将竹编技艺作为一种生态产品去传承和保护」

过去在开化县各个乡镇村落，都有代代相传的竹编人家，但现在这项民间工艺也面临着失传的局面。

国家公园的建设和发展，正在致力于挖掘这些传统手工艺在新时代背景下的新生，作为代表当地历史和传统生活方式的手工艺品也许有机会成为国家公园的特色生态产品。

注意事项

- 解说重点放在竹编技艺体现了当地山民因地取材的传统智慧与朴素审美，以及传统技艺的传承与保护。

辅助道具推荐

- 农户家的竹制品或者随身携带的一些小型竹制品

拓展活动推荐

- 设计一堂竹编体验课程，邀请当地的竹编匠人，为大家展示竹编技艺，并教大家编制一个简单的竹艺作品。

General Interpretation | Village and Traditional Culture Section

Interpretation No. TY-CL-10

10 Inheritance of bamboo weaving skills

● Key points

1. Bamboo weaving skills and traditional simple life wisdom
2. Inherit and protect bamboo weaving skills as an ecological product

● Observation places recommended

Taihuishan Natural Village of Changhong Town

"Bamboo weaving skills and traditional and simple life wisdom"

The mountains of Qianjiangyuan National Park are rich in bamboos, among which the moso bamboos is the most common and has a diameter as big as a bowl. The bamboo is full of treasure. In addition to providing fresh bamboo shoots for cooking, bamboo is also an important building material for houses, as well as an indispensable raw material for making furniture and various daily necessities.

You can see various household appliances made of bamboo everywhere in the backyard of the local residents. For example, bamboo sieves for drying vegetables, bamboo baskets for lifting, bamboo buckets, dustpans, brooms, cups for holding water, fishing and hunting tools, *etc*.. Bamboo has transformed into a variety of practical objects under the ingenuity of local craftsmen with a natural and simple beauty.

Please look at this bamboo tube, it is actually a tool for pickles. The thick bamboo tube is polished, which is not only convenient and easy to use, but also the pickles will have a light bamboo fragrance. And this shoulder basket looks ordinary, in fact it is very carefully weaved. The size of the pores should be big enough for shaking off the mud when holding vegetables, and meanwhile it needs to be small enough not to miss any small fruits or leaves. In addition, there are a lot of details on this basket, such as the pattern style, the difference in the materials used in the basket body and the handle, and the craftsmanship of the edging.

Bamboo can't be directly used for weaving. It has to undergo various processes such as splitting, cutting, polishing, scraping, *etc*. to make bamboo sticks with the same width and thickness as raw materials.

Raw materials alone are not enough, the bamboo weaving skills are also complicated. In addition to the traditional warp and weft

weaving methods, the locals have also developed many new techniques, such as sparse weaving, inserting, threading, cutting, locking, nailing, tying, covering, *etc*., so that we have more patterns and richer shapes.

"Inherit and protect bamboo weaving skills as an ecological product"

In the past, in all towns and villages in Kaihua County, bamboo weaving skills were handed down from generation to generation, but now this folk craft is also facing the situation of being lost.

The construction and development of national parks are dedicating to the rejuvenation of these traditional handicrafts in the new era. As handicrafts representing local history and traditional lifestyles, they may have the opportunity to become special ecological products of national parks.

Attention

- The focus of the interpretation is on the fact that bamboo weaving skills reflects the traditional wisdom and simple aesthetics of local mountain people based on local materials, as well as on the inheritance and protection of traditional skills.

Recommended auxiliary facilities and toolkits

- Bamboo products in farmers' households or some small bamboo products carried you by

Extra information

- Design a bamboo weaving experience course, invite local bamboo weaving craftsmen to show you the skills of bamboo weaving, and teach you to compose a simple bamboo work.

通用解说 | 村落与传统文化解说　　解说编号：TY-CL-11

11 依山傍水的传统村落

● 解说要点

1. 公园传统村落的布局特点
2. 传统民居的生态格局

● 观察地点推荐

1. 长虹乡高田坑自然村、台回山自然村、库坑村、河滩自然村
2. 苏庄镇古田村、横中村
3. 齐溪镇仁宗坑村

解说方式 Interpretation Style

☑ 人员解说 Staff Interpretation　☐ 教育活动 Educational Activity
☑ 解说牌 Interpretation Signage　☐ 场馆解说 Hall Interpretation

解说季节 Interpretation Season

☑ 春季 Spring　☑ 夏季 Summer
☑ 秋季 Autumn　☑ 冬季 Winter

解说词

钱江源山多地少的地势特征决定了这里不同于平原地区的村庄和房屋布局。国家公园内 72 个自然村落大都依山傍水而建，根据山水走势的不同，有的呈条带状，有的呈树叶状，有的则呈团块状……尽管每个村落的布局各有特色，但我们仍不难发现其中的一些共同特征。

首先，有村子的地方一定有水。这里的村庄通常沿着河流或者溪流而建，房屋沿着水流两侧排开，距离紧凑。村庄的房屋建筑受到徽派建筑影响较大，多黄土黛瓦。从山顶俯瞰村落，你会发现这些房屋高矮相近，瓦片错落，紧贴在一起，几乎看不到房屋之间的空隙。村中的小路也是就地取材，多用青石板铺成，但因为土地资源稀缺，这里的路大都比较狭窄，仅一米多宽。

除了溪流之外，古树几乎也是每个村庄的标配。古树中常见的品种有香樟、南方红豆杉和枫香等，它们的年纪动辄上百年，是村里最为古老的居民。

走进村庄，我们会发现，当地民居的建造也是人们利用自然、改造自然，与自然和谐共生的智慧结晶。钱江源地区多雨，为了排掉多余的雨水，这里的屋顶一般倾斜度都比较大，以便雨水落到屋顶后自然落下而不至于积累。同时，为了让落下的雨水及时排掉，村民们还会在墙角挖凿排水的小沟渠，将雨水引至溪流中，以免造成院落积水。

除了排水之外，村民还非常善于因势利导，引用溪流活水进行清水养鱼，或者在自己院子里，用石板搭置一个桌台，用当地产的竹子做成一个自动引水装置，引流山上的溪水，作为日常清洗之用。

由于山间交通不便，村民们的房屋也大都就地取材，以二层的土木结构为主，呈中轴对称排列。每个房屋和家具的布置也都十分讲究，自有一套逻辑。同时，如果你留心观察的话，会发现二楼的窗户一般都是长方形没有窗户，且从窗口常常伸出几根长杆。这是做什么用的呢？没错，是用来晾晒谷物、玉米、辣椒等农作物的。由于山里平地较少，且气候阴湿，阳光稀缺，用这种方法进行晾晒不仅可以节省空间，同时也使农作物可以最大限度地暴露在阳光下。

注意事项

- 可以结合村落特点，选择合适的视角进行解说。

辅助道具推荐

- 谷歌地图上截取的村庄整体布局
- 现场无法看到的实景照片补充

General Interpretation | Village and Traditional Culture Section

Interpretation No. TY-CL-11

11 Traditional villages by mountains and rivers

Key points

1. The layout characteristics of traditional villages in the park
2. The ecological pattern of traditional dwellings

Observation places recommended

1. Gaotiankeng Natural Village, Takhuishan Natural Village, Kukeng Village and Hetan Natural Village of Changhong Town
2. Gutian Village and Henghzong Village of Suzhuang Town
3. Renzongkeng Village of Qixi Town

The topography of Qianjiangyuan Mountain with more mountains and less land determines the layout of villages and houses that is different from that in plain areas. There are 72 natural villages in the national park, most of which are built by mountains and rivers. According to the different trend of the landscape, some are banded, some are leaf-shaped, and some are clumpy··· Although the layout of each village has its own characteristics, we can still find some common features.

First of all, where there is a village, there must be water. Villages here are often built along rivers or streams, with houses spaced close apart along either side of the current. The houses and buildings in the village are greatly influenced by Hui-style buildings, which mostly have loess and black tiles. Overlooking the village from the top of the mountain, you will find that these houses are similar in height, with scattered tiles and close to each other, and you can hardly see the gaps between the houses. The paths in the village are also made of local materials, mostly paved with blue slate. However, due to the scarcity of land resources, the roads here are mostly narrow, only more than one meter wide.

In addition to streams, ancient trees are standard in almost every village. Common species of ancient trees include camphor, *Taxus chinensis* and maple fragrance. They are often more than a hundred years old and the oldest residents in the village.

Walking into the village, we will find that the construction of every residential building is also the wisdom crystallization of people's utilization of nature, transformation of nature and harmonious coexistence with nature.

QianJiangyuan area is rainy, in order to drain the excess rain, the roof here is generally inclined to be large, so that the rain will fall naturally after falling on the roof. At the same time, in order to allow the rain to drain away in a timely manner, the villagers would also dig a small ditch in the corner to drain the rainwater into the stream, so as not to cause ponded water in the yard.

In addition to drainage, villagers are also very good at taking advantage of the situation, using fresh water from streams to raise fish, or setting up a table in their yard with flagstones and using locally grown bamboo as an automatic diversion device to drain the mountain stream for daily cleaning.

Due to the inconvenience of mountain transportation, the houses of the villagers are mostly made of local materials, with the civil structure of the second floor as the main structure, in the axisymmetric arrangement.The layout of each house and furniture are also very exquisite, with a set of logic. And if you look closely, you'll notice that the second-floor windows are usually rectangular and often a few long rods are extended out of the window. What are these rods to do with? That's right. They are used for drying crops like grains, corn and peppers. Since there are few flat areas in the mountains and the weather is damp and sunlight is scarce, drying in this way not only saves space but also maximizes exposure to the sun's rays.

Attentions

- It can be interpreted from an appropriate perspective based on the characteristics of villages.

Recommended auxiliary facilities and toolkits

- The overall layout of the village from the Google map
- Live photos that can't be seen at the scene

通用解说 | 村落与传统文化解说 解说编号：TY-CL-12

12 村子里最古老的居民

解说要点

1. 古树所蕴含的家族与家乡的情感凝结
2. “风水林”的由来、象征意义
3. 植树节中体现的对自然资源永续利用的恪守与践行

观察地点推荐

1. 何田乡田畈村、陆联村
2. 长虹乡高田坑自然村
3. 苏庄镇古田村、唐头村、横中村
4. 齐溪镇仁宗坑村

解说方式 Interpretation Style

☑ 人员解说 Staff Interpretation ☐ 教育活动 Educational Activity
☑ 解说牌 Interpretation Signage ☐ 场馆解说 Hall Interpretation

解说季节 Interpretation Season

☑ 春季 Spring ☑ 夏季 Summer
☑ 秋季 Autumn ☑ 冬季 Winter

钱江源的村落为我们展示了人与自然、人与山水是如何在空间上和谐共生的，其中蕴含着经过岁月沉淀、铅华洗尽后依然保留的原真味道。而这一切的见证者，可以说就是几乎每个村落里都有的矗立数百年的古树。

「古树所蕴含的家族与家乡的情感凝结」

村落的入口处、溪流边、祠堂外、古墓旁，常常都能看到带有文化符号意义的高大古树甚至古树群：比如，苏庄镇平坑村口的百年香樟、长虹乡高田坑村 500 多岁高龄的南方红豆杉，何田乡田畈村河畔的古树群......

「“风水林”的由来、象征意义」

这些古树大多是遵照传统，以宗族或村坊为单位种植而形成的风景林、水口林、祖墓林和宗族共有林等，也被统称为“风水树”或“风水林”，一般以香樟、南方红豆杉、枫香、苦槠等树种为主，且都具有寿命长、不落叶、易生长，树形挺拔雄伟，枝叶茂盛，能显示威巍的气派等特点。这些古树千百年来无论晴雨旱涝，战争或饥荒，一直繁茂生长，它们本身就象征着一个村落的持续繁荣和兴盛。

孟浩然有诗云：“绿树村边合，青山郭外斜。开轩面场圃，把酒话桑麻。”正如诗中所言，古树就像村落迎接远方来客的标志一样，同时也是村民们聚会议事的场所。每逢村子里有重大事情需要集体商议，大家便会不约而同地聚集到古树下讨论、裁决。而平日里，偌大的树冠为村民提供遮阳避雨的休憩空间，是老人闲聊，孩子玩耍的好去处。

「植树节中体现的对自然资源的永续利用与恪守」

古树的守候承载了一代代人的家族和血缘记忆，植树节则是当地居民对于自然资源特别是木材资源永续利用的践行和恪守。节日中仪式的庄重也体现了当地人对自然的感怀敬重与对未来的祈愿。我们今天在各个村落所见的几百年、上千年的老树，如樟树、枫树、红豆杉、银杏等不少古树名木，都是历史上历年植树节种植而来的。

注意事项

- 钱江源几乎所有的村落都有古树的存在，它们一般都在数百岁以上，它们的存在印证了村落的古老。
- 古树一般位于村口或者村中央比较显著的位置，是村中风水好的象征，也是人们休闲聚会的好去处。

辅助道具推荐

- 在推荐地点的古树处解说

General Interpretation | Village and Traditional Culture Section

Interpretation No. TY-CL-12

12 The oldest "residents" in villages

Key points

1. Emotions of family and homeland embedded in ancient trees
2. The origin and symbolic meaning of "Feng Shui Forest"
3. Practice of and abidance by sustainable use of natural resources reflected in the Tree Planting Day

Observation places recommended

1. Tianfan Village and Lulian Village of Hetian Town
2. Gaotiankeng Natural Village of Changhong Town
3. Gutian Village, Tangtou Village and Hengzhong Village of Suzhuang Town
4. Renzongkeng Village of Qixi Town

Qianjiangyuan's villages show us how humans and nature live harmoniously, which contain the originality that has been preserved after the baptism of years. And who witness all this are ancient trees that have stood for hundreds of years in almost every village.

"The emotions of family and home land embedded in ancient trees"

You can see tall ancient trees and even ancient tree groups with cultural symbols at the entrance of the village, by the stream, outside the ancestral temple, and next to the ancient tomb, such as the 100-year-old camphor tree at the entrance of Pingkeng Village of Suzhuang Town, 500 years Chinese yew of Gaotian Pit Village in Changhong, the ancient trees on the banks of Tianfan Village in Hetian Town.

"The origin and symbolic meaning of 'Feng Shui Forest'"

Most of these ancient trees follow the tradition and are planted in units of clan or village to form landscape forests, shuikou forests, ancestral tomb forests, and clan-shared forests. They are also referred to as "feng shui trees" or "feng shui forests". It is generally dominated by camphor trees, Chinese yew, Chinese sweet gum, sweet oachestnut, and other tree species. They are all long-lived, no deciduous and easy to grow, having tall and majestic tree shape, lush foliage and imposing manner. These ancient trees have been flourishing for thousands of years, regardless of rain, drought, floods, wars or famine. It itself symbolizes the continued prosperity of a village.

Meng Haoran has a poem: "There are woods and mountains surrounding the village. Open the window there is the vegetable farm, villagers drink wine and talk about their day." As the poem says, the ancient tree is like a sign that the village welcomes visitors from afar, and it is also the villagers' meeting place. Every time there is a major event in the village, everyone will gather under the ancient tree to discuss and adjudicate. On weekdays, the large canopy provides villagers with shade and shelter from the rain and is a good place for the elderly to chat and children to play.

"Practice of and abidance by sustainable use of natural resources reflected in the Tree Planting Day"

The shield of ancient trees bears the memory of family and blood from generation to generation. The "Tree planting Day" is the practice and abidance by the local residents for the sustainable use of natural resources, especially woods. The solemn ceremony of the festival also reflects local people's respect for nature and pray for the future. The old trees of hundreds or thousands of years that we have seen in various villages today, such as camphor trees, maple trees, yew ginkgo, and many other famous trees, were planted in the Tree Planting Day in history.

Attentions

- There are ancient trees in almost all villages in Qianjiangyuan. They are generally over hundreds of years old. Their existence confirms the old age of villages.
- Ancient trees are generally located in the entrance of the village or the prominent place in the center of the village. They are symbols of good feng shui and also great places for people to socialize.

Recommended auxiliary facilities and toolkits

- Interpretation at the ancient tree at the recommended location

通用解说 | 村落与传统文化解说　　　　解说编号：TY-CL-13

13 封山育林的传统习俗

● 解说要点

1. 封山育林的传统习俗
2. 封山节：人地关系的演进与文化传承

● 观察地点推荐

1. 苏庄镇唐头村
2. 长虹乡库坑村、河滩村
3. 何田乡田畈村、陆联村

解说方式 Interpretation Style

☑ 人员解说 Staff Interpretation　☐ 教育活动 Educational Activity
☑ 解说牌 Interpretation Signage　☐ 场馆解说 Hall Interpretation

解说季节 Interpretation Season

☑ 春季 Spring　☑ 夏季 Summer
☑ 秋季 Autumn　☑ 冬季 Winter

中国有句古话叫作："取之有度，用之有节，则长足。"生活在钱江源的山民们或者并不知道这个理论，但是这种热爱自然、亲近自然和崇尚自然的天性却早已融进他们的血液里。现今保存的众多百年古碑中，就记录了当地人在百年以前就有以制度保护生态的优良传统，如封山育林、禁止伐木、伏季休渔等，充分体现了古人的环境保护意识。

「封山育林的传统习俗」

开化县是林业大县，森林资源十分丰富，历代政府和民间为保护森林都采取了许多有效的措施，其中，立碑禁山护林是其中最常见的一类。在长虹乡星河村莘田自然村榨油厂的黄泥墙上，就嵌有一块 200 多年历史的石碑，碑额横刻"荫木禁碑"四个大字。碑文主要记载了当时族人立碑的缘由以及严禁盗砍山林等约定。历经岁月的洗涤，石碑上的字迹虽然已经模糊，但是这项传统却依然传承至今。

古代没有专门的森林保护法，禁伐林木靠的更多的是自觉和乡风民俗的制约，这处禁山碑就成了早期成文版的村规民约，其中不乏"重申严禁"以及"禀官判处，决不徇私"这样严厉的惩戒。这些规定为保护森林的一草一木、一石一水与山林的永续发展作出了了不起的贡献。

「封山节：人地关系的演进与文化传承」

现代人做事喜欢讲求仪式感，但古人的仪式感可不是刻意追求，而是他们生活中必不可少的元素。以封山育林为例，一般要经过以下程序：首先要选好日子，在特定的时节，由村里德高望重的族长或里长主持，根据族规或村规民约，划定村前屋后、入水口、溪河两旁的护林地界，并用稻草扎在周围的树木上作为标记。仪式后提供大锅饭，召集全村人聚餐。吃过封山仪式的大锅饭，就意味着认可遵守保护山林的规定，此后如有侵犯封山区的一草一木，一经发现，便处以与仪式中全村封山饭价值相等的罚金。

随着历史的变迁和社会的发展，有些"封山"的传统仪式逐渐发展为封山节这样的文化活动。碑文的规定也逐渐被纸质的村规民约所取代。但无论时代如何变迁，传统习俗如何演化，人们与这片土地早已经紧紧绑定在一起。祖辈们亲手种下的一棵幼苗，不仅成长为未来的护村林、风景林，同时也保护了当地丰富的原生物种资源，创建了一条纽带，连接了古人与今人，成为追溯今天生态系统保护的源头。

除了从石碑上的铭文之外，我们还可以从典籍、宗祠等途径找到更多当地传统文化中闪光的生态智慧。据开化当地的历史典籍记载，这里还有禁猎、禁伐、禁渔、禁采矿等各种传统，它们有的继续着，有的消失了，有的摇身一变成了新的村规民约，随着时代的步伐，继续谱写着人与自然互相适应、互相守候的新篇章。

General Interpretation | Village and Traditional Culture Section

Interpretation No. TY-CL-13

13 Tradition customs of mountain closure for afforestation

● Key points

1. Traditional customs of mountain closure for afforestation

2. Mountain Closure Festival: the evolution and cultural inheritance of the relationship between man and land

● Observation places recommended

1. Tangtou Village of Suzhuang Town

2. Kukeng Village and Hetan Natural Village of Changhong Town

3. Tianfan Village and Lulian Village of Hetian Town

There is an old saying in China called: "Take the right amount, have something to spare." The mountain people living in Qianjiangyuan may not know this theory, but loving nature, being close to nature and advocating nature has embedded into their blood. Among the many hundred-year-old monuments preserved today, it is recorded that the local people have had a fine tradition of protecting the ecology since 100 years ago, such as mountain closure for afforestation, logging prohibition, and seasonal fishing ban. The complete local laws and regulations of environmental protection fully reflect the environmental protection awareness of the ancients.

"Traditional customs of mountain closure for afforestation"

On the yellow mud wall of Xintian Village Oil Extraction Plant, Xinghe Village, Changhong Town, a century-old stone monument is embedded, and the forehead is engraved with the four characters "Ban in the Forest". The inscription mainly records the reasons why the clans established the monument at that time and the agreement that prohibiting the theft of forests resources in mountains is prohibited. It has a history of more than 200 years. After the baptism of years, although the handwriting on the stele is getting blurry, this tradition is still inherited to this day.

In ancient times, there was no special forest protection law, and the prohibition of logging forests relied more on consciousness and folk customs. This mountain monument is an early written version of the village regulations and civil agreements. These regulations have made great contributions to the protection of the grass, wood, stone, water and sustainable development of forests.

"Mountain Closure Festival: evolution and cultural inheritance of human–earth relationship"

Modern people like to emphasize the sense of ritual, but the sense of ritual of the ancients is much stronger than ours. Taking mountain closure for afforestation as an example, the following procedures are generally required: First, at a specific time of a good day, the village's prestigious patriarch presides over to delineate the boundary of the forest protection area in the front of the village, at the back of the house and the water inlet and on both sides of the river according to the national regulations or village regulations, and use straw tieing the surrounding trees as a mark. After the ceremony, a meal will be offered to all the villagers. Having eaten the meal of the mountain closure ceremony means to abide by the regulations for protecting the forests. After that, if anyone infringe the mountain closure regulations, a fine equal to the value of the mountain closure meal in the ceremony will be imposed.

With the change of history and the development of society, some of the traditional ceremonies of "mountain closure" gradually developed into cultural activities such as Mountain Closure Festival. No matter how the traditional customs evolve, it is actually the deepening and evolution of the local people's understanding of the new relationship between man and land. Either way, it is for the protection of mountain forests. Seedlings planted by the ancestors not only grow into a future village-protection and landscape forest, but also protects the local rich native species resources, creating a link that connects the ancients and the present, and evolving into the ecosystem of today.

通用解说｜村落与传统文化解说　　解说编号：TY-CL-14

14 香火草龙：颂赞自然的盛会

解说要点

1. 香火草龙的制作工艺与美好寓意
2. 香火草龙体现的人与自然和谐共生的自然观

观察地点推荐

苏庄镇古田村、横中村、唐头村

解说方式 Interpretation Style

- ☑ 人员解说 Staff Interpretation
- ☐ 教育活动 Educational Activity
- ☑ 解说牌 Interpretation Signage
- ☐ 场馆解说 Hall Interpretation

解说季节 Interpretation Season

- ☑ 春季 Spring
- ☑ 夏季 Summer
- ☑ 秋季 Autumn
- ☐ 冬季 Winter

解说词

每年的八月十五是我国传统的中秋节。对于生活在钱江源的苏庄人来说，这一天除了家人团聚，一起吃月饼、拜月神之外，还有一项必不可少的节目——舞香火草龙。这是开化当地一项极具地方特色的传统习俗，始于唐宋，在明朝达到鼎盛，至今已有六七百年的历史。

「香火草龙的制作工艺与美好寓意」

舞龙作为一种传统的庆祝表演，大家并不陌生。但是和大家印象中用竹纸扎制的龙不同，开化的龙是用稻草做的。每年稻谷丰收过后，村民们会挑选上好的稻草，由多人历经数日精心编制成草龙，在他们的巧手匠心之下，一根根普通的稻草变成层次分明的龙身、龙角、龙须、龙鳞，活灵活现、栩栩如生。编好后的草龙会先在宗族祠堂中陈列着，在中秋前夕，村民们在龙身上遍插寄托心愿的香支，香火草龙就这样做好了。

中秋节当天，人们会将香火草龙全部点燃后举行热烈华美的舞龙仪式。稻草扎制的草龙在香火中绽放耀眼夺目的光彩，也同时倾诉了乡民对来年风调雨顺和丰收平安的虔诚祈愿，在仪式的尾声，草龙随香火焚烧殆尽，并被村民们送入村中的河流，寓指一切来源于自然，也必将最终回归自然的哲学思想。

「香火草龙体现的人与自然和谐共生的自然观」

凝聚村民们朴素情感与精巧手艺的香火草龙，是人们在土地上辛勤躬耕一年，在秋天品尝到丰收硕果之时，对自然馈赠所奉上的最淳朴的感怀之心。在热闹的仪式背后，更呈现了当地人善用自然、尊重自然规律的朴素自然观。

开化香火草龙于 2011 年被列为第三批国家级非物质文化遗产，与海宁观潮、西湖赏月并称为我国中秋节的标志性传统习俗。

注意事项

- 重点在于香火草龙背后透出的节俗文化、尊重自然的情感。

辅助道具推荐

- 照片，视频或在中秋节期间现场展示香草火龙

拓展知识点

- 舞香火草龙活动一般在八月中秋晚间举行，主要分3段式。

1. 起山月。舞龙前，村民要在祠堂摆起香案“接龙”，祭祀祖先，主持训示。
2. 满山月。这是舞龙的高潮时刻。皓月当空，夜色朦胧，只见一条长长的火龙在刚收割后的田野里飞奔狂舞，上下翻腾，吐云驾雾，气势磅簿。
3. 落山月。这是舞龙的尾声，人们将欢送草龙顺着河流回归自然。

Interpretation No. TY-CL-14

14 Incense stick grass dragon dance: a celebration of nature

Key points

1. Manufacture process and beautiful meaning of the incense stick grass dragon reflected by the incense stick grass dragon dance
2. The natural view of harmony between man and nature

Observation places recommended

Gutian Village , Hengzhong Village and Tangtou Village of Suzhuang Town

Every year on August 15th is the traditional Mid-Autumn Festival in China. For Suzhuang people living in Qianjiangyuan, in addition to family reunion, eating moon cakes and worshiping the moon god, there is another essential show – the incense stick grass dragon dance. This is a traditional custom with great local characteristics in Kaihua. It started in the Tang and Song dynasties and reached its peak in the Ming Dynasty. It has a history of six or seven hundred years.

"The craftsmanship and beautiful meaning of the incense stick grass dragon"

The dragon dance is no stranger to everyone as a traditional celebration performance, But unlike the dragons made of bamboo paper that everyone is familiar with, the Kaihua dragons are made of straw. After the harvest of rice every year, the villagers will choose the best straw, and it will be carefully crafted into grass dragons by many people over several days. Under their ingenuity, ordinary straws will become layered dragon horns, dragon beards, and dragon scales, coming alive and vivid. The finished grass dragon will be displayed in the patriarchal ancestral temple first. On the eve of Mid-autumn Festival, the villagers put the incense sticks that they use for making a fish on the dragon, so the incense stick grass dragon is finished.

On the day of the Mid-autumn Festival, people will ignite the incense stick grass dragons to hold a warm and gorgeous dragon dance ceremony. The straw dragon blooms dazzlingly telling the pious wishes of the villagers for the weather and the harvest of the coming year. At the end of the ceremony, the dragon will be burned with the incense sticks and sent by the villagers into the river that flows through the village, which implies that everything comes from nature, and eventually returns to nature.

"The natural view of harmony between man and nature reflected by the incense stick grass dragon dance"

The incense stick grass dragon, which gathers the emotions and exquisite craftsmanship of the villagers, is the simplest feeling that people have to nature's gifts after they work hard for a year and taste the rich fruits in autumn. Behind the lively ceremony, there is a simple natural view of locals making good use of nature and respecting the laws of nature.

Kaihua incense stick grass dragon was listed as the third batch of national intangible cultural heritage in 2011. It together with the tide-watching in Haining and moon-watching in the West Lake are regarded as the iconic traditional customs of our country's Mid-autumn Festival.

Attentions

- The focus is on the festival culture and the emotions of respecting nature refflected by the incense stick grass dragon dance.

Recommended auxiliary facilities and toolkits

- Photos, videos, or live shows of the incense stick grass dragon during the Mid-autumn Festival

Extra information

- The incense stick grass dragon dance event is generally held at the mid-autumn night in August, which is divided into three stages.

1. The starting moon. Before the dragon dance, the villagers had to set up the incense case to "welcome the dragon" in the ancestral temple, memorizing their ancestors, and presiding over the instructions.
2. The full moon. This is the climax of the dragon dance. The bright moon is in the sky and the night was hazy.You can see a long fire dragon flying in the field that has just been harvested, tumbling up and down, looking gorgeous.
3. The falling moon. This is the end of the dragon dance. People will send the grass dragon back to the sea along the river.

通用解说 | 村落与传统文化解说　　　　解说编号：TY-CL-15

15 保苗节：田间的尊重与爱护

● 解说要点

保苗节体现人们对美好生活的向往、对自然的尊重与爱护

● 观察地点推荐

苏庄镇古田村

解说方式 Interpretation Style

- ☑ 人员解说 Staff Interpretation
- ☐ 教育活动 Educational Activity
- ☑ 解说牌 Interpretation Signage
- ☐ 场馆解说 Hall Interpretation

解说季节 Interpretation Season

- ☐ 春季 Spring
- ☑ 夏季 Summer
- ☑ 秋季 Autumn
- ☐ 冬季 Winter

解说词

钱江源地区历史久远，很早就有人在这里定居。人类聚落形成之后，为保证四季食物充足，自然就会开荒拓田，于是村庄周边多见水稻田、玉米地、红薯地、蔬菜地……这些开放的农田经常会有一些野生动物邻居来拜访，比如，野猪。

从人类驯化家猪开始，人与野猪的博弈就开始上演，时至今日，这样的局面并没有减轻，因为生态保护和野生物保护的卓著成效，野猪的野外种群数量得到恢复，导致它们出现在农田的概率这几年还提高了。

不过，即使在国家公园建设之前，钱江源地区的人与野生动物从来就不是对立的存在，人们有更智慧的方式去平衡这种关系。苏庄镇长久以来逐渐形成的保苗节传统，就用一种特殊的警示方式很好地平衡了农业生产与自然保护的关系，在确保辛勤耕耘的农作物能够丰产和自足的同时，又不伤害和我们共同生活在这片山林中的瑞兽灵禽。

保苗节在每年的农历六月的插秧时节举行。这一天，村民会自发组织起来，从关帝庙出发，抬着明太祖、关公的塑像走进田陌，在田间地头插满各种红的、蓝的、黄的等颜色醒目的小三角旗作为警示，同时组成巡游队伍，通过敲打锣鼓、吹奏唢呐等形式，沿着田间开始巡游。

没有陷阱，没有捕猎。村民们通过这种温和但有效的方式，提醒山上的野兽不要在插秧育苗和后续的收获季节来犯，为保持人与自然的和谐共处，划定了合理的距离。没有过度的干扰和恫吓，更没有不必要的侵犯和伤害，整个保苗节充满古老的传统文化色彩，也体现了村民对丰收和美好生活的向往，以及对自然的尊重与爱护。

拓展知识点

「溪流中的世代守护：敬鱼节」

- 钱江源地区的传统村庄中，往往都有清澈的溪流穿村而过。都说水至清则无鱼，然而在这清澈的溪流中，仔细观察，我们会发现游弋着成群结队的石斑鱼等原生鱼种，这还要得益于延续千年的适度捕捞和定期禁渔的传统村规，即“敬鱼节”。当地村规民约规定，任何人和单位不能以任何方式在实施护鱼规定的河道进行捕捞、采砂挖沙等破坏河道自然生态的活动，否则就要依规罚以敬鱼节仪式花费同等金额的罚金。

「山间的生产号子：满山唱」

- 满山唱是开化山区居民在山上劳作时，为了消除劳累，驱散寂寞，根据劳动情景和生活趣事自编自唱的一种自由式歌谣。歌唱时，此山与那山相互呼应，由此得名“满山唱”。满山唱具有强烈的地域性、原生性，歌谣的题材来源于居民的生活，题材广泛，有反映劳动的“采茶山歌”，有表达爱情的“九娘歌”，有反映尊敬老人的“孝敬歌”，有反映妇女生活的“绣荷包”，等等。

General Interpretation | Village and Traditional Culture Section

Interpretation No. TY-CL-15

15 Seeding Conservation Festival: respect and love for nature among the field

● Key points

Seeding Conservation Festival reflects people' s longing for a better life, respect and love for nature

● Observation places recommended

Gutian Village of Suzhuang Town

The Qianjiangyuan region has a long history. Some people settled here very early. After the formation of human settlements, in order to ensure adequate food for the four seasons, people started to develop wasteland and expand fields. There are rice fields, corn fields, sweet potato fields, vegetable fields around the village···These open farmlands often have neighbors from the wild, such as wild boars.

Since the domestication of pigs, the game between humans and wild boars has started. Today, this situation has not been alleviated. The population of wild boars has increased because of the outstanding results of ecological conservation and wild animal protection. The probability that they appear on farmlands has also increased in recent years.

But even before the establishment of the national park, people and wildlife in the Qianjiangyuan region were never opposed, since people have a smarter way to balance this relationship. The long-established tradition of the Seeding Conservation Festival in Suzhuang Town has used a special warning method to balance the relationship between agricultural production and nature protection. While ensuring that the hard-working crops can be productive and self-sufficient, it will not harm the precious animals and birds that live with us in this mountain forest.

The Seeding Conservation Festival is held every year during the seedling transplanting season in the six month of the lunar year. On this day, the villagers will organize themselves, to carry the statues of Ming Taizu and Guan Gong from the Guandi Temple into the farmland, filling the fields with red, blue, yellow and other small flags as warnings. At the same time, villagers form a parade team to parade along the field, while hitting gongs and drums, playing suona and so on.

In this gentle but effective way, the villagers reminded the animals on the mountain to

behave themselves during the planting season and the subsequent harvest season. In order to maintain mutual respect between man and nature, a reasonable distance was established without excessive interference, intimidation, unnecessary infringement and injury.

The entire Seeding Conservation Festival is full of mystery of ancient culture and tradition, which also reflects the villagers' hope for harvest and a better life, as well as respect and love for nature.

Extra information

"The guardian of generations in the stream: Fish Respect Festival"

- In traditional villages in the Qianjiangyuan region, clear streams often pass through the village. It is said that there is no fish when the water is clear. However, in this clear stream, if you look closely, we will find groupers and other native fish species are cruising, thanks to the tradition of moderate fishing and village regulations that lasts for thousands of years. The local village regulations stipulate that no person or unit may in any way engage in activities such as fishing, sand digging, damaging the natural ecology of the river channel in the river channel where fish protection regulations are implemented. Otherwise, a fine equal to the value of the ceremony will be imposed.

"The production trumpet among the mountains: the mountain song"

- The mountain song is a kind of free-style song that is written and sung by the residents of the Kaihua mountain area in order to eliminate fatigue and dispel loneliness. When singing, there are echoes among the mountains, hence the name "Mountain Song". The mountain song has a strong regional and original character. The theme of the song comes from the life of the residents. There are "tea picking folk songs" reflecting labor, "nine girls songs" expressing love, and "song of filial piety" respecting the elderly, as well as "embroidered purses" reflecting women's lives, *etc*..

通用解说 | 村落与传统文化解说　　解说编号：TY-CL-16

16 守护暗夜星空

● 解说要点

1. 城市化与环境污染等对星空观测的影响
2. 钱江源优越的观星资源
3. 暗夜星空的标准与保护价值

● 观察地点推荐

长虹乡高田坑自然村、台回山自然村

解说方式 Interpretation Style

☑ 人员解说 Staff Interpretation　☐ 教育活动 Educational Activity
☑ 解说牌 Interpretation Signage　☐ 场馆解说 Hall Interpretation

解说季节 Interpretation Season

☑ 春季 Spring　☑ 夏季 Summer
☑ 秋季 Autumn　☑ 冬季 Winter

解说词

星空自古以来就承载了人类对于浩瀚宇宙、未知世界的无限想象，而成为一种神秘又浪漫的象征。大家一定很熟悉古代中国神话传说中《牛郎织女》《董永与七仙女》《嫦娥奔月》的故事，还有古希腊罗马神话中的各种星座故事。

但是各位访客，你们中又有几位在自己生活的城市有机会看见星空呢？随着人口增加、工业化与城市化进程加快，我们的夜晚已经被人造灯光所改造，城市越来越亮，星空却越来越暗淡。从美国国家航空航天局（NASA）拍摄的全球夜间灯光分布图中可以看出，中国东部沿海，尤其是长江三角洲地区像是被点燃了一样，分外明亮，而钱江源国家公园正好位于这块亮斑的西部边缘。这里虽然紧邻繁华，但却依然保留了澄净的夜空。

「钱江源国家公园优越的观星资源」

没有什么能够阻挡人们对于美的向往和追求，为了追逐暗夜星空，人们不惜翻山越岭、跨越山海，去到高海拔地区的旷野或无人打扰的天涯海角。但是对于生活在钱江源地区的人们来说，星空就在自家门口。

要想看到最美的星空，天空的通透度、天光高度、海拔与实时天气等客观条件缺一不可。钱江源国家公园最好的观星地是位于长虹乡的高田坑村，这个被称为华东地区“暗夜公园”的地方，每年都会吸引大量天文爱好者。

高田坑村也很珍惜自己的这份资源禀赋，每天晚上十点，村里的路灯就会准时熄灭，将夜晚让位给星空。这对于合理利用资源、减少光污染并保护暗夜行动都有着重要的意义。

注意事项

- 从夜间照明分布图、既定印象中的观星场所引入，说明光污染带来的影响。
- 钱江源国家公园观星与背后良好的环境及守护的行动。

辅助道具推荐

- 全球夜间灯光分布图

拓展知识点

- 国际上，国际暗夜协会(IDA)为呼吁治理全球光污染，保护可被观察的星空资源，提出了“暗夜公园”的概念，用以肯定在全球范围内评选出的一些暗夜条件极佳的区域。“暗夜公园”或“暗夜保护区”，从某种意义上来说可以被认为是星空的避难所，也是我们人类应该珍惜的自然馈赠。

General Interpretation | Village and Traditional Culture Section

Interpretation No. TY-CL-16

16 Guarding the stars and dark sky

Key points

1. The influence of urbanization and environmental pollution on starry sky observation
2. Qianjiangyuan's superior stargazing resources
3. Standard and protection value of the dark night sky

Observation places recommended

Gaotiankeng Natural village and Taihuishan Natural village of Changhong Town

The starry sky has carried the infinite imagination of the vast universe and unknown world since ancient times, and has become a mysterious and romantic symbol. Everyone must be familiar with the stories of *Cowboy and Weaver Girl*, *Dong Yong and the Seven Fairies* and *Chang'e Run to the Moon* in ancient Chinese mythology, as well as the stories of various constellations in ancient Greek and Roman mythologies.

But tourists, how many of you have the chance to see the starry sky in the city where you live? With the increase in population, the acceleration of industrialization and urbanization, our night has been reshaped by artificial lights, the city is getting brighter, but the star sky is getting darker. Let's take a look at this picture. This is the global night light distribution map taken by the National Aeronautics and Space Administration (NASA). Have you found that the eastern coast of China, especially the Yangtze River Delta, is extremely bright as if it was lit? Qianjiangyuan National Park is located exactly on the western edge of this bright spot. Although it is close to the urban area, it still retains the clear night sky.

"Supreme stargazing resources of Qianjiangyuan National Park"

Nothing can stop people's pursuit of beauty. In order to chase the dark night sky, people are willing to cross mountains and seas to the wilderness of high altitude areas, or undisturbed horizons. But for people living in the Qianjiangyuan area, the starry sky is at their doorstep.

To see the most beautiful starry sky, objective conditions such as sky transparency, sky light height, altitude and real-time weather are indispensable. The best stargazing place in Qianjiangyuan National Park is Gaotian Pit Village in Changhong Town. This place, known as the "Dark Night Park" in East China, attracts a large number of astronomy

lovers every year.

Gaotian Pit Village also cherishes its natural endowment. The street lights in the village will turn off on time every night at 10 o'clock, giving way to the night sky. This has important significance for the rational use of resources, the reduction of light pollution and the protection of dark night.

Attentions

- Start the interpretation from the map of night lighting distribution and the stargazing place in the established impression to illustrate the impact of light pollution.
- Stargazing in Qianjiangyuan National Park and the good environment and guardian actions behind.

Recommended auxiliary facilities and toolkits

- Photos of nighttime lighting distribution across the country

Extra information

- Internationally, the International Dark Night Association (IDA) has called for the treatment of global light pollution and the protection of observable starry sky resources. It has proposed the concept of "dark night park" to affirm some of the areas with excellent dark night conditions worldwide. "Dark Night Park" or "Dark Night Reserve" can be regarded as a refuge of the starry sky in a sense, and it is also a natural gift that we humans should cherish.

通用解说 | 村落与传统文化解说　　　　解说编号：TY-CL-17

17 在地居民的社区参与与角色转型

● 解说要点

公园在处理人地关系方面的一些尝试：生态补偿、特许经营、产业富民、巡护员管理等

● 观察地点推荐

1. 齐溪镇里秧田村、仁宗坑村
2. 苏庄镇古田村、横中村
3. 何田乡田畈村

解说方式 Interpretation Style

- ☑ 人员解说 Staff Interpretation
- ☐ 教育活动 Educational Activity
- ☑ 解说牌 Interpretation Signage
- ☐ 场馆解说 Hall Interpretation

解说季节 Interpretation Season

- ☑ 春季 Spring
- ☑ 夏季 Summer
- ☑ 秋季 Autumn
- ☑ 冬季 Winter

解说词

钱江源国家公园的面积为 252km²，仅为三江源国家公园总面积的 1/500，是神农架国家公园面积的 1/4 不到，但却涵盖 4 个乡镇 72 个自然村，生活有近万名居民。这是其他国家公园所不具备的。作为一个以实现最严格保护为目标的国家公园，如何妥善处理好本地的人地关系，是摆在公园面前一道切实的难题。

为了让村民更广泛地参与到公园的建设与保护工作中来，钱江源国家公园尝试了通过生态补偿的方式对集体林地役权进行改革，并取得了一定的成效。什么是生态补偿呢？简单来说就是对土地进行有偿使用。根据规定，国家公园内要实行最严格的保护，这样就必定会触及当地原住民的一些利益。为了弥补他们的损失，实现保护和发展的双赢，公园会通过租赁、置换等方式补偿村民的损失，进行生态补偿。

不同区域生态补偿程度也不一样。在国家公园的核心保护区，人为活动是被禁止的，而在一般控制区可以允许适当开展生产、生活活动，但禁止使用有害农药、化肥等。公园内已有些地方和乡村约定，如果村民可以做到禁用农药化肥、禁止焚烧秸秆、禁止干扰野生动物等，村集体和村民每亩每年享受国家公园提供的专项补贴。

生态补偿是钱江源国家公园在兼顾最严格保护、统一管理、当地社区发展和居民利益四者平衡的一次有效尝试，也为其他国家公园的地役权改革提供了示范。除了补贴之外，公园还协助村民和相关企业签订了特许经营合同，约定以较高的价格收购这些农产品，同时允许使用国家公园相关品牌提高产品附加值。

此外，公园还鼓励村民利用本地资源优势，发展和经营与生态旅游和生态农业相关的特色产业，构建以龙顶茶、山茶油和清水鱼等为主导的休闲农业特色产业体系，科学开展游憩活动，让居民通过自己的力量富起来。

居民不仅是公园的参与者、建设者，也是公园的管理者和守护者。国家公园建成以后，招募村民组建了一支巡护员队伍，他们每天身着工作服在公园巡游，给予公园最悉心的照料。

注意事项

- 政策会跟着时间而变化，解说前可以查找阅读关于钱江源国家公园的最新报道资料进行解说。

辅助道具推荐

- 钱江源国家公园地形和植被分布图
- 钱江源国家公园功能分区

General Interpretation | Village and Traditional Culture Section

Interpretation No. TY-CL-17

17 Local residents engagement and role transformation

Key points

Some attempts of the park to deal with the relationship between people and land: ecological compensation, franchise, industrial enrichment, rangers management, *etc*.

Observation places recommended

1. Liyangtian Village and Renzongkeng Village of Qixi Town
2. Gutian Village and Hengzhong Village of Suzhuang Town
3. Tianfan Village of Hetian Town

Qianjiangyuan National Park covers an area of 252 square kilometers, which is only one-five hundredth of the total area of Sanjiangyuan National Park and less than a quarter of the area of Shennongjia National Park. However, it covers four towns and 72 natural villages with nearly 10,000 residents. This is something that no other national park has. As a national park whose goal is to achieve the strictest protection, how to properly handle the relationship between local people and land is a real problem facing the park.

In order to make the villagers participate more widely in the construction and protection of the park, the park has tried to reform the easement of collective forest by means of ecological compensation, and has achieved some results. What is ecological compensation? To put it simply, it is the paid use of land. According to the regulations, the most stringent protection must be conducted in the national park, which is bound to touch on the interests of the local indigenous people. In order to make up for their losses and realize a win-win situation of protection and development, the park will compensate the villagers for their losses and make ecological compensation through leasing and replacement.

The degree of ecological compensation is also different in different regions. In the core protection area of the national park, human activities are prohibited, while in the general control area, production and living activities can be carried out appropriately, but the use of harmful pesticides and fertilizers is prohibited. Some places in the park have agreed with the villagers that if the villagers can ban pesticides and fertilizers, burning straw and interfering with wild animals, the village collective and the villagers can receive a subsidy from the National Park (1*mu*=1/15ha).

Ecological compensation is an effective attempt to balance the strictest protection, unified management, local community development and residents' interests in Qianjiangyuan National Park. It also provides a model for easement reform in other national parks. In addition to subsidies, the park has helped villagers sign franchise contracts with related businesses. The franchise contracts stipulate that these agricultural products will be bought at a higher price, while the products are allowed to use the national park-related brands for their added value.

In addition, the park also encourages villagers in taking advantage of local resources to develop and manage characteristic industries related to ecological tourism and ecological agriculture, build a characteristic industrial system of leisure agriculture dominated by Longding tea, camellia oil and freshwater fish, and carry out recreational activities scientifically, so that residents can get rich through their own efforts.

Residents are not only the participants and builders of the park, but also the managers and guardians of the park. After the park was built, villagers were recruited to form a team of rangers who prowled the park every day in uniforms, giving it the most attentive care.

Attentions

- The policy will change with time. Before interpretation, you can look up and read the latest reports about Qianjiangyuan National Park.

Recommended auxiliary facilities and toolkits

- Topography and vegetation map of Qianjiangyuan National Park
- Functional zoning map of Qianjiangyuan National Park

04
方法与技巧
Skills and Principles

4.1 行前准备

（1）了解目标听众

对听众了解得越多，越有助于你准备节目。他们多大年纪？他们来自哪里？他们想要什么？不管存在多少差异，所有听众都有一些共同的喜好。尽你所能了解听众以后，与听众的经历联系起来，将有助于为你的解说选择更好的素材。如果可以让听众参与进来，相比单纯的解说而言，可以让听众获得更有价值的体验。注意一些特殊听众的需求。（特殊听众包括：老人、小孩、少数民族、视觉或听觉受损者、行动不便者、家庭群体等。）

（2）了解所在的场域、季节、天气状况、访客的行程时间

对解说场地和天气情况的了解，可以让你有更加充分的准备，以便更灵活地应对随时有可能出现的突发状况，保障访客安全和解说活动的顺利开展。

（3）规划特定的解说路线

沿着特定的路线，布设一些相对固定的解说点。每个路线都应该有一个以资源特色为基础制定的主题。这些解说点即是根据路线的资源和设施情况筛选出的能具体阐释主题内涵的解说素材。相对稳定的解说路线能使你提前对将要解说的场域和资源有比较充分的把握，更好地组织解说内容，达成解说目标。

（4）组织解说的逻辑结构和语言

解说的开头要能引起访客兴趣，抓住听众注意力。可以结合访客的兴趣点，采用提问等方式引入主题内容。围绕主题，尽量通过案例或者故事的形式，将能够支持主题的观点和契合主题的各类事实埋藏在故事中，讲述给你的听众。解说要有一个结尾，可以是对观点的重申或者总结，也可以是一种行为倡议。

（5）准备解说工具

每一个户外解说员都需要一个可随身携带的解说工具包，并尽可能穿一件口袋较多的衣服，可以盛放各种各样的工具和小物件，比如标本、地图、照片、掉落的松果等。这些小东西有助于解说员在解说的过程中创造“惊喜”，吸引访客的注意力，与访客进行互动，激发他们的好奇心，最终这样的解说会更容易被访客接受和理解。

下面列出的一些物件可以作为解说员准备随身背包携带工具的参考。

【常用的解说辅助工具】

望远镜

钱江源国家公园的天空、森林、溪流边经常有各种各样的鸟类飞过或停留，望远镜可以帮助你认清楚那是什么，在哪里可以找到它。建议使用双筒望远镜。

照相机及其配件

如果解说员认为目标物体十分奇特，值得记录和探索，建议直接拍摄下来，比如，动物的脚印。

放大镜

20 倍的手持放大镜是推荐的工具，用于做植物鉴定以及观察微小的目标。

防水笔记本

可用于记录所见所感，绘制自然笔记或示意图。

电筒或头灯

用于探索树洞或者岩洞，或者夜观时使用。一般应准备有红光光源，或用红布蒙住光源，以减少对夜行性生物的影响。

地图及解说工具书

在活动介绍或特定地点，可辅助展所在位置等信息。

图卡

常见物种、景观的卡片，辅助解说提供访客生动直观的感受，建议塑封后重复多次使用。

小样品袋和瓶子

用于收集或展示样品的临时道具（用完后要放回原处）。

卷尺

可以测量树木的胸径、高度等信息。

麻绳

可以用来开展相关活动，还可用于紧急情况。

观星笔

可在夜晚观星时使用。

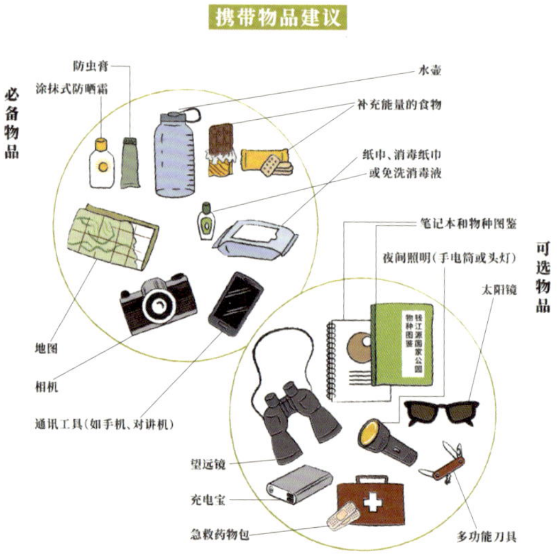

【野外工作工具】

急救药箱和生存套件

可以用于必要时的急救，也可以用于解说野外活动的相关生存常识。

水壶

在行程中补充必要的水分，或者用水冲洗以便于观察某些对象。

补充能量的食物

长时间的徒步行走耗费体能时，尤其在寒冷天气时，需要食物帮助迅速补充能量。

手机和充电宝

沿途解说的过程中并不需要手机，但当我们遇到陌生的动植物，手机可帮我们查询或记录。更重要的是，保持通讯畅通，电力充足，以防万一。

手表

计算并管理好徒步游览体验的速度。记住，你应该准备有指南针的手表，这样不仅可以知道时间，还可以指引方向。

湿巾 / 纸巾

如果团队成员观察大自然时，身上湿了或者脏了，可用来清理脏物。

垃圾袋

准备好垃圾袋，做好环境保护的榜样或者在雷暴雨时避免游客淋湿。

【小道具使你的解说更有趣】

照片

稀有的动植物照片、航拍图、有关季节转换的照片或者其他访客无法亲眼看到的事物的照片。

地图

地形图、植被图、水系图和历史地图等。

各类岩石等自然物样本

如果你不能保证随时都能找到你需要的样本，提前准备一个是很有必要的。

凋落物和标本

有些物种只能在特定的地点或者季节才能观察到，为了以防万一，可以提前准备好。

野外指导和描写当地特征的关键性材料

包括鸟类、哺乳类、昆虫类、爬行类和两栖类动物，动物的足迹和迹象，野花，岩石和矿产，气候等。

注：可以携带的物品还有很多，很难穷尽。但并非所有的物品都适合放在包里或者写进这本手册。解说员应该决定自己的背包里放什么。这取决于个人的兴趣、擅长的领域和解说活动的需求。

4.1 Preparations

(1) Know your target audience

The more you know about your audience, the more it will help you prepare for the show. How old are they? Where are they from? What do they want? No matter how many differences there are, all listeners have some common preferences. After trying your best to understand the audience, connecting with the audience's experience will help you choose better material for your interpretation. If the audience can be involved, it can give the audience a more valuable experience than a simple interpretation. Pay attention to the needs of some special audiences. (Special audiences include: the elderly, children, ethnic minorities, people with visual or hearing impairment, people with reduced mobility, family groups, *etc.*)

(2) Understand where you are, the season, weather conditions, and the planned travel time of visitors

The understanding of the interpretation site and weather conditions can make you more fully prepared to respond more flexibly to emergencies that may occur at any time, to ensure the safety of visitors and the smooth development of the interpretation activities.

(3) Plan a specific interpretation route

Set up some relatively fixed interpretation locations along a specific route. Each route should have a theme based on resource characteristics. These interpretation points are selected according to the route's resources and facilities, and can explain the content of the theme in detail. A relatively stable interpretation route enables you to have a full grasp of the field and resources to be interpreted in advance, better organize the interpretation content, and achieve the interpretation goal.

(4) Organize the logical structure and language of interpretation

The beginning of the interpretation should arouse the interest of visitors and grab the attention of the audience. The main content can be introduced in combination with visitors' points of interest and using questions. Try to use cases or stories to bury various facts that support the theme in the story and tell your audience. The interpretation should have an end, which can be a reiteration or summary of the interpretation, or a behavioral suggestion.

(5) Prepare props

Every outdoor interpreter needs an interpretation kit that can be carried with him, and wear the clothes that has as many pockets as possible. It can hold a variety of tools and small objects, such as specimens, maps, photos, fallen pine cones, *etc.*. These small objects help the interpreter create "surprises" in the process of interpretation and attract the attention of visitors, interact with visitors and stimulate their curiosity, and eventually your interpretation will be more easily accepted and understood by visitors.

Some of the items listed below can provide a reference for preparing your personal interpretation backpack.

[Essential interpretation tools]

Binoculars or general telescopes
There are often various birds hovering or staying in the sky, forests, and streams of Qianjiangyuan National Park. A telescope can help you identify what it is and where to find it.

Camera and its accessories
If the interpreter thinks that the target object is very peculiar and worthy of being photographed, it may be worth recording and exploring, such as animal footprints.

Hand-held lens and magnifying glass
The 20x handheld lens is the best tool for plant identification and to see tiny objects.

Waterproof notebooks
Used for recording good thoughts, thoughts and sketches.

Charcoals
Used for marking on trees and stones.

Flashlights
Used for exploring tree caves or rock caves, or used at night.

Stargazing pens
Used when stargazing at night.

Tape measures
Used for measuring the circumference of trees.

Hemp ropes
Used for carrying out related activities and also for emergency repairs.

Sample bags and bottles
Temporary tools for collecting or displaying samples (to be put back after being used).

[Tools for field work/trail]

First aid kits
Safety is the first duty of a team leader and interpreter. Moreover, these tools can also be used to explain reminding and knowledge of field work.

Kettles
Replenish water during the trip, or rinse with water so that the cleaned objects can be observed.

Energy food
When a long hike takes its toll, especially in cold weather, you need food to help you refuel quickly.

Cell phones and power banks
You don't need to use your cell phone during the journey, but when you come across strange plants and animals, your cell phone may help you look them up or record them. More importantly, keep your communications available and power on, just in case of any accident.

Watches
Calculate and manage the speed of the hiking experience. Remember, you should have a watch with a compass to know the time and guide you.

Wipes/paper towels
If team members get wet or dirty while watching nature, clean up the mess.

Garbage bags
Bring the bags with you. Pilot the environmental protection behavior or keep visitors from getting wet during thunderstorms.

[Small props for more interesting interpretation]

Photos

Rare photos of flora and fauna, aerial images, photos of seasonal changes, or photos of other things that visitors cannot see with their own eyes.

Maps

Topographic maps, vegetation maps, water system maps, historical maps, *etc*..

Samples of natural objects

If you can't guarantee that you will find the samples you need at any time, it is necessary to prepare one in advance.

Litter and specimens

Some species can only be observed in certain places or seasons. Just in case, you can prepare them in advance.

Field guidance books and key materials for description of local characteristics

Including birds, mammals, insects, reptiles, amphibians, animals' footprints and signs, wildflowers, rocks and minerals, climate, *etc*..

Notes: There are so many items to carry that it is hard to run out. But not all items fit in a bag or in this manual. The commentator gets to decide what you put in your backpack? It depends on your personal interests, your area of expertise, and your interpretative needs.

4.2 面向访客的行前说明

（1）最高等级自然保护地属性的特殊性

不同于传统的自然环境中的旅游体验，我们的国家公园，是以加强自然生态系统原真性、完整性保护为基础，以实现国家所有、全民共享、世代传承为目标的最高等级的自然保护地。要理顺管理体制，创新运营机制，健全法治保障，强化监督管理，构建统一规范高效的中国特色国家公园体制，并建立分类科学、保护有力的自然保护地体系。

因此，将现有各类自然保护地进行系统整合，是构建中国国家公园的核心任务之一。在我们的国家公园，面向公众的体验主要为科学研究、自然教育、游憩体验三个部分，借由这些面向公众的活动，来加深我们对国家公园的认知，通过了解走向守护。守护是第一要义，也是我们作为负责任的访客来国家公园旅行的核心原则。

（2）转变我们的身份

钱江源国家公园是我国现有国家公园体制试点单位中可达性最高、对普通公众到访最便捷的国家公园。过去，我们通常只是以普通访客的身份在公园中体验自然。国家公园的设立，除了提醒公众转变意识，也让我们有了更多参与到保护工作中的可能性，实现从旁观者到参与者的转变——参与公民科学家的科学活动，探秘原真自然里的精彩纷呈，了解国家公园的生态系统如何作为生态屏障润泽华东；参与在地的环境解说与环境教育活动，对国家公园的山川草木与人文生活有更深的认知，与国家公园建立更紧密的情感联结。

（3）走访人文——尊重本土传统文化

守护国家公园，不仅仅是保护整个公园原始的生态环境，更有世世代代生活于此的居民与他们在此衍生的传统民俗文化、劳作技艺以及因地制宜的民居营造，都值得我们访客去尊重。

留心饱经风霜的拙朴民居

钱江源国家公园内的村落民居，往往用黄泥、毛石、原木所建，饱经风霜之下，既有拙朴之美，也有很多需要修缮的地方，应当留意垒砌用的石块、容易脱落的外墙面以及构造所用的木材，等等。依山而建的村落对访客而言是不同寻常的美，更是这里居民们的家园，减少不必要的打扰才是访客应有的素质。

静心拜访世代凝结的宗族祠堂

在钱江源国家公园，几乎每个村落都有自己的祠堂（详见“通用人员解说方案”的“村庄与传统文化”部分），有的成为村里公共活动的空间，有的还延续着宗族祭奠、牌位拜访的功能。在拜访这些祠堂的时候，应当不喧哗、不擅自走入私人的阁楼，静心去体会祠堂所讲述的历代故事。

全心感受与自然相依的节庆民俗

一年里的时节中，有许多钱江源国家公园内本土村落特有的节庆民俗活动，如保苗节、香火草龙，等（详见“通用人员解说方案”的“村庄与传统文化”部分）。倘若我们有幸遇到当地的节庆民俗，应当尊重节庆习俗中的每一个环节，全心去参与、去感受，了解这些活动背后的历史沿袭，以及它们与自然紧密结合的关系。

珍惜主人们的劳作果实

在“九山半水半分田”的钱江源国家公园，农耕文化下衍生出不少劳作的传统与果实，如清水养鱼、土蜂蜜酿造、油茶种植与古法榨油等，热情的居民也会邀请我们体验和品尝他们的收获。不浪费粮食，珍惜源自山间自然恩惠的劳作成果，是我们对当地居民最大的尊重。

4.2 Pre-departure instructions for visitors

(1) The peculiarities of the highest level nature protected area

Different from the traditional tourist experience in the natural environment, our national park is based on strengthening the authenticity and integrity protection of the natural ecosystem, with the goal of realizing national ownership, sharing by the whole people, and passing on from generation to generation. It is necessary to straighten out the management system, innovate the operation mechanism, improve the guarantee of the rule of law, strengthen the supervision and management, build a unified, standardized and efficient national park system with Chinese characteristics, and establish a scientifically classified and powerfully protected nature protected area system.

Therefore, the systematic integration of various existing nature protected area is one of the core tasks of building China's national parks. In our national parks, the public-oriented experience mainly consists of three parts: scientific research, nature education, and recreational experience. Through these public-oriented activities, we can deepen our understanding of the national park and move towards protection through understanding. Guarding is the first priority, and it is also the core principle of our travels to national parks as responsible visitors.

(2) Transform our identity

Qianjiangyuan is the national park with the highest accessibility and the most convenient access to the general public among the existing national park system in our country, giving us more opportunities to visit and experience. In the past, we usually only experienced nature in the park as ordinary visitors. The establishment of national parks not only reminds the public to change their consciousness, but also gives us more possibilities to participate in conservation work, so as to realize the transformation from bystanders to participants—participate in the scientific activities as citizen scientists to explore the secrets and the splendor in true nature, understand how the ecosystem of the national park serves as an ecological barrier to moisturize East China; and participate in local environmental interpretation and environmental education activities to have a deeper understanding of the mountains and plants and human life of the national park, and establish a closer emotional connection with the national park.

(3) Visit the human culture — respect local traditional culture

Protecting the national park not only protects the original ecological environment, but also the residents who have lived here for generations and the traditional folk culture, labor skills, and the construction of dwellings tailored to local conditions are worthy of our visitors' respect.

Pay attention to the ancient houses

The village dwellings in Qianjiangyuan National Park are often built with yellow mud, rubble, and logs. Under the baptism of time, there is not only simple beauty, but also many places that need to be repaired. You should pay attention to the stones used for the construction, which are easy to fall off, the outer wall of the building, the wood used in the construction, *etc*.. The village built on the hill is unusually beautiful for visitors, and it is also the home of the residents here. In this case, please reduce unnecessary disturbance.

Visit the clan ancestral temple passed down for generations

In Qianjiangyuan National Park, almost every village has its own ancestral temple (see the "General Interpretation of Village and Traditional Culture" for details). Some have become a space for public activities in the village, and some continue to function as clan memorials and spirits tablet visits. When visiting these ancestral temples, please don't make noise or walk into a private attic without authorization, and experience the stories told by the ancestral temples with your heart.

Wholeheartedly feel the festive folk customs that depend on nature

During the seasons of the year, there are many festivals and folklore activities unique to the native villages in Qianjiangyuan National Park, such as the Seeding Conservation Festival and the incense stick grass dragon (see the "General Interpretation of Village and Traditional Culture" for details). If we are fortunate enough to encounter local festivals and folklore, we should respect every aspect of festival customs, participate and feel wholeheartedly, and understand the historical lineage behind these activities and the close relationship between these activities and nature.

Cherish the fruits of the local residents work

In the Qianjiangyuan National Park with of 90% Mountains, 5% of water and 5% of fields, many labor traditions and fruits have been derived from the farming culture, such as fish farming in clean water, honey brewing, camellia planting and oil squeezing, *etc*.. Enthusiastic residents will also invite us to experience and taste their harvest. It is our greatest respect for the local residents by not wasting food, and cherishing the fruits of the work derived from the natural blessings of the mountains.

4.3 解说过程中的技巧与注意事项

（1）准确的时间观念

提前到达

至少提前 15min 到达指定出发地点，以便第一次来参观的访客能够确定这个地方正是出发的地点。

按时出发

你应当给予那些按时到达的访客以回报。同时还需要在还能看清出发点的位置上，安排队伍进行一次停顿，以便给迟到的访客提供跟上队伍的机会。

按时折返

按时回到出发点，既是为了安全，也是为了建立游客对你的信任。

（2）领导你的团队

提前了解你的访客

在出发前或者行走过程中，非正式的谈话是联系你与访客之前的桥梁。对访客的了解越多，越有助于你的解说活动的展开。

成为团队的带领者

全程引导参访过程，让访客感觉和你在一起比自己在小路上独自前行更加激动人心。

始终位于团队的核心位置

经过要介绍的目标时，返回到队伍中间去，确保所有人能看到并听到你。这是为团队解说的关键技巧。

（3）照顾好你的访客

给访客预设期待

告知访客本次游线有多长，徒步难易程度以及游览时间。提前预告或者埋一些伏笔，告诉访客可以看见什么，让访客心中有所期待。对于事物的确定感能让人感到愉快。

使访客感到舒适

尽量避免访客遭受风吹日晒和雨淋，确保游览过程舒适愉快。

确保访客安全

确保游览过程安全，并且在结束前，不允许访客独自离开。

（4）解说实践技巧

让访客参与其中

行动比语言更有意义，让游客参与其中，融入感情，多提问和讨论。

大声说话但注意语调的变化

户外环境下，你的声音很难传到很远，因此注意要保证所有访客都能听到你的声音，但不要吼叫。

携带百宝箱

好的道具可以让你的解说更加生动有趣，同时还有助于让潜在的“制造麻烦”的访客忙于协助你展示物件，从而没有捣乱的机会。

把握好解说的时机

如果在溪流中遇到燕尾，及时停止讨论青蛙，开始聚焦于眼前这幕动物的一举一动。

重视总结

好的总结不仅是解说的圆满结束，也会提升教育的有效性。

（5）其他注意事项

确保活动不会破坏自然和人文资源，注意提醒访客不要乱丢垃圾。

4.3 Skills and principles during interpretation

(1) Be punctual

Arrive early

Arrive at the designated departure place at least 15 minutes in advance so that first-time visitors can be sure that this place is the place of departure.

Start on time

You should reward those visitors who arrive on time. However, give late visitors a chance to keep up with the team at a location where the starting point can still be seen.

Return on time

Return to the starting point on time, both geographically and thematically.

(2) Lead your team

Getting to know your visitors in advance

Before departure or during walking, casual conversation is the bridge between you and the visitor. The more you know about the visitor, the more helpful your interpretation activities are.

Become a leader

Guide the whole visit to Make visitors feel that being with you is more exciting than walking alone on the trail.

Staying on the central position

When go past the goal you want to introduce, then return to the middle of the team to make sure everyone can see you. This is the key to a rhythmic tour for large groups.

(3) Take care of your team visitors

Set expectations for visitors

Tell visitors how long is the tour route, how easy it is to hike, and how long to visit. Foretell or bury some foreshadowing in advance, tell visitors what they can see so as to let visitors have expectations. The certainty of things can make people feel happy.

Make visitors comfortable

Try to avoid them being exposed to wind, sun, and rain.

Ensure your visitors are safe

Ensure the trip is safe. Before it is ended, do not allow visitors to leave by themselves.

(4) Interpret practical skills

Involve visitors

Actions are more meaningful than words, so involue visitors in interpretation, get into feelings and ask more questions and discuss.

Speak loudly but pay attention to changes in intonation

In an outdoor environment, your voice is difficult to reach very far.

Carry necessary toolkits

Propriate tools can make your interpretation more lively and interesting, and at the same time it can make potential "trouble-making" visitors in helping you display the object so that they have no chance to make trouble.

Make good use of proper timing

If you encounter forktails in a stream, stop discussing frogs and start focusing on the animal in front of you.

Pay attention to the summary

A good summary will not only serve as a perfect ending of the interpretation, but also improve the effectiveness of education for visitors.

(5) Other considerations

Ensure that activities will not have negative impacts on natural and human resources, and remind the visitors not to litter.

4.4 反思与评价

对每一个解说员来说，对解说活动的反思是解说员自觉地把自己的每一次解说实践作为认识对象而进行全面而深入的冷静思考和总结。这是解说员用于提升自身解说技能，改进解说策划的学习方式。

你怎么知道你的解说活动是成功的？你的直觉一定会首先告诉你答案。访客的注意力、倾听的时间长度、参与度等，都是这个活动感染力的评价标准。但为了使评价更加客观，你可以考虑采用以下形式重复地检查你的“直觉”是否准确可靠。

（1）自我评价

为了了解解说活动是否成功，你必须有一个衡量“成功”的标准。在解说之前，为自己树立解说目标，并且设定一些切实可行的具体的评价方法，比如，测试访客是否回答出你提出的问题。在解说的过程中，你可以一直以目标驱动自己，保持在追求目标的轨道上。

解说主题 ______________________________

地点 ____________________ 日期 ____________________

体验时长 ____________________ 团队人数 ____________________

解说员 ____________________ 天气 ____________________

自我评价

1. 本次解说的目标是什么？
2. 本次解说过程中，有多少访客与我进行了互动？
3. 本次解说过程中，访客学到了什么？
4. 本次解说过程中，我的哪些设计引起了访客较高的共鸣？
5. 本次解说过程中，我的哪些设计对访客有所启迪？我是怎么知道的？
6. 本次解说过程中，我做得最棒的一件事是什么？
7. 如果再给我一次机会，我还可以做得更好的地方有什么？
8. 如果再给我一次机会，我绝对不会再犯的错误是什么？

（2）访客评价

试着使用书面或者口头询问的方式，邀请访客对本次活动进行评价。询问他们此行的收获，他们最喜欢哪一个解说点，他们对国家公园有没有新的认识，以及他们对于此次解说活动有没有其他建议。

访客评价

感谢您参加此次活动，请您花几分钟时间，帮助我们改进活动。

游览地点 ______________________ 游览日期 ______________________

您来自 _________ 省 ___________ 市 您的年龄 ______________________

1. 此行印象最深刻或者新学习到的三个知识或内容：

2. 本次活动过程中，我最喜欢的两个解说点是：

3. 回到生活的地方，我会这样给家人或朋友介绍钱江源国家公园：

4. 我想对解说员说：

注：你可以用视频或者音频的形式将你的现场解说记录下来。稍后可以回看，用于自我评价；或者倾听他人的意见。

4.4 Reflection and evaluation

For every interpreter, reflection on the interpretation activity is that the interpreter consciously takes each of his own interpretation practices as an object of understanding to conduct comprehensive and in-depth calm thinking and summary. This is a learning method used by interpreters to improve their interpretation skills and improve interpretation planning.

How do you know your interpretation activity is successful? Your intuition will tell you the answer first. Visitors' attention, length of listening time, participation, *etc.* are all evaluation criteria for the appeal of this activity. But in order to make the evaluation more objective, you can consider repeatedly checking your "intuition" in the following form.

(1) Self-evaluation

In order to understand whether the interpretation is successful, you must have a measure of "success". Before the interpretation, set up an interpretation goal for yourself and set some practical and specific evaluation methods. For example, test whether the visitor answered the question you asked. In the process of interpreting, you can always drive yourself with goals so as to keep on track of pursuing goals.

Self-evaluation

interpretation theme________________________________

location____________________ date____________________

length of experience ________________________________

Number of team's people______________________________________

interpreter ____

weather ____

1. What is the goal of this interpretation?
2. During this interpretation, how many visitors interacted with me?
3. What did the visitors learn during this interpretation?
4. During this interpretation, which of my designs resonated with visitors?
5. During this interpretation process, which of my designs have inspired visitors? How did I know?
6. What was the best thing I did during this interpretation?
7. If you give me another chance, what can I do better?
8. If you give me another chance, what is the mistake I will never make again?

(2) Visitor evaluation

Try to use written or oral inquiries to invite visitors to comment on this activity. Ask them what they learned from the trip, what are their favorite interpretation points, whether they have new understanding of national parks, and whether they have any other suggestions for this interpretation.

Visitor evaluation

Thank you for participating in this event, please take a few minutes to help us improve our event.

Place to visit________________ Tour date____________________

You are from __________________Province_________________ City

Your age_____________________

1. Three most impressive things/knowledges I learned from national parks:

2. During this event, my two favorite interpretation points are:

3. Back to where I live, I will introduce Qianjiangyuan National Park to my family or friends like this:

4. One thing I want to tell the interpreter:

Suggestions: You can record your live interpretation in the form of video or audio. You can look back later, use it for self-evaluation; or you can listen to the opinions of others.

05

自然教育活动学习单

Nature Education Learning Sheet

01 中国国家公园巡礼

经过这一节的介绍，你对我国国家公园的概念、设立目的和整体分布情况是否有了一个全面的了解？让我们一起来回顾一下吧！请你用连线的方式将国家公园和所在的省市连起来！

云南省	三江源国家公园
四川省	祁连山国家公园
黑龙江省	大熊猫国家公园
甘肃省	东北虎豹国家公园
青海省	神农架国家公园
陕西省	普达措国家公园
湖北省	钱江源国家公园
吉林省	南山国家公园
海南省	武夷山国家公园
福建省	热带雨林国家公园
湖南省	
浙江省	

2021 年 10 月 12 日，习近平总书记代表中国政府宣布中国正式设立首批五个国家公园，你能查一查是哪五家么？请在它们的名称后面标上☆。

02 钱江源国家公园打卡地图

面积广阔的国家公园，有着丰富的目的地等着你探索。下面这些地点你去过哪些呢？用线将去过的地点连起来，标记一张你自己的国家公园旅游路线地图吧！

1 古田山

2 古田飞瀑

3 古田村

4 瞭望台

5 源头碑

6 大峡谷

7 台回山

8 田贩村

9 高田坑

03 古田山常见植物打卡

钱江源国家公园被比作“亚热带常绿阔叶林之窗”，这里拥有长江三角洲非常珍稀的原真森林。在古田山上行走的你，遇见了哪些植物呢？请将你遇见的植物的名称后面的框涂黑。

甜槠（zhū） □

古田山最常见的常绿阔叶树种，叶子革质，卵圆形

木荷 □

高大乔木，叶革质或薄革质，椭圆形，在荒山灌丛是耐火的先锋树种。

枳（zhǐ）椇（jǔ） □

俗称“鸡爪梨”或“拐枣”，又因果实形态酷似楷书“万”字，又称为万寿果树。

马尾松 □

钱江源海拔较高地区最常见的针叶树种，树皮红褐色，一簇簇针叶好像马尾。

黄山松 □

常见的针叶树种，在山顶和峭壁上傲然挺立。

树参 □

叶片看起来很厚，叶形变异很大，椭圆形、椭圆状披针形，甚至还有分裂的倒三角形叶片。

香果树 □

古老的孑遗植物，被植物学家誉为“中国森林中最美丽动人的树”。

04 叶子的形态收集

叶子是植物进行光合作用的重要器官。为了适应不同的环境，不同的植物各自长出不同形状和特征的叶子，它们是辨认植物的重要方式。在钱江源国家公园，你观察到什么类型的叶子呢？在下面的空白处画下它们吧！

植物名称	分类	叶片形态	自然笔记速写

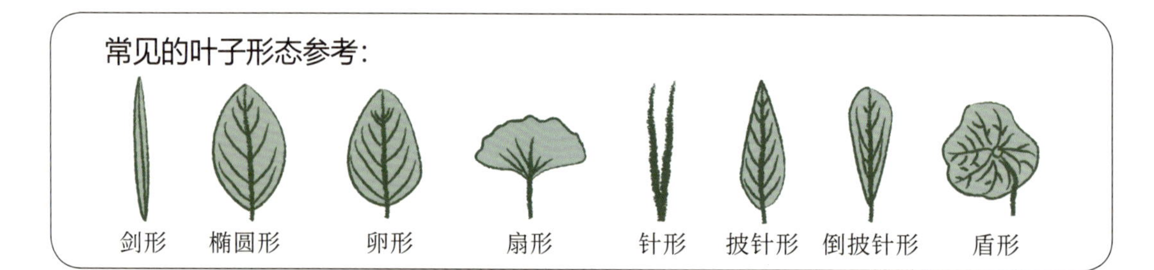

05 花朵的小小收藏家

美丽的花朵是植物的繁殖器官，不同的植物演化出不同的花朵形态，用来吸引授粉者的注意。在钱江源国家公园，你观察到什么类型的花朵呢？在下面的空白处画下它们吧！

植物名称	分类	花序形态	自然笔记速写

常见的花序形态参考：

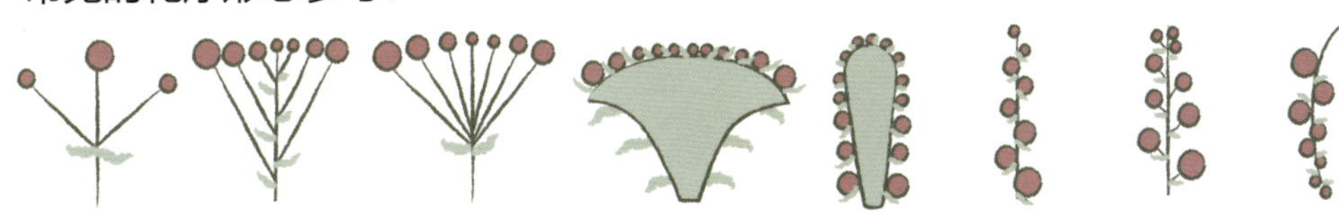

聚伞花序　伞房花序　伞形花序　头状花序　肉穗花序　穗状花序　总状花序　柔荑花序

06 千奇百怪的果实

果实担负着植物种子传播和繁衍下一代的重任。不同植物果实的类型多种多样，形态也千奇百怪。在钱江源国家公园，你发现了哪些奇怪的果实呢？在下面空白处画下它们吧！

植物名称	分类	果实形态	自然笔记速写

常见的果实形态参考：

坚果（单果）

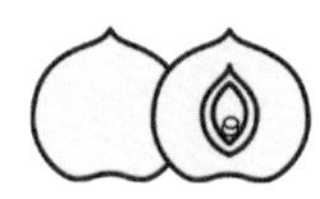
核果（单果）

浆果（单果）

荚果（单果）

翅果（单果）

聚合果

聚花果

07 钱江源常见植物打卡

沿着莲花塘的步道漫步，随着海拔升高，你有发现身边植物的变化么？留心观察，有没有遇到这些植物呢？请将你遇见的植物的名称后面的框涂黑。

青钱柳 □

高大乔木，果实像一串串绿色的铜钱。

阔叶箬竹 □

虽然是竹属的一种，却是低矮的灌木丛，拥有宽大叶子，是包粽子的典型叶片。

小沼兰 □

兰花一种，地生小草本，叶片铺地生长，生于林下或阴湿处的岩石上。

马尾松 □

钱江源海拔较高地区最常见的针叶树种，树皮红褐色，一簇簇针叶好像马尾。

芒萁 □

真蕨类植物，多年生草本，高30～60cm。根状茎细长横走，叶片疏生，叶轴一至二回或多回分叉。

杉木 □

钱江源最为常见的针叶树种，树冠圆锥形，树皮灰褐色

地衣 □

路边岩石表面经常生长的一种叶片状地衣，其实是藻类和真菌的共生复合体。

长柄双花木 □

落叶灌木，古老的孑遗物种，心形叶片是辨认它的重要标志之一。

08 小雨滴的旅行

你在钱江源遇到雨天了么？你知道每一滴落下的雨滴在自然界有着什么样的经历吗？试着回答下面的问题，按照答案的提示向前走。你走到终点了么？

1. 雨季

钱江源国家公园的春季和夏季雨水充沛，小雨滴和伙伴们开心极了。

2. 林冠层

小雨滴落在常绿阔叶林的林冠上，茂密的枝叶像是给每颗大树戴上一顶帽子。

恭喜你！完成了旅程！

12. 生命循环

还有一些小雨滴，被植物吸收或动物饮用后变成它们身体的一部分。当然，经过生命循环，它们还是会回到空气中。

恭喜你！完成了旅程！

11. 蒸发

汇入自然水体中的小雨滴，还有那些一直留在叶片和枝干表面的雨水，通过自然蒸发也回到了空气中。

10. 溪流河流

这些地表径流进一步汇聚，加速流淌，最终汇入地面的溪流、河流、湖泊和各种水体中，钱塘江的源头也是这么汇聚而成的呢。

游戏结束了，你是否发现，小雨滴的旅行并没有唯一的终点。原来并不是所有的雨滴都能汇入钱塘江。那么请你再想一想，为什么我们说国家公园的森林可以涵养水源呢？

3. 枝叶

有的小雨滴顺着叶片倾斜的角度向下流向树枝，又顺着枝条向下向树干的方向流去。

4. 树干

小雨滴顺着粗壮结实的树干一路向下流淌直至根部，突然被一层松软的“垫子”接住了。

5. 枯落物

这层“垫子”原来是枯叶堆积形成的枯落物。有些小雨滴干脆留了下来。这里湿润又松软，非常的舒服。

6. 土壤

更多小雨滴穿过了枯落物，渗入到了地面的土壤中。这里的土壤常年不受干扰，肥沃而带着淡淡的自然清香。

7. 植物根系

在土壤里，小雨滴遇到了各种植物的根系。这些根系密密地分布，钻到土壤深处，努力地为植物吸收着水和养分。

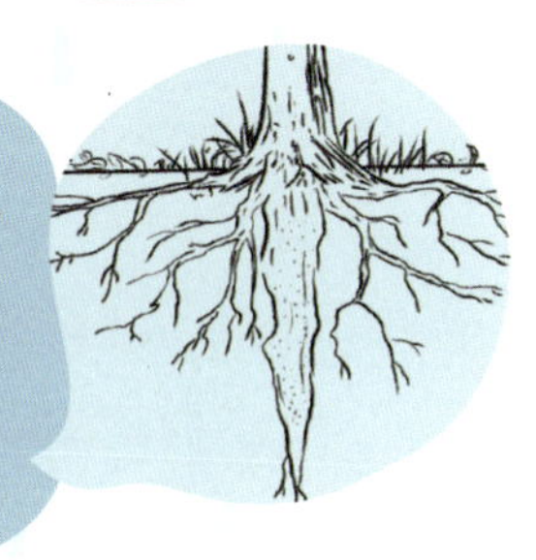

8. 蒸腾作用

有些小雨滴被努力工作的根系抓住，被吸收到植物体内，又顺着枝干回到叶片，通过气孔的蒸腾作用回到空气中。

9. 地表径流

大雨时，有些小雨滴迅速地流到地面，还没来得及被土壤吸收，就汇成水流，向低处流淌。

恭喜你！
完成了旅程！

09 折纸游戏

回忆一下，你在钱江源的参访中听到青蛙的叫声了么？用白纸按下面的步骤折一只蛙，为它涂上颜色，试着说一个关于青蛙的故事吧！

你在钱江源的参访中听到青蛙的叫声了吗？用白纸按下面的步骤折一只蛙，为它涂上颜色，试着说一个关于青蛙的故事吧！

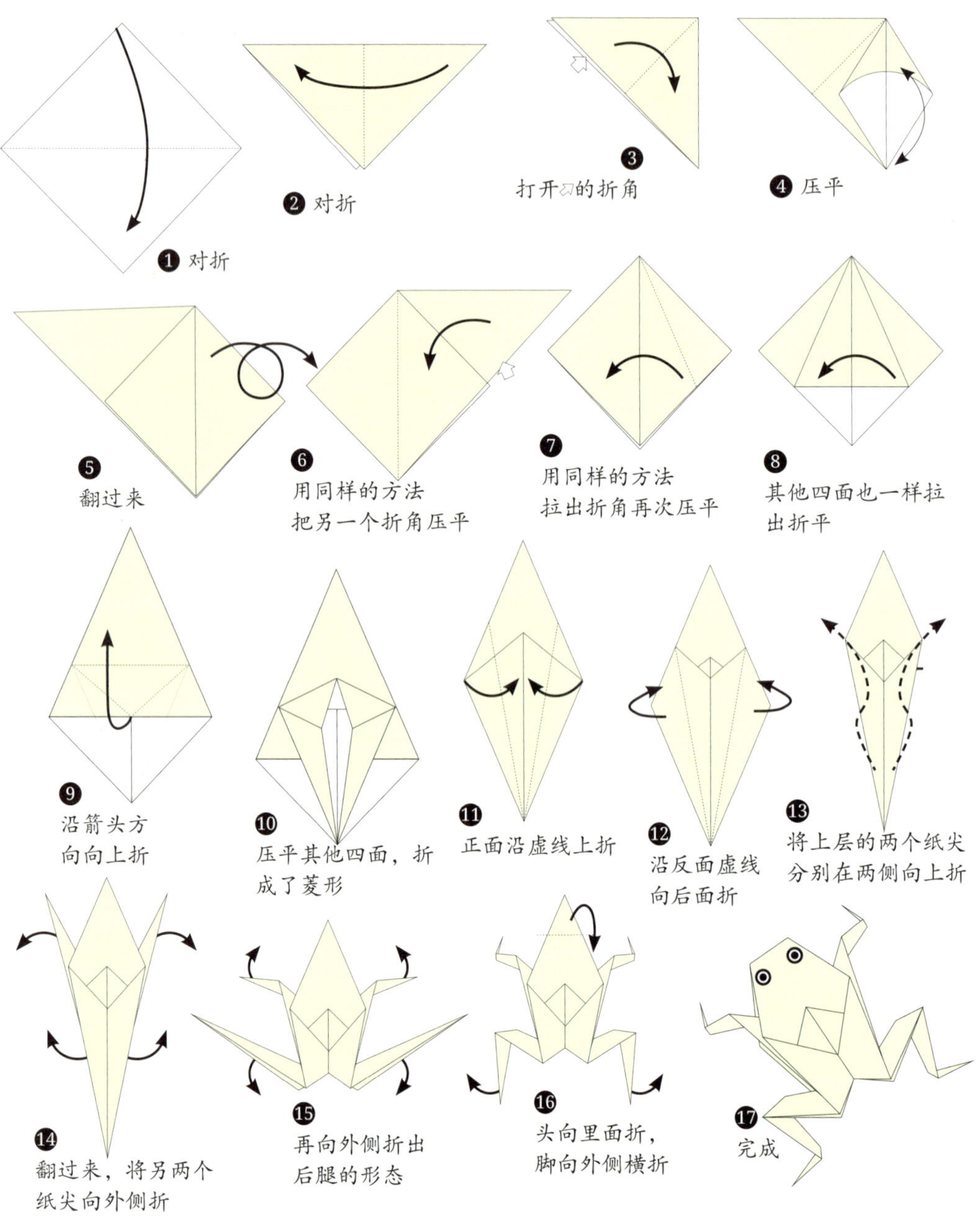

注：仿照日本折纸师新宫文明（Fumiaki Shingu）的免费折纸网站（Origami Club）资料绘制。

10 填色大作战

在钱江源，你遇到了哪些让你印象深刻的生物呢？参考提供的物种照片，为下面的植物填色，看一看，你有仔细观察它们吗？

11 填色大作战

在钱江源，你遇到了哪些让你印象深刻的生物呢？参考提供的物种照片，为下面的动物填色，看一看，你有没有仔细地观察它们呢？

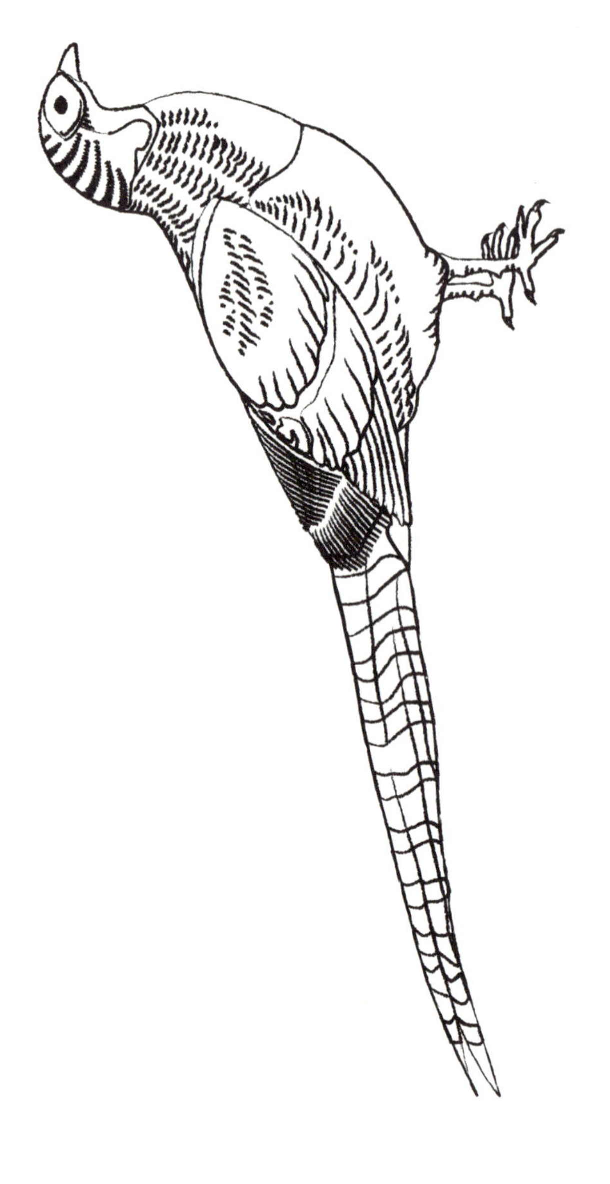

572

12 昆虫连连看

昆虫是自然界种类数量最多的一类生物。不同的昆虫为了适应环境，特化出不同形式的口器，你了解哪些呢？试着将下面不同昆虫和相应的进食特点联系起来。

蝴 蝶

虹吸式口器，偏爱吸食花距底部香甜的花蜜。

刺吸式口器，可以刺穿人和动物的表皮组织，吸食血液。

蚊 子

纺织娘

咀嚼式口器，具有可动的大颚，可以切割、磨碎植物。

嚼吸式口器，既可吸食花蜜，也可咬食花粉。

蜜 蜂

人的皮肤

13 种间关系填空

大自然的演化中，不同生物之间演化出形形色色的种间关系和适应策略，如捕食、竞争、拟态等。你能将下的关系对应起来么？

【　　　　】

崖壁上的地衣是真菌和藻类共同构成的复合体，一般藻类负责提供营养，真菌决定地衣的形态。

【　　　　】

长喙天蛾的长喙能够伸入大彗星风兰长长的花距采食花蜜，这是二者在进化中互相适应的结果。

【　　　　】

作为顶级捕食者，蛇雕曾经是中国南方最常见的猛禽之一。

【　　　　】

芒萁的枯叶含有抑制其他草本植物生长的酚类物质，从而在与其他草本植物的竞争中获得优势。

捕食 a

竞争 b

协同进化 c

互利共生 d

14 来自动物的启发——仿生技术连连看

动物仿生学是人们研究动物适应环境的独特结构，进而应用到人类的产品设计中去的一门学问。下面这些技术与产品，你知道都来自于哪些植物或动物么？

鱼的浮沉

鱼类通过控制腹中鱼鳔的空气调整在水中的位置。

蝴蝶身上的鳞片会随阳光的照射方向自动变换角度调节体温。

蝴蝶翅膀的微观结构

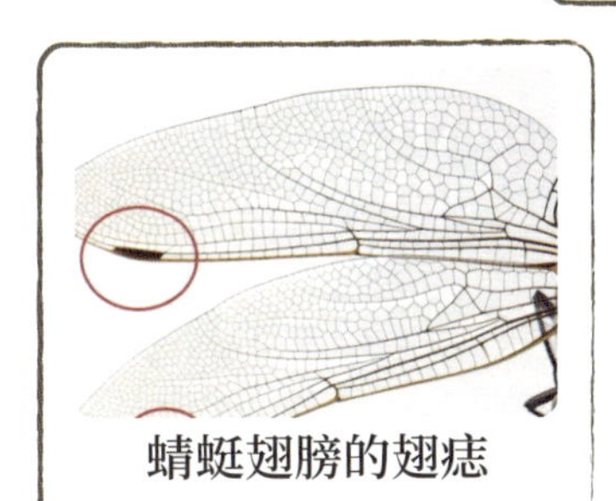
蜻蜓翅膀的翅痣

蜻蜓仅靠两对翅膀不停拍打进行高速飞行，为了维持稳定，依靠加重的翅痣在高速飞行时确保平衡。

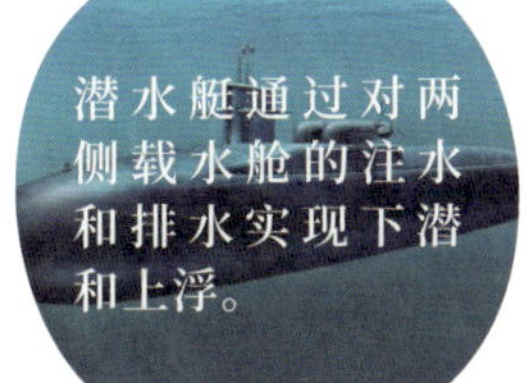

萤火虫发出的光不产生热，是一种“冷光”，一般很柔和，很适合人类的眼睛，光的强度也比较高。

萤火虫的腹部光源

15 子遗植物连连看

钱江源国家公园的延绵山脉为众多古老的子遗植物提供了避难所。这些子遗植物各具特征，试着将它们与右侧的特征连起来吧！

落叶灌木，又称为并蒂花。

果实像一串串绿色的铜钱。

叶形如马褂，又被叫作马褂木。

断开的叶片有明显的“叶断丝连”现象。

叶片条形柔软，在长枝上成螺旋状散生。

16 森林的食物链

森林里生活着形形色色的生物，它们之间有着什么样的关系呢？试着用食物箭头将下面的不同物种连接起来，绘制一份森林里的食物链关系图吧。

想一想，如果食物链中的某种物种消失了，对这个食物链会有什么样的影响呢？食物链还会安全吗？

17 描绘星空

钱江源是长江三角洲地区难得的适宜观星的地区，这里的高田坑村被称为华东地区最佳的观星点。图为星空下的梯田和古村落，试着为下面的画面添上色彩吧。

18 我为古树做铭牌

每个古树都是钱江源地区历史的见证者和守护者。回忆你在这里看到的一棵古树，试着为它设计一个独特的古树解说牌吧！设计的时候想一想，要表达什么，又要注意什么呢？

古田村古樟树

高田坑古红豆杉

古树群

19 村舍访谈

访问一位村里的长者，请他讲讲以前的生活、农耕方式，谈谈有哪些现在已经消失了，又有哪些值得我们传承，写一篇访谈笔记。

访谈对象: 日期: 地点:

20 发现传统生活的智慧

在钱江源的传统村庄里，有很多建筑、工具甚至传统习俗都体现了当地人巧妙利用自然、因地制宜的生活智慧，你观察到了么？选择你最感兴趣的一项画在下面吧！

清水养鱼

当地人利用山区的溪流和地形高差摸索出挖池蓄水养鱼的独特方式。

传统农居

独特的民居和空间利用方式。

水车石磨（传统农具）

利用水车带动的石磨是当地人加工粮食的重要工具。

21 “香火草龙”连线

请从数字“1”开始，将下面的点位按照标记的数字顺序从小到大进行连线，看一看，你发现了什么？

22 $1m^2$ 的植物观察

$1m^2$ 的空间，对植物来说已经可能成为生机盎然的世界。在户外指定的场地划一个 1mx1m 的方格，把观察到的植物记录到下面的笔记中。

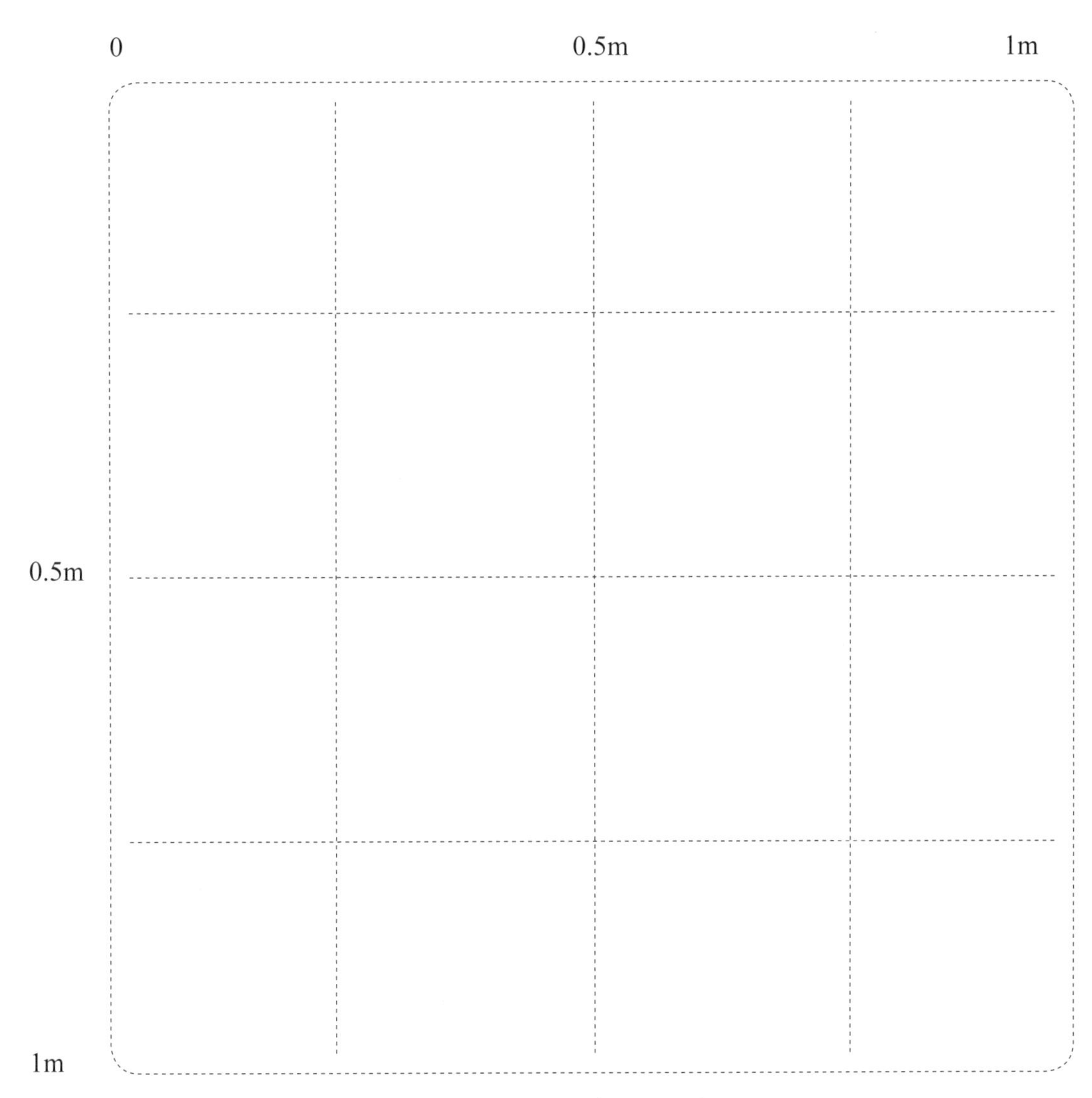

注：除了植物，在调研中，你发现了一些动物吗？试着也记录下来。

地点：　　　　日期：　　　　记录人：

23 种子雨收集器 DIY 与运用

种子雨收集器（seedtrap），是科学家们通过收集森林里的凋落物，进而研究森林生长变化和植被更替的工具。试着在老师的指导下完成小型种子雨收集器的组装，想一想，过一段时间，会收集到什么呢？

形形色色的种子雨收集器:

钱江源国家公园的种子雨收集器。你有发现它们么？

科学家将收集器收集的森林凋落物进行分类整理。

小型种子雨收集器（澳大利亚自然保护区）。

针叶林里的种子雨收集器（日本北海道）。

森林中大型种子收集器（墨西哥）。

24 塔吊设计图 / 模型制作

古田山的森林塔吊是全世界最大的林冠层观察和研究装置之一，高达60m。你知道塔吊是怎么工作的么？参照下面的图纸，试着制作一个塔吊的简易模型吧！

延伸设计：想一想，除了书中的纸质模型，你还有什么好的方法或利用其他材料，制作一个更逼真的塔吊模型吗？

参考文献
References

Andrew T. Smith, 解焱 . 中国兽类野外手册 [M]. 长沙：湖南教育出版社 , 2009.

蔡君 . 公园作为学习场所——国家公园解说和环境教育发展探讨 [J]. Landscape Architecture Journal, 2019. 26 (6)：91-96.

彩万志 , 李淑娟 , 米青山 . 昆虫拟态的多样性 [J]. 应用昆虫学报 , 2002, 39(5) 390-396.

陈晨 , 王民 , 蔚东英 . 环境解说历史及其理论基础的研究 [J]. 环境教育 , 2005.7:15-17.

陈建华 , 冯志坚 . 浙江古田山种子植物区系的地理成分研究 [J]. 华东师范大学学报：自然科学版 , 2002. 1：104-111.

陈声文 , 余建平 , 陈小南 , 等 . 利用红外相机网络调查古田山自然保护区的兽类及雉类多样性 [J]. 兽类学报 , 2016, 36(3)：292-301.

陈帅奇 . 浙西开化地区燕山期岩体特征及成矿 背景研究 [D]. 北京：中国地质大学 , 2011.

陈维立 , 罗惠方 , 王逸欣 , 等 . 与解说大师聊聊天：环境传播之启示与实践 [J]. 环境教育研究 , 2008, 5 (2)：33-87.

陈征海 , 刘安兴 , 李根有 , 等 . 浙江天然湿地类型研究 [J]. 浙江大学学报 (农业与生命 科学版), 2002, 28(2)：156-160.

丁炳扬 , 曾汉元 , 方腾 , 等 . 浙江省古田山自然保护区蕨类植物区系的研究 [J]. 浙江大学学报 (农业与生命科学版), 2001, 27(4)：370-374.

丁平 , 诸葛阳 , 姜仕仁 . 浙江古田山自然保护区鸟类群落生态研究 [J]. 生态时报 (2 期)：121-127.

杜彦君 , 马克平 . 浙江古田山自然保护区常绿阔叶林种子雨的时空变异 [J]. 植物生态学报 , 2012, 36(8)：717.

方小斌 , 周奕琳 , 杨璐伊 , 等 . 浙江省古田山国 家级自然保护区两栖爬行动物资源现状 [J]. 四川动物 , 2013, 32(1)：125-130.

国家林业局湿地保护管理中心 . 生机湿地 [M]. 北京 : 中国环境出版社 , 2017.

郭育任 , 林姗妮 , 郑琬平 . 阳明山径 : 阳明山国家公园步道导览手册 [M]. 台湾：阳明山国家公园管理处 , 2014.

贺琼 . 地衣的新概念 : 地衣是三种生物共生的生命复合体 [J]. 生物学教学 , 2017, 42(3)：74-75.

胡正华 , 于明坚 . 浙江古田山常绿阔叶林演替序列研究：群落物种多样性 [J]. 生态学杂志 , 2006, 25(6)：603-606.

胡正华 , 于明坚 , 彭传正 , 等 . 古田山国家自然保护区黄山松林主要种群生态位研究 [J]. 生态环境 , 2004, 13(4)：619-621.

胡正华 , 于明坚 , 索福喜 , 等 . 古田山国家自然保护区针阔叶混交林植物物种多样性特征 [J]. 生态环境 , 2008, 17(5)：1961-1964.

胡正华 , 于明坚 , 涂红 , 等 . 浙江古田山自然保护区甜槠林群落特征研究 [J]. 生态学杂志 , 2004, 24(2)：15-18.

胡正华 , 于明坚 , 张腾 , 等 . 浙江古田山自然保护区常绿阔叶林群落特征 [J]. 南京气象学报 , 2003,

26(1) 63-69.
开化县地方志编纂委员会 . 开化年鉴 . 2017[M]. 北京 : 中国文史出版社 , 2018.
李博 , 杨持 , 林鹏 . 生态学 [M]. 北京 : 高等教育出版社 , 2010.
刘睿杰 , 章祎雯 , 刘家顺 . 自然地理要素对浙江省方言空间分布的影响 [J]. 语言应用研究 , 2015, 0 (6) :127-129.
马克平 , 陈彬 , 米湘成 , 等 . 浙江古田山森林 : 树种及其分布格局 [M]. 北京 : 中国林业出版社 , 2009.
马克平 , 方腾 , 陈建华 . 中国常见植物野外识别手册 : 古田山册 [M]. 北京 : 高等教育出版社 , 2013.
齐忠伟 . 开化香火草龙 [M]. 杭州 : 浙江摄影出版社 , 2015.
衢州市开化县水利局 . 开化水利志 [M]. 北京 : 中国文史出版社 , 2006.
孙孝平 , 李双 , 余建平 , 等 . 基于土地利用变化情景的生态系统服务价值评估 : 以钱江源国家公园体制试点区为例 [J]. 生物多样性 , 2019, 27(01): 51-63.
汪长林 . 钱江源国家公园 [M]. 金华 : 西泠印社出版社 , 2018.
汪长林 , 钱海源 , 余建平 . 钱江源国家公园鸟类图鉴 [M]. 杭州 : 浙江大学出版社 , 2019.
王建华 . 开化 : 钱塘江之源 [M]. 杭州 : 浙江人民出版社 , 2007.
汪劲武 . 常见野花 [M]. 北京 : 中国林业出版社 , 2009.
Tuden F. 解说我们的遗产 [M]. 许世璋 , 高思明译 . 台北 : 五南书局 , 2006.
雍怡 . 江源古田 : 钱江源国家公园环境解说 [M]. 北京 : 中国林业出版社 , 2021.
雍怡 . 原野之窗——生物多样性教育课程 [M]. 北京 : 中国林业出版社 , 2020.
于明坚 , 胡正华 , 余建平 , 等 . 浙江古田山自然保护区森林植被类型 [J]. 浙江大学学报 (农业与生命科学版), 2001, 27(4): 375-380.
于海燕 . 钱塘江流域生态功能区划研究 [D]. 杭州 : 浙江大学 , 2008.
凯瑟琳· 雷尼尔 , 迈克尔· 格罗斯 , 罗恩· 齐默尔曼 . 解说人员指导手册 : 环境解说设计和展示技巧 [M]. 赵金凌 , 张岚译 . 北京 : 中国环境出版社 , 2013.
赵欣如 . 中国鸟类图鉴 [M]. 北京 : 商务印书馆 , 2018.
李立 , 陈建华 , 任海保 , 等 . 古田山常绿阔叶林优势树种甜槠和木荷的空间格局分析 [J]. 植物生态学报 , 2010, 34 (3) : 241.
浙江动物志委员会 . 浙江动物志· 淡水鱼类 [M]. 杭州 : 浙江科学技术出版社 , 1991.
浙江动物志委员会 . 浙江动物志· 两栖类 , 爬行类 [M]. 杭州 : 浙江科学技术出版社 , 1990.
浙江动物志委员会 . 浙江动物志· 兽类 [M]. 杭州 : 浙江科学技术出版社 , 1989.
祝燕 , 赵谷风 , 张俪文 , 等 . 古田山中亚 热带常绿阔叶林动态监测样地——群落组成与结 构 [J]. 植物生态学报 , 2008, 32(2) 262-273.
胡人亮 . 苔藓植物学 [M]. 北京 : 高等教育出版社 , 1987.
Freeman Tilden. Interpreting Our Heritage [M]. 世璋 , 高思明 , 译 . 台北 : 五南文化事业机构 , 2007.
HU Zheng-Hua, QIAN Hai-Yuan, YU Ming- Jian. The niche of dominant species populations in *Castanopsis eyrei* forest in Gutian Mountain National Nature Reserve[J]. Acta Ecologica Sinica, 2009, 29 (7) : 3670-3677.
Tilden F. Interpreting our heritage[M]. University of North Carolina Press, 2009.

附　录
Appendix

钱江源国家公园的自然条件

钱江源国家公园属于我国中亚热带湿润季风气候区。夏季，从太平洋吹来的暖湿气流经过白际山脉时受到阻挡，在山地迎风坡产生丰沛的降水；冬季，来自内陆的西伯利亚地区的冷干气流南下时同样会遇到山脉的阻挡，此时山脉在某种程度上对公园起到保温作用。

公园受夏季风影响较大，湿润多雨，全年四季分明。根据多年降水资料统计，公园年平均降水天数约在 140 天以上，年降水总量为 1963mm。其中，6 月降水天数最多，10 月降水天数最少，降水量也最低（图 1）。公园年平均气温为 15.3℃，其中 1 月和 7 月平均气温分别为 5.1℃和 26.6℃，年温差相对周边城镇变化小，冬暖夏凉（图 2）。此外，由于公园内多山，地形复杂，形成了丰富多样的局部山区小气候。行走其中，要特别注意天气的变化。

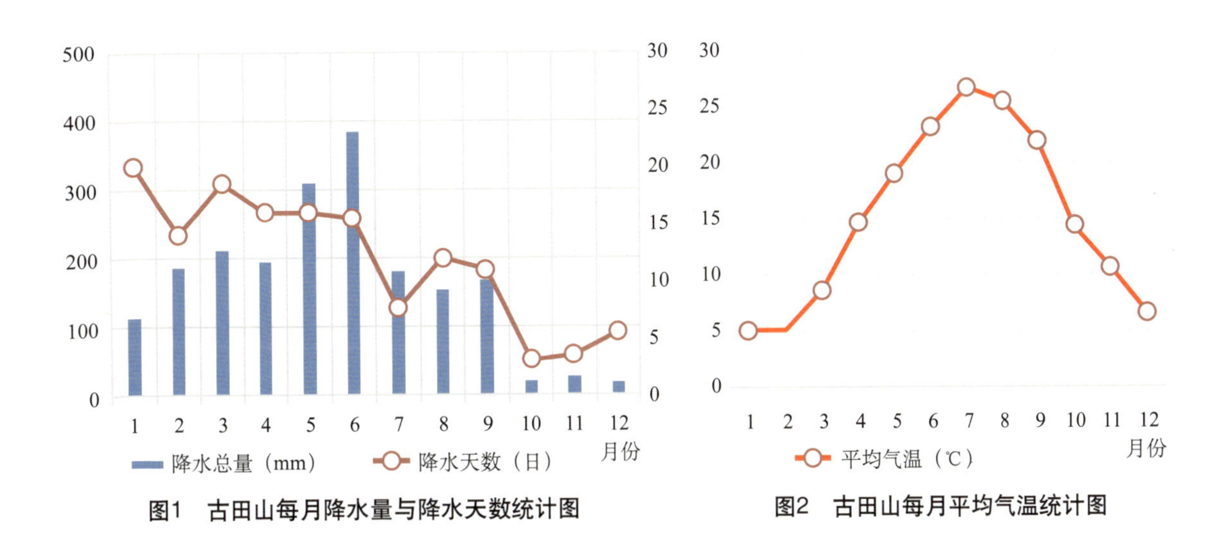

图1　古田山每月降水量与降水天数统计图

图2　古田山每月平均气温统计图

各季节气候状况与注意事项

3 ~ 6 月为雨季，山区经常阴雨绵绵，行走山间请携带雨具，并留意您的脚下，避免滑倒；

7 ~ 8 月为伏旱期，山区天气炎热，蚊虫较多，行走山间需注意防晒伤、防蚊虫叮咬；

8 月底至 9 月为台风多发季节，气候较不稳定，且山区午后常有雷阵雨，请随时注意天气预报；如遇汛期，请不要从事登山徒步等活动；

夜晚山区视线不佳，请尽早下山，避免摸黑时迷路。